U0168668

《中国周边海域海底地理实体图集丛书》

Atlas Series of Undersea Features of China's Surrounding Seas

吴自银◎主编

南海北部海域
海底地理实体图集

Atlas of Undersea Features of Northern South China Sea

吴自银　赵荻能　等◎著

By Wu Ziyin, Zhao Dineng, et al.

海洋出版社

2022年・北京

图书在版编目(CIP)数据

南海北部海域海底地理实体图集：汉、英 / 吴自银等著. — 北京：海洋出版社, 2022.11
（中国周边海域海底地理实体图集丛书 / 吴自银主编）
ISBN 978-7-5210-1008-4

Ⅰ. ①南… Ⅱ. ①吴… Ⅲ. ①南海－海底地貌－图集
Ⅳ. ①P722.7-64

中国版本图书馆CIP数据核字(2022)第176677号

审图号：GS(2022)2671 号

南海北部海域海底地理实体图集
NANHAI BEIBU HAIYU HAIDI DILI SHITI TUJI

丛书策划：王 溪　　责任编辑：屠 强 王 溪
助理编辑：李世燕　　责任印制：安 森

海洋出版社 出版发行
http://www.oceanpress.com.cn
北京市海淀区大慧寺路 8 号　邮编：100081
北京顶佳世纪印刷有限公司印刷
2022年11月第1版　2022年11月第1次印刷
开本：880mm × 1230mm　1 / 16　印张：23
字数：547千字　图幅数：127幅　定价：560.00元
发行部：010-62100090　总编室：010-62100034
海洋版图书印、装错误可随时退换

指导委员会

金翔龙　李家彪　陈丹红　相文玺　李四海

从　书

主　　编： 吴自银

从书编委会： 徐贺云　吴自银　朱本铎　邢　喆

图　集

作者名单：

吴自银　赵荻能　周洁琼　朱　宝　许津德
张经纬　朱本铎　刘丽强　杨安秀　罗　宇
廖定海　姜军毅　金涛勇　沈义俊　葛骑岐

技术支撑人员：

赵荻能　朱　宝　周洁琼　朱　超　刘　洋
崔丙浩　王明伟　罗　宇　刘时桥　汪九尧
秦晓铭　王嘉翀　姜军毅　刘志豪　王家豪
钟　皓　孙　恺　王　强

序 言

海洋覆盖面积约占地球表面积的71%，海水覆盖之下的海底地形地貌多姿多彩，既有平缓的海底平原，也有连绵起伏的海岭、高耸的海山和深邃的海沟，其复杂程度不亚于陆地山川地貌。如同陆地的自然地理实体需要赋予名称一样，位于海底的地理实体也应该赋予名称，可用于航海制图、海洋科学研究、海洋管理等，其重要性不言而喻。

国际海底地名分委会（SCUFN）是由国际海道测量组织（IHO）和政府间海洋学委员会（IOC）联合领导下的大洋水深制图委员会的下属专业组织，成立于1975年，是当今海底地理实体研究与命名领域具有较高权威性和影响力的国际组织，致力于全球海底地理实体命名的指导方针、原则以及相关标准规范的研究和制定工作，审议沿海国提交的海底地理实体命名提案。经SCUFN审议采纳的海底地理实体名称将写入世界海底地理实体名录（IHO-IOC publication B-8）中，主要用于世界通用大洋制图。

国际上关于海底地理实体的命名工作，最早可追溯到1899年，在第七届国际地理大会上，由两位德国科学家提出建立海洋地理实体命名国际协议的建议。当前，包括美国、德国、俄罗斯、日本、韩国、法国等很多世界沿海发达国家高度重视海底地理实体命名工作，并成立了专门的海底命名委员会。

我国学者早在20世纪80年代就开展了海底地理实体命名工作。中国于2011年9月在北京承办了SCUFN第24次会议，在这次会议上审议通过了我国提交的7个位于太平洋海域的海底地理实体命名提案，这批以《诗经》和中国美好传说中的词命名的海底地理实体，不仅打开了我国参与国际海底地理实体命名的新局面，更弘扬了中国传统文化，将中国文化符号永久镌刻在海底。我国向SCUFN提交的位于南海、太平洋、印度洋、大西洋和南大洋的上百个海底地名提案已经获得审议通过，并收录到世界海底地理实体名录中。

中国周边海域位于亚洲大陆东缘，总海域面积约4.80×10^6 km^2，包括“四海一洋”，其中有属于内陆海的渤海、属于陆缘海的黄海、属于陆架海和边缘海的东海与南海，以及台湾以东的太平洋海域。中国周边海域不仅包含广阔的大陆架，还有地形复杂的大陆坡、深海盆和深海槽，四大海域从北向南面积越来越大，水深越来越深，地形特征也越来越复杂。

准确规范的海底地理实体名称是进行科学交流的基础，但在20世纪90年代以前，受限于当时的单波束调查技术手段，我们对海底精细地形地貌特征了解的很不够，对于海底地理实体命名的研究工作开展的不多。自20世纪90年代我国大规模引进多波束测深系统以来，在中国周边海域实施了大规模的多波束海底地形地貌调查，获取了一批全新的高分辨和高精度的多波束水深数据资料，可精细地刻画海底地形地貌特征，大幅度提升了对海底地形地貌的科学认知，为我国周边海域的海底地理实体命名研究奠定了扎实的数据资料基础。

自中国地名委员会海底地名分委会成立以来，我国相关单位的科学家，依据我国地名管理法律法规和国际海底地理实体命名规则，结合中国文化特色，对中国周边海域的海底地理实体进行了长期、系统、深入的研究，先后形成了一批规范化的海底地理实体名称，并经主管部门审核后报国务院批准。为尽快向全社会分享该项最新研究成果，我们全面启动了《中国周边海域海底地形与地名图》（以下简称《地名图》）及《中国周边海域海底地理实体图集丛书》（以下简称《地名图集》）的编纂工作。《地名图》的成图比例尺为1：200万，包括《渤海 黄海 东海海底地形与地名图》及《南海海底地形与地名图》两幅。《地名图集》按照海区分布，结合海底地理实体的实际数量设计了三册图集丛书，包括：《渤海 黄海 东海和台湾以东海域海底地理实体图集》《南海北部海域海底地理实体图集》和《南海南部海域海底地理实体图集》，其中南海北部和南部以14°N为界划分为两册。《地名图》从空间上直观、全面地展示了我国海底地名的新成果，《地名图集》则采用平面图、立体图和信息表三者相结合的模式详细地展示每个海底地理实体的精细特征。《地名图》和《地名图集》二者相辅相成，有利于中国周边海域最新海底地理实体命名成果的传播和准确使用。

为了高质量地完成本次编图工作，我们系统地收集了20多年来在中国周边海域获取的最新多波束测深数据，并采用先进的技术方法、统一的软件平台，对来自不同项目、不同年代、不同设备和不同格式的多波束测深数据进行了统一和规范化的处理，使其达到统一的精度标准。在其他非多波束勘测区域，我们还收集了大量出版的海图和海底地形地貌图件及部分公开数据，并采用多源数据融合处理方法对复杂来源的地形地貌数据进行了有机融合处理，从而形成覆盖了中国周边海域的高分辨率、高精度的数字水深模型，并在此基础上研编了《地名图》和《地名图集》。

《地名图集》共收录了769项新命名的海底地理实体名称，并以“图－表”相结合的方式全面展示中国周边海域的海底地理实体命名成果。命名标准参考了国际海底地名分委会《海底地名命名标准》（B-6出版物）和中国国家标准《GB 29432—2012 海底地名命名》，海底地形编图和研究参考了中国国家标准《GB/T 12763.10—2007 海洋调查规范 第10部分：海底地形地貌调查》。在遵循国际海底地理实体命名基本规则的基础上，本次海底地理实体名称的分级基于地貌形态与成因相结合的原则，按照地貌形态、规模大小和主从关系，先宏观后微观、先群体后个体进行分级与分类，确保海底地理实体名称的层次性和实用性。在此指导原则基础上，将中国周边海域海底地理实体名称分为四级类别，下级类别海底地理实体名称一般由上级类别海底地理实体名称派生而来。第一级是依据海域的大地构造特征划分的大型海底地理实体名称，专名一般以所在的海域名称或依靠的岛屿名称命名；第二级是依据区域大地构造特征和地貌形态划分的海底地理实体名称，其专名一般以就近的陆地和海岛、所在海区的名称、邻近陆地上规模较大的地名和描述实体形态的词命名；第三级是依据地貌形态划分的海底地理实体名称，是本次命名的主体；第四级是构成第三级海底地理实体名称的最小一级的海底地理实体名称。第三级和第四级的专名通常结合该海域已有的海底地名进行命名，以指示该海底地理实体的位置命名，或使用同类事物命名群组出现的海底地理实体，或根据地貌的几何形态，或根据第二级海底地理实体名称进行团组化和序列化命名。

《地名图集》由自然资源部第二海洋研究所、国家海洋信息中心、中国地质调查局广州海洋地质调查局、中国航海图书出版社和中国地质调查局海口海洋地质调查中心等多家单位的科学家共同完成（见作者列表），自然资源部第二海洋研究所吴自银研究员负责牵头《地名图》和《地名图集》的策划、设计、制作和出版。《地名图》和《地名图集》可供海洋科研、海洋调查、地图制作、教学等使用参考。

《地名图》和《地名图集》的研编受到国家自然科学基金（资助编号：41830540、41906069、42006073），国家重点研发计划（资助编号：2022YFC2806600），浙江省自然科学基金（资助编号：LY21D060002、LY23D060007），上海交通大学“深蓝计划”基金（资助编号：SL2020ZD204、SL2022ZD205、SL2004），东海实验室开放基金项目（资助编号：DH-2022KF01005、DH-2022KF01001），自然资源部海洋测绘重点实验室开放基金项目（资助编号：2021B05），水声技术重点实验室稳定支持项目（资助编号：JCKYS2021604SSJS018）等多个项目的联合资助。

鉴于作者拥有的资料程度和研究水平的限制，《地名图》和《地名图集》难免存在诸多不足之处，为了更好地促进我国海洋科学的发展，敬请各位同行、专家批评指正。

吴自银 等

2021 年 9 月

Foreword

The ocean covers about 71% of the earth's surface area. The topography of the seafloor is as diversified and complicate as its continental counterparts, ranging from flat and gentle submarine plain, undulating ridges, and towering seamounts to deep trenches. Just as the continental features need names, so do undersea features. The undersea feature names can be used in nautical charting, marine scientific research, and marine management. Their importance cannot be overstated.

The International Hydrographic Organization – Intergovernmental Oceanographic Commission General Bathymetric Chart of the Oceans (IHO-IOC GEBCO) Sub-Committee on Undersea Feature Names (SCUFN), established in 1975 as the Sub-Committee on Geographical Names and Nomenclature of Ocean Bottom Features (SCGN), and its name changed in 1993. SCUFN is an international organization with high authority and influence in the field of undersea feature names. SCUFN is dedicated to the research and development of guidelines and principles of undersea feature names as well as related standards, and considers proposals submitted by coastal States for the naming of undersea features. All undersea feature names selected by SCUFN are contained in the IHO-IOC GEBCO Gazetteer of Undersea Feature Names (IHO-IOC publication B-8), and select those names of undersea features in the world ocean appropriate for use on the GEBCO graphical and digital products, on the IHO small-scale international chart series, and on the regional International Bathymetric Chart (IBC) series.

The international work on undersea feature names was traced back to as early as 1899, when two German scientists proposed to appoint a committee on the "Nomenclature of Sub-oceanic Features" at the 7th International Geographical Congress held in Berlin, Germany. So far, many developed coastal states, including the United States, Germany, Russia, Japan, South Korea and France, have attached great importance to undersea feature naming and have set up special committee on undersea feature names.

Chinese scholars have been working on undersea feature naming early in 1980s. China hosted the 24th SCUFN Meeting in Beijing in September, 2011, in which seven proposals for undersea feature names in the Pacific Ocean submitted by China were considered and approved. These names, which were originated from *The Book of Songs* and beautiful Chinese legends, not only initiated and promoted China's participation in the international undersea feature naming, but also spread Chinese traditional culture and permanently engraved Chinese cultural symbols on the seabed. So far, hundreds of proposals for undersea feature names located in the South China Sea, Pacific Ocean, Indian Ocean, Atlantic Ocean and Southern Ocean submitted by China to SCUFN have been approved and contained in the IHO-IOC GEBCO Gazetteer of Undersea Feature Names.

China's surrounding seas are located on the eastern Asian continental margin with a total sea area of about 4.80×10^6 km^2, including the Bo Hai that falls into the category of inland sea, the Yellow Sea that falls into the category of epicontinental sea, the East China Sea and the South China Sea that fall into the category of shelf sea and marginal sea respectively, as well as the Pacific Ocean to the east of Taiwan Dao. China's surrounding seas not only include the extensive continental shelves, but also include the continental slopes, deep-sea basins and deep-sea troughs with complicate topography. The four seas increase in area and water depth with more and more complicated topographic features from north to south.

Uniform undersea feature names are the basis for scientific communication. However, before the 1990s, limited by the single-beam survey technique, our knowledge of the detailed topographic and geomorphological features was far from enough and our research on undersea feature names was insufficient. Since the large-scale introduction of multi-beam echosounder system in China in 1990s, a large-scale multi-beam survey of seafloor topography and geomorphology has been carried out in China's surrounding seas, and a new batch of high-resolution and high-precision multi-beam bathymetric data has been obtained to characterize seafloor topographic and geomorphological features in details. This has greatly improved our knowledge of seafloor topography and geomorphology and laid a solid database for the study of undersea feature names in China's surrounding seas.

Since the establishment of Sub-Committee on Undersea Feature Names of China Committee on Geographical Names, scientists of relevant organizations have made long-time, systematic and in-depth research of the undersea features in China's surrounding seas in combination with China's cultural characteristics and in accordance with Chinese laws and regulations on the administration of place names and the international rules for undersea feature names. We have successively named a series of standardized undersea feature names, which had been submitted to the State Council for approval upon verification of the competent departments. In order to share the latest research findings to the whole society as soon as possible, we have carried out the compilation of the *Maps of Submarine Topography and Undersea Feature Names of China's Surrounding Seas* ("*Maps of Undersea Features*" for short) and *Atlas Series of Undersea Features of China's Surrounding Seas* ("*Atlases of Undersea Features*" for short) .The *Maps of Undersea Features*, with a scale of 1 : 2 000 000, consists of two parts, namely *Map of Submarine Topography and Undersea Feature Names of Bo Hai, Yellow Sea and East China Sea* and *Map of Submarine Topography and Undersea Feature Names of South China Sea*. The *Atlases of Undersea Features* is divided into three volumes based on the distribution of the sea areas and the actual quantity of undersea features, namely, *Atlas of Undersea Features of Bo Hai, Yellow Sea, East China Sea and East of Taiwan Dao*, *Atlas of Undersea Features of Northern South China Sea* and *Atlas of Undersea Features of Southern South China*

Sea. The northern part and the southern part of South China Sea are divided into two volumes bounded by 14°N. The *Maps of Undersea Features* visually and comprehensively presents China's new achievement in undersea feature names, while the *Atlases of Undersea Features* presents the details of each undersea feature names with combination of information table, plan and three-dimensional map. The *Maps of Undersea Features* and the *Atlases of Undersea Features* are complementary and are conducive to the spreading and accurate use of the latest achievements in undersea feature names of China's surrounding seas.

In order to ensure the quality of the compilation, we systematically collected the latest multi-beam bathymetric data obtained in China's surrounding seas in the past two decades, and applied advanced technologies and methods and unified software platform to make unified and standardized processing of the multi-beam bathymetric data of different projects, different eras, different equipment and different formats, so as to achieve a unified accuracy standard. For other areas without available multi-beam bathymetric data, we collected a large number of published nautical charts, submarine topographic and geomorphological maps and some public data and use data fusion processing methods to integrate the topographic and geomorphologic data from complicated sources. In this way, we developed high-resolution and high-precision digital bathymetric model of the China's surrounding seas and compiled *Maps of Undersea Features* and *Atlases of Undersea Features* on this basis.

Atlases of Undersea Features includes 769 new undersea feature names and comprehensively displays China's achievement in undersea feature names of surrounding seas with combination of maps and tables. We referred to *Standardization of Undersea Feature Names* by the IHO-IOC GEBCO Sub-Committee on Undersea Feature Names (IHO-IOC Publication B-6) and the Chinese standard of *Nomenclature of Undersea Feature Names* (GB 29432—2012) for the standards of names, and referred to the Chinese standard of *Specifications for Oceanographic Survey Part* 10: *Submarine Topography and Geomorphology* (GB/T 12763.10—2007) in the compiling of submarine topographic maps. On the basis of the international principles for undersea feature names, the grading and classification of undersea feature names were made first for groups in macroscopic scale and then for individuals in microscopic scale based on the principle of combining submarine geomorphological feature with cause of formation and according to the topographic form, scales and different levels, to ensure the hierarchy and practicability of undersea feature names. On the basis of this guideline, we divided the undersea feature names in China's surrounding seas into four levels. The lower order undersea feature names are generally derived form upper order undersea feature names. The first order is the large-scale undersea feature names determined based on the geotectonic features of the sea area and their specific terms are named after the sea area where the features are located and the islands they relied upon. The second order is the undersea feature names determined based on the geotectonic characteristics and geomorphological features of the region and their specific terms are named after the nearby continents and islands, the names of the sea areas, the names of the large-scale landward places and the terms that describe the morphology of the features. The third order is the undersea feature names determined by the topographic features and is the main part of the mentioned feature naming for this time. The fourth order is the lowest level of the third level undersea feature names. The specific terms of the third and fourth order are generally named after the existing names of the sea area to designate the location of the undersea features, or after the undersea features appeared in the group names for similar features, or after geometric shapes of the geomorphology, or adopt group names and serial names based on the second-order undersea feature names.

Atlases of Undersea Features was jointly completed by the Second Institute of Oceanography of MNR, the National Marine Data and Information Service, the Guangzhou Marine Geological Survey of China Geological Survey, the China Navigation Publications Press, the Haikou Marine Geological Survey Center of China Geological Survey and other organizations (see list of the authors). Wu Ziyin, a researcher from the Second Institute of Oceanology of Ministry of Natural Resources, was in charge of the organizing, design, production and publishing of the *Maps of Undersea Features* and *Atlases of Undersea Feature Names*, which can be used for reference in marine scientific research, oceanographic survey, cartography and teaching.

The *Maps of Undersea Features* and *Atlases of Undersea Feature Names* are supported by the National Natural Science Foundation of China (41830540, 41906069, 42006073), the National Key Research and Development Program of China (2022YFC2806600), the Natural Science Foundation of Zhejiang province (LY21D060002, LY23D060007), the Oceanic Interdisciplinary Program of Shanghai Jiao Tong University (SL2020ZD204, SL2022ZD205, SL2004), the Science Foundation of Donghai Laboratory (DH-2022KF01005, DH-2022KF01001), the Key Laboratory of Ocean Geomatics, Ministry of Natural Resources, China (2021B05), the Key Laboratory of Acoustic Science and Technology, Harbin Engineering University (JCKYS2021604SSJS018), and several other projects.

Due to the limitation of data and research level of the authors, it is inevitable that the *Maps of Undersea Features* and *Atlases of Undersea Features* are inadequate in many respects. Criticisms and corrections from peers and experts are welcomed so that we can more effectively promote the development of marine science in China.

Wu Ziyin et al.

09/2021

前　言

《南海北部海域海底地理实体图集》收录海底地理实体名称共计 319 项，采用信息表、平面图和立体图三者相结合的模式，全面展示了位于南海北部海域的海底地理实体的中、英文标准名称、精细地形地貌特征、界限范围以及专名由来等信息。

海底地理实体名称包含专名和通名两部分，其中的通名部分反映了海底地理实体名称的地貌类型，可按照海底地貌学的分类原则，以地貌形态、规模大小和主从关系，将海底地理实体名称划分为四个级别。本图集共收录一级海底地理实体名称 6 项，二级海底地理实体名称 26 项，三级海底地理实体名称 279 项，四级海底地理实体名称 8 项，各海底地理实体名称的级别信息详见各章开头处的海底地理实体列表。

按照海底地理实体名称所在的海域和专名的命名由来，可将本图集收录的海底地理实体名称划分为六大海区包括：（Ⅰ）南海北部陆架和陆坡海区，包含海底地理实体名称共计 59 项，专名主要取词自历史习惯名称、我国古代科学家的名字、周边已有的海底地理实体名称或邻近陆地或岛屿上的地名；（Ⅱ）西沙群岛周边海区，包含海底地理实体名称共计 55 项，专名主要取词自西沙群岛的岛礁暗沙名称、周边已有的海底地理实体名称或中国古代神话人物的名字；（Ⅲ）中沙北海隆周边海区，包含海底地理实体名称共计 62 项，专名主要取词自中沙大环礁中的暗沙浅滩名称、周边已有的海底地理实体名称或唐朝诗人李白的《梦游天姥吟留别》中的词；（Ⅳ）盆西海岭海区，包含海底地理实体名称共计 25 项，专名主要取词自周边已有的海底地理实体名称、唐朝诗人王之涣的《凉州词》、王勃的《滕王阁序》或李白的《关山月》中的词；（Ⅴ）南海海盆北部海区，包含海底地理实体名称共计 64 项，专名主要取词自历史习惯名称、周边已有的海底地理实体名称、我国古代科学家、医学家和文人的名字、海洋软体动物的名字、毛泽东诗词作品中的词、我国古代司法官吏名或我国古代文学作品中的词；（Ⅵ）南海海盆南部和西部海区，包含海底地理实体名称共计 54 项，专名主要取词自周边已有的海底地理实体名称、海洋软体动物的名字、我国古代文学作品中的词、我国唐宋文人的名字、宝石矿物的名称或与“玉”相关的词。

本图集由自然资源部第二海洋研究所完成，主要作者为吴自银、赵荻能、周洁琼等，其他主要作者见作者列表。

Preface

Atlas of Undersea Features of Northern South China Sea contains a total of 319 undersea feature names. It uses information table, plan and three-dimensional map to comprehensively display the standard Chinese and English names, the detailed topographical and geomorphological features, the boundaries and scopes, and the origins of the specific terms of undersea feature names located in the northern South China Sea.

An undersea feature name consists of a specific term and a generic term. The generic term reflects the geomorphological types of the undersea feature names and can be classified into four levels according to the classification principle of submarine geomorphology, by topographic form, scales and principal-subordinate relationship of the undersea feature names. This atlas contains 6 first order, 26 second order, 279 third order, and 8 fourth order undersea feature names, and the order information of each names is shown in the list of undersea features at the beginning of each chapter.

Six major sea areas were further delimited according to the sea areas where the undersea feature names are located, as well as the origins of specific terms of names. (Ⅰ) 59 undersea feature names in the Nanhaibeibu Lujia and Nanhaibeibu Lupo. The specific terms of these names are mainly taken from historical names, the names of famous ancient Chinese scientists, the existing undersea feature names around the area or geographic names of neighboring lands and islands. (Ⅱ) 55 undersea feature names in the sea area off Xisha Qundao. The specific terms of these names are mainly taken from the names of the islands and reefs of Xisha Islands, the existing undersea feature names around the area or the names of ancient Chinese mythological figures. (Ⅲ) 62 undersea feature names in the sea area off Zhongshabei Hailong. The specific terms of these names are mainly taken from the names of the shoals and banks in Zhongsha atoll, the existing undersea feature names around the area, or the phrases in the poem *A Visit to Mount Tianmu in Dream* by Li Bai (701–762 A.D.) in the Tang Dynasty (618–907 A.D.). (Ⅳ) 25 undersea feature names in the Penxi Hailing. The specific terms of these names are mainly taken from the existing undersea feature names around the area, the phrases in the poem *Out of the Great Wall* by Wang Zhihuan (688–742 A.D.), the phrases in the poem *A Tribute to King Teng's Tower* by Wang Bo (650–676 A.D.), or the phrases in the poem *Moon over Fortified Pass* by Li Bai (701–762 A.D.) in the Tang Dynasty (618–907 A.D.). (Ⅴ) 64 undersea feature names in the north of Nanhai Haipen. The specific terms of these names are mainly taken from historical names, the existing undersea feature names around the area, the names of ancient Chinese scientists, medical scientists and literati, the names of marine mollusks, the words in Mao Zedong's poems, the titles of ancient Chinese judicial officials, or the phrases in ancient Chinese literatures. (Ⅵ) 54 undersea feature names in the south and west of Nanhai Haipen. The specific terms of these names are mainly taken from the existing undersea feature names around the area, the names of marine mollusks, the phrases in ancient Chinese literatures, the names of famous men of letter of the Tang and Song Dynasties (618–1279 A.D.), the names of gems and minerals, or the words related to jade ("Yu" in Chinese).

This atlas was completed by the Second Institute of Oceanography, Ministry of Natural Resources of China, with the main authors Wu Ziyin, Zhao Dineng, Zhou Jieqiong, and other main authors listed in the author list.

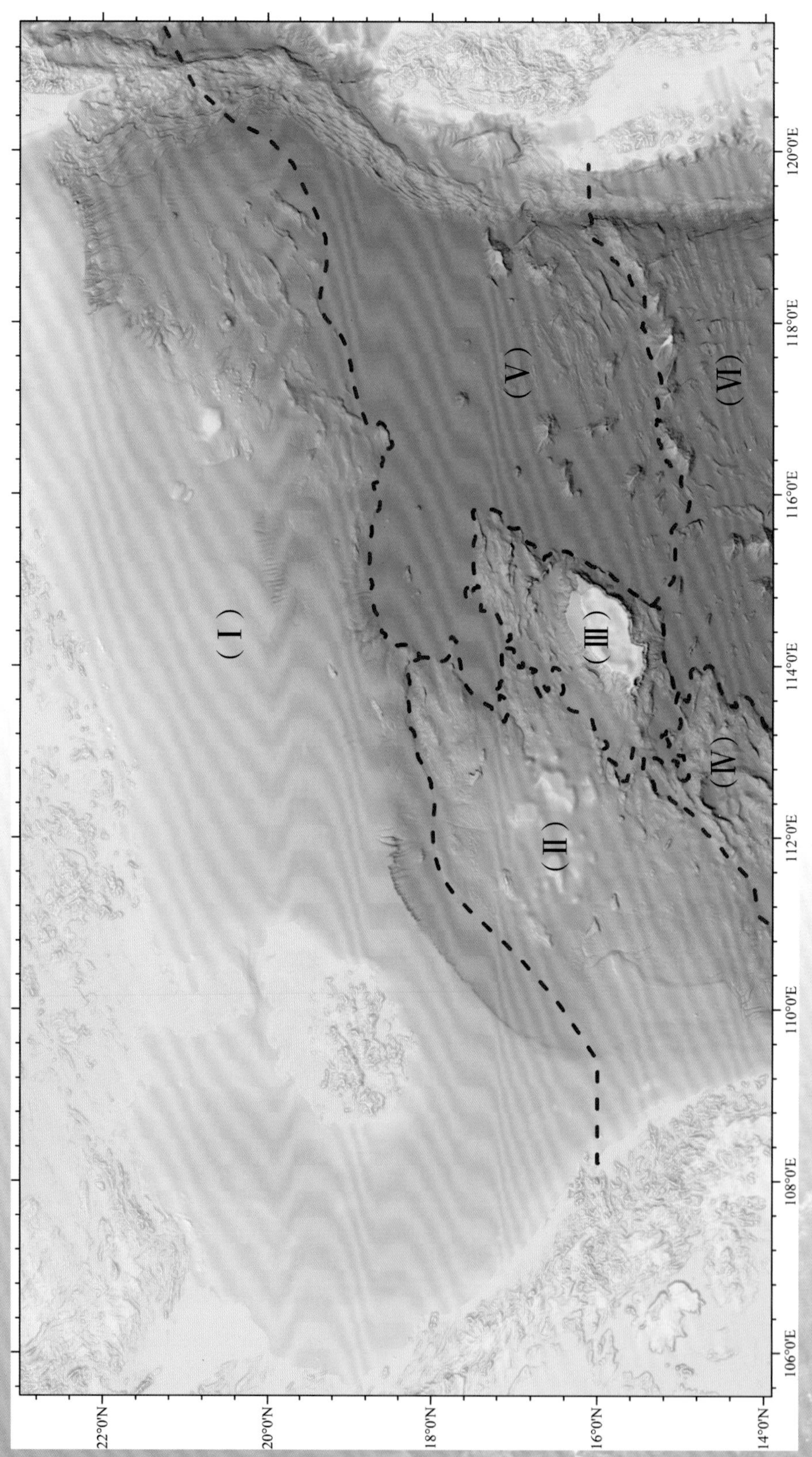

南海北部海域海底地理实体命名分区示意图
The Naming Region of Undersea Features of Northern South China Sea

Contents

目　录

1. 南海北部陆架陆坡海区海底地理实体

2. 西沙群岛周边海区海底地理实体

3. 中沙北海隆周边海区海底地理实体

4. 盆西海岭海区海底地理实体

5. 南海海盆北部海区海底地理实体

6. 南海海盆南部和西部海区海底地理实体

Contents

1. Undersea features in the Nanhaibeibu Lujia and Nanhaibeibu Lupo

2. Undersea features in the sea area off Xisha Qundao

3. Undersea features in the sea area off Zhongshabei Hailong

4. Undersea features in the Penxi Hailing

5. Undersea features in the north of Nanhai Haipen

6. Undersea features in the south and west of Nanhai Haipen

1

南海北部陆架陆坡海区海底地理实体

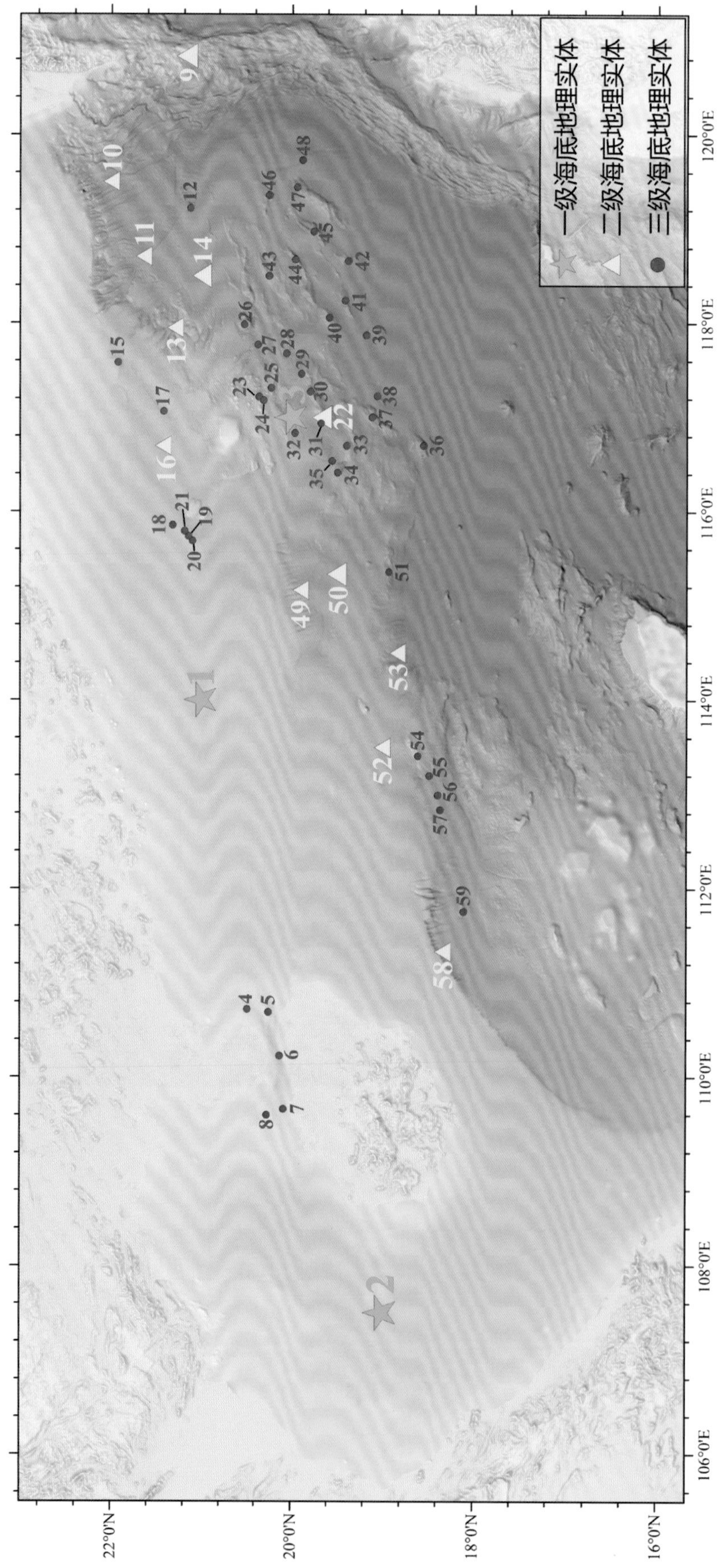

图 1-1 南海北部陆架和陆坡海区海底地理实体中心点位置示意图，序号含义见表 1-1

Fig.1-1 Location of center coordinates of undersea features in the Nanhaibeibu Lujia and Nanhaibeibu Lupo, with the meanings of the serial numbers shown in Tab. 1-1

表 1–1 南海北部陆架和陆坡海区海底地理实体列表

Tab.1–1 List of undersea features in the Nanhaibeibu Lujia and Nanhaibeibu Lupo

序号 No.	标准名称 Standard Name	汉语拼音 Chinese Phonetic Alphabet	类别 Generic Term	中心点坐标 Center Coordinates		实体等级 Order
				纬度 Latitude	经度 Longitude	
1	南海北部陆架 Nanhaibeibu Lujia	Nánhǎiběibù Lùjià	大陆架 Shelf	21°00.0'N	114°00.0'E	1
2	南海北部陆坡 Nanhaibeibu Lupo	Nánhǎiběibù Lùpō	大陆坡 Slope	20°00.0'N	117°00.0'E	1
3	北部湾陆架 Beibuwan Lujia	Běibùwān Lùjià	大陆架 Shelf	19°00.0'N	107°30.0'E	1
4	徐闻东浅滩 Xuwendong Qiantan	Xúwéndōng Qiǎntān	浅滩 Shoal	20°28.9'N	110°43.3'E	3
5	徐闻东水道 Xuwendong Shuidao	Xúwéndōng Shuǐdào	水道 Channel	20°14.8'N	110°41.4'E	3
6	徐闻南海槽 Xuwennan Haicao	Xúwénnán Hǎicáo	海槽 Trough	20°07.5'N	110°13.5'E	3
7	徐闻西水道 Xuwenxi Shuidao	Xúwénxī Shuǐdào	水道 Channel	20°05.0'N	109°39.7'E	3
8	徐闻西沙脊群 Xuwenxi Shajiqun	Xúwénxī Shājǐqún	海底沙脊群 Sand Ridges	20°15.9'N	109°35.9'E	3
9	恒春海脊 Hengchun Haiji	Héngchūn Hǎijǐ	海脊 Ridge	21°10.0'N	120°50.0'E	2
10	澎湖海底峡谷群 Penghu Haidixiaguqun	Pénghú Hǎidǐxiágǔqún	海底峡谷群 Canyons	22°00.0'N	119°30.0'E	2
11	台湾海底峡谷 Taiwan Haidixiagu	Táiwān Hǎidǐxiágǔ	海底峡谷 Canyon	21°38.5'N	118°42.3'E	2
12	浦元海山 Puyuan Haishan	Pǔyuán Hǎishān	海山 Seamount	21°07.7'N	119°13.0'E	3
13	笔架海底峡谷群 Bijia Haidixiaguqun	Bǐjià Hǎidǐxiágǔqún	海底峡谷群 Canyons	21°17.7'N	117°56.7'E	2
14	东沙斜坡 Dongsha Xiepo	Dōngshā Xiépō	海底斜坡 Slope	21°00.0'N	118°30.0'E	2
15	潮汕海底崖 Chaoshan Haidiya	Cháoshàn Hǎidǐyá	海底崖 Escarpment	21°55.5'N	117°34.8'E	3
16	东沙北斜坡 Dongshabei Xiepo	Dōngshāběi Xiépō	海底斜坡 Slope	21°25.6'N	116°41.6'E	2
17	东沙北海底崖 Dongshabei Haidiya	Dōngshāběi Hǎidǐyá	海底崖 Escarpment	21°25.2'N	117°03.7'E	3
18	北卫海丘 Beiwei Haiqiu	Běiwèi Hǎiqiū	海丘 Hill	21°19.3'N	115°51.3'E	3
19	北卫西台地 Beiweixi Taidi	Běiwèixī Táidì	海台 Plateau	21°08.9'N	115°44.4'E	3

续表

序号 No.	标准名称 Standard Name	汉语拼音 Chinese Phonetic Alphabet	类别 Generic Term	中心点坐标 Center Coordinates		实体等级 Order
				纬度 Latitude	经度 Longitude	
20	南卫西台地 Nanweixi Taidi	Nánwèixī Táidì	海台 Plateau	21°06.3'N	115°41.4'E	3
21	东沙海底崖 Dongsha Haidiya	Dōngshā Hǎidǐyá	海底崖 Escarpment	21°11.4'N	115°47.6'E	3
22	尖峰斜坡 Jianfeng Xiepo	Jiānfēng Xiépō	海底斜坡 Slope	19°40.0'N	117°00.0'E	2
23	西樵海丘 Xiqiao Haiqiu	Xīqiáo Hǎiqiū	海丘 Hill	20°21.5'N	117°13.0'E	3
24	丹霞海丘 Danxia Haiqiu	Dānxiá Hǎiqiū	海丘 Hill	20°18.4'N	117°10.8'E	3
25	罗浮海丘 Luofu Haiqiu	Luófú Hǎiqiū	海丘 Hill	20°13.5'N	117°18.7'E	3
26	北坡海山 Beipo Haishan	Běipō Hǎishān	海山 Seamount	20°31.1'N	117°59.0'E	3
27	北坡南海底峡谷群 Beiponan Haidixiaguqun	Běipōnán Hǎidǐxiágǔqún	海底峡谷群 Canyons	20°22.1'N	117°46.2'E	3
28	北坡南海丘 Beiponan Haiqiu	Běipōnán Hǎiqiū	海丘 Hill	20°03.6'N	117°40.6'E	3
29	李春北海底峡谷群 Lichunbei Haidixiaguqun	Lǐchūnběi Hǎidǐxiágǔqún	海底峡谷群 Canyons	19°53.7'N	117°27.7'E	3
30	李春海山 Lichun Haishan	Lǐchūn Hǎishān	海山 Seamount	19°47.7'N	117°16.3'E	3
31	李春西海底峡谷 Lichunxi Haidixiagu	Lǐchūnxī Hǎidǐxiágǔ	海底峡谷 Canyon	19°40.5'N	116°56.1'E	3
32	东沙南海底峡谷 Dongshanan Haidixiagu	Dōngshānán Hǎidǐxiágǔ	海底峡谷 Canyon	19°58.1'N	116°49.9'E	3
33	沈括海山 Shenkuo Haishan	Shěnkuò Hǎishān	海山 Seamount	19°23.5'N	116°41.8'E	3
34	尖峰圆丘 Jianfeng Yuanqiu	Jiānfēng Yuánqiū	圆丘 Knoll	19°29.6'N	116°24.7'E	3
35	尖峰北海底峡谷 Jianfengbei Haidixiagu	Jiānfēngběi Hǎidǐxiágǔ	海底峡谷 Canyon	19°33.2'N	116°32.1'E	3
36	石申海山 Shishen Haishan	Shíshēn Hǎishān	海山 Seamount	18°32.9'N	116°41.9'E	3
37	杜诗海山 Dushi Haishan	Dùshī Hǎishān	海山 Seamount	19°06.8'N	117°00.1'E	3
38	秦九韶海丘 Qinjiushao Haiqiu	Qínjiǔsháo Hǎiqiū	海丘 Hill	19°03.5'N	117°13.5'E	3
39	徐光启海丘 Xuguangqi Haiqiu	Xúguāngqǐ Hǎiqiū	海丘 Hill	19°10.8'N	117°52.2'E	3

续表

序号 No.	标准名称 Standard Name	汉语拼音 Chinese Phonetic Alphabet	类别 Generic Term	中心点坐标 Center Coordinates		实体等级 Order
				纬度 Latitude	经度 Longitude	
40	宋应星海丘 Songyingxing Haiqiu	Sòngyìngxīng Hǎiqiū	海丘 Hill	19°35.0'N	118°03.3'E	3
41	裴秀海丘 Peixiu Haiqiu	Péixiù Hǎiqiū	海丘 Hill	19°24.7'N	118°14.3'E	3
42	郭守敬海丘 Guoshoujing Haiqiu	Guōshǒujìng Hǎiqiū	海丘 Hill	19°22.6'N	118°39.7'E	3
43	蔡伦海山 Cailun Haishan	Càilún Hǎishān	海山 Seamount	19°45.4'N	118°58.3'E	3
44	笔架海丘 Bijia Haiqiu	Bǐjià Hǎiqiū	海丘 Hill	20°14.9'N	118°30.0'E	3
45	丁缓海山 Dinghuan Haishan	Dīnghuǎn Hǎishān	海山 Seamount	19°58.0'N	118°40.4'E	3
46	马钧海丘 Majun Haiqiu	Mǎjūn Hǎiqiū	海丘 Hill	20°14.9'N	119°21.3'E	3
47	墨子海山 Mozi Haishan	Mòzǐ Hǎishān	海山 Seamount	19°56.5'N	119°26.3'E	3
48	刘焯海谷 Liuzhuo Haigu	Liúzhuō Hǎigǔ	海谷 Valley	19°52.8'N	119°43.8'E	3
49	神狐海底峡谷特征区 Shenhu Haidixiagutezhengqu	Shénhú Hǎidǐxiágǔtèzhēngqū	海底峡谷特征区 Canyon Province	19°54.0'N	115°10.0'E	2
50	珠江海谷 Zhujiang Haigu	Zhūjiāng Hǎigǔ	海谷 Valley	19°30.0'N	115°20.0'E	2
51	毕昇海丘 Bisheng Haiqiu	Bìshēng Hǎiqiū	海丘 Hill	18°55.7'N	115°21.5'E	3
52	一统斜坡 Yitong Xiepo	Yītǒng Xiépō	海底斜坡 Slope	19°00.0'N	113°30.0'E	2
53	一统海底峡谷群 Yitong Haidixiaguqun	Yītǒng Hǎidǐxiágǔqún	海底峡谷群 Canyons	18°50.0'N	114°30.0'E	2
54	张衡海丘 Zhangheng Haiqiu	Zhānghéng Hǎiqiū	海丘 Hill	18°36.4'N	113°24.4'E	3
55	万泉海底峡谷 Wanquan Haidixiagu	Wànquán Hǎidǐxiágǔ	海底峡谷 Canyon	18°28.9'N	113°12.0'E	3
56	昌化海底峡谷 Changhua Haidixiagu	Chānghuà Hǎidǐxiágǔ	海底峡谷 Canyon	18°23.4'N	112°59.3'E	3
57	南渡海底峡谷 Nandu Haidixiagu	Nándù Hǎidǐxiágǔ	海底峡谷 Canyon	18°21.9'N	112°50.1'E	3
58	西沙北海底峡谷群 Xishabei Haidixiaguqun	Xīshāběi Hǎidǐxiágǔqún	海底峡谷群 Canyons	18°20.0'N	111°20.0'E	2
59	万户海山群 Wanhu Haishanqun	Wànhù Hǎishānqún	海山群 Seamounts	18°06.5'N	111°45.4'E	3

1.1 南海北部陆架

标准名称 Standard Name	南海北部陆架 Nanhaibeibu Lujia	类别 Generic Term	大陆架 Shelf
中心点坐标 Center Coordinates	21°00.0′N, 114°00.0′E	规模（千米 × 千米） Dimension（km × km）	1200 × 250
最小水深（米） Min Depth (m)	0	最大水深（米） Max Depth (m)	360
地理实体描述 Feature Description	南海北部陆架宽广，东西向长约 1200 千米，南北向宽约 250 千米，呈东窄西宽之势。东部最窄处约 150 千米，西部较宽，靠近海南岛处可达 300 千米，等深线总体呈平行于岸线排列（图 1–2）。 Nanhaibeibu Lujia is expansive, with a length of about 1200 km from east to west, and a width of about 250 km from north to south. The Shelf is narrower in east and wider in the west, with the narrowest part of about 150 km in the east. The western part is relatively wider, with a width of up to 300 km near Hainan Dao. Its isobaths are generally arranged in parallel to the coastline (Fig.1–2).		
命名由来 Origin of Name	该大陆架位于南海北部，因此得名。 The Slope is located in the northern part of the South China Sea, and “Beibu” means northern part in Chinese, so the word “Nanhaibeibu” was used to name the Shelf.		

1.2 南海北部陆坡

标准名称 Standard Name	南海北部陆坡 Nanhaibeibu Lupo	类别 Generic Term	大陆坡 Slope
中心点坐标 Center Coordinates	20°00.0′N, 117°00.0′E	规模（千米 × 千米） Dimension（km × km）	1400 × 400
最小水深（米） Min Depth (m)	200	最大水深（米） Max Depth (m)	3900
地理实体描述 Feature Description	南海北部陆坡位于南海北部陆架和南海海盆之间的过渡带，西起海南岛南部的西沙海槽，东至台湾岛南侧的恒春海脊，东西向长达 1400 千米，总体呈东西两侧狭窄陡峭、中部宽缓的地形特征（图 1–2）。 Nanhaibeibu Lupo is located at the transitional zone between Nanhaibeibu Lujia and Nanhai Haipen, starting from Xisha Haicao on the south of Hainan Dao in the west and ending at Hengchun Haiji on the south of Taiwan Dao in the east, with a length of 1400 km from east to west. Generally, it shows a topographic characteristics of being narrow and steep in the east, west sides and wide and gentle in the middle part (Fig.1–2).		
命名由来 Origin of Name	该大陆坡位于南海北部，因此得名。 The Slope is located in the northern part of the South China Sea, and “Beibu” means northern part in Chinese, so the word “Nanhaibeibu” was used to name the Slope.		

(a)

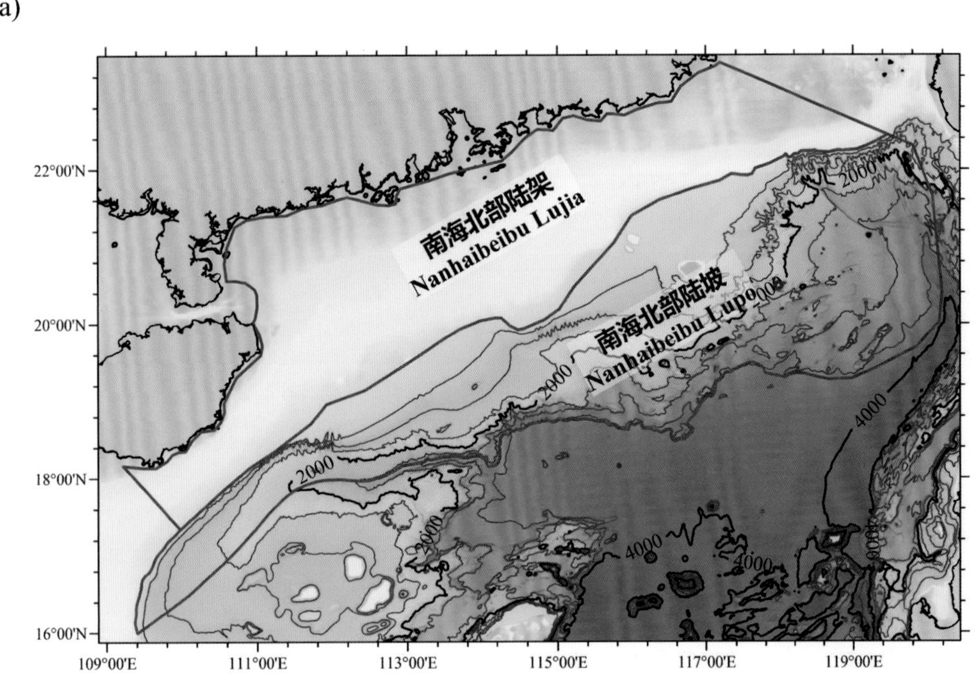

(b)

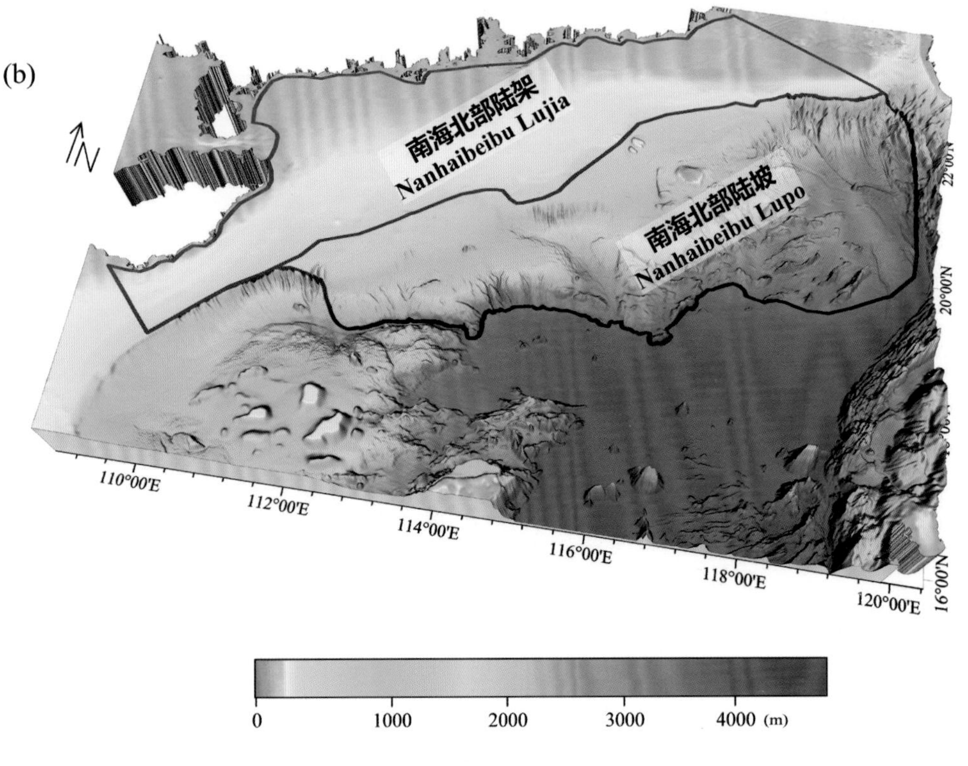

图 1-2 南海北部陆架、南海北部陆坡

(a) 海底地形图（等深线间隔 500 米）；(b) 三维海底地形图

Fig.1-2 Nanhaibeibu Lujia, Nanhaibeibu Lupo

(a) Seafloor topographic map (with contour interval of 500 m); (b) 3-D seafloor topographic map

1.3 北部湾陆架

标准名称 Standard Name	北部湾陆架 Beibuwan Lujia	类别 Generic Term	大陆架 Shelf
中心点坐标 Center Coordinates	19°00.0′N, 107°30.0′E	规模（千米 × 千米） Dimension（km × km）	600 × 300
最小水深（米） Min Depth (m)	0	最大水深（米） Max Depth (m)	200
地理实体描述 Feature Description	北部湾陆架位于南海北部西侧，其北、东、西三面被陆地环绕，海底地形受海岸制约明显，等深线呈顺岸线排列，其中北部等深线为北东—南西向展布，南部等深线转为北西—南东向展布（图 1-3）。 Beibuwan Lujia is located on the west of the northern part of the South China Sea and is surrounded by land on the north, east and west sides. The submarine topography is obviously restricted by the coastline. The isobaths are arranged in the direction of the coastline, of which, the isobaths in north central are arranged in the direction of NE−SW, while the isobaths in the south are arranged in the direction of NW−SE (Fig.1−3).		
命名由来 Origin of Name	该大陆架位于南海北部湾，因此得名。 The Shelf is located in the Beibu Wan of the South China Sea, so the word "Beibuwan" was used to name the Shelf.		

(a)

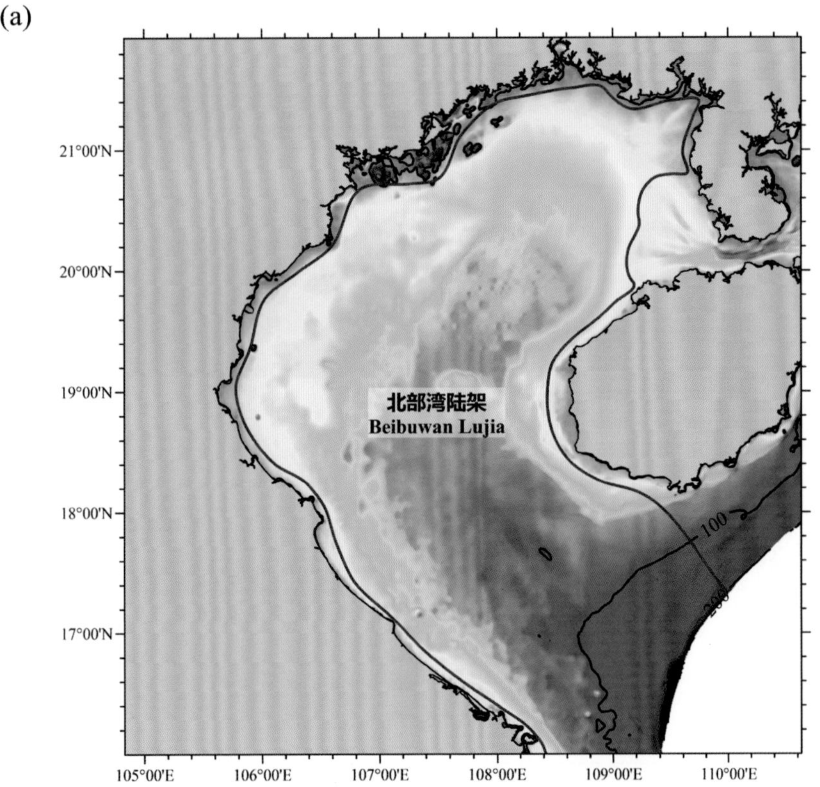

(b)

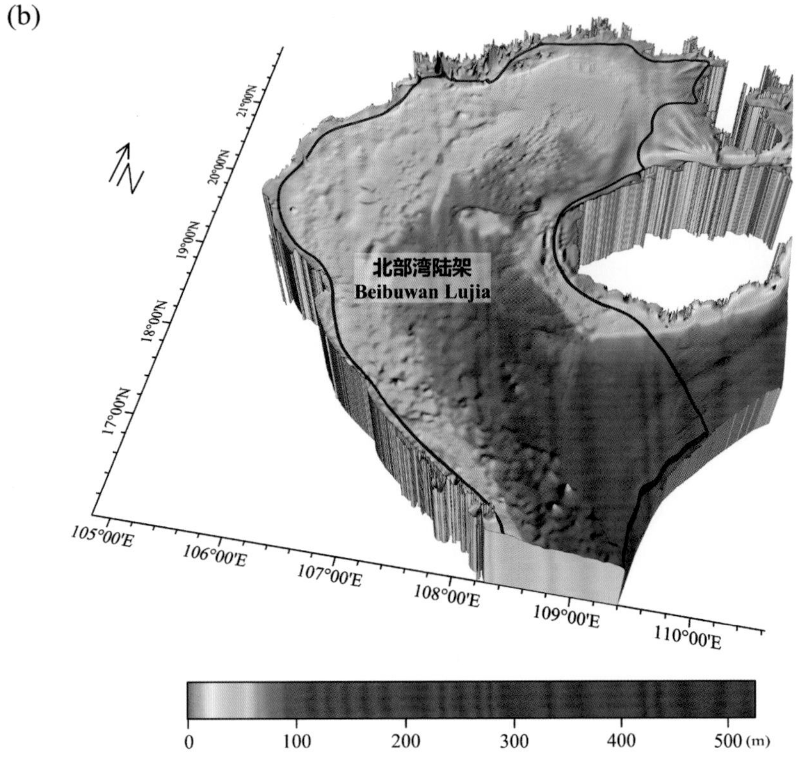

图 1-3　北部湾陆架

(a) 海底地形图（等深线间隔 100 米）；(b) 三维海底地形图

Fig.1-3　Beibuwan Lujia

(a) Seafloor topographic map (with contour interval of 100 m); (b) 3-D seafloor topographic map

1.4 徐闻东浅滩

标准名称 Standard Name	徐闻东浅滩 Xuwendong Qiantan	类别 Generic Term	浅滩 Shoal
中心点坐标 Center Coordinates	20°28.9'N, 110°43.3'E	规模（千米 × 千米） Dimension（km × km）	40 × 45
最小水深（米） Min Depth (m)	0	最大水深（米） Max Depth (m)	40
地理实体描述 Feature Description	徐闻东浅滩位于雷州半岛徐闻县以东海域，平均水深约 15 米，其上发育数条北东—南西向的潮流沙脊（图 1-4）。 Xuwendong Qiantan is located in the sea area to the east of Xuwen county in Leizhou Bandao, with an average depth of about 15 m. Several tidal sand ridges have been developed on it in the direction of NE−SW (Fig.1−4).		
命名由来 Origin of Name	该浅滩位于广东省湛江市徐闻县以东，因此得名。 The Shoal is located to the east of Xuwen county in Zhanjiang, Guangdong province, and “Dong” means east in Chinese, so the word “Xuwendong” was used to name the Shoal.		

1.5 徐闻东水道

标准名称 Standard Name	徐闻东水道 Xuwendong Shuidao	类别 Generic Term	水道 Channel
中心点坐标 Center Coordinates	20°14.8'N, 110°41.4'E	规模（千米 × 千米） Dimension（km × km）	45 × 25
最小水深（米） Min Depth (m)	30	最大水深（米） Max Depth (m)	80
地理实体描述 Feature Description	徐闻东水道位于海南岛以北、徐闻县以东海域的琼州海峡东出口处，由数条东—西向的海底水道组成（图 1-4）。 Xuwendong Shuidao is located at the east entrance of the Qiongzhou Haixia in the sea area to the north of the Hainan Dao and to the east of Xuwen county and is composed of several channels in the direction of E−W (Fig.1−4).		
命名由来 Origin of Name	该水道位于广东省湛江市徐闻县以东，因此得名。 The Channel is located to the east of Xuwen county in Zhanjiang, Guangdong province, and “Dong” means east in Chinese, so the word “Xuwendong” was used to name the Channel.		

1.6 徐闻南海槽

标准名称 Standard Name	徐闻南海槽 Xuwennan Haicao	类别 Generic Term	海槽 Trough
中心点坐标 Center Coordinates	20°07.5'N, 110°13.5'E	规模（千米 × 千米） Dimension（km × km）	70 × 10
最小水深（米） Min Depth (m)	60	最大水深（米） Max Depth (m)	110
地理实体描述 Feature Description	徐闻南海槽位于海南岛以北、徐闻县以南的琼州海峡中部海域，平面形态呈东—西向的长条形（图 1–4）。 The Xuwennan Haicao is located in the central area of Qiongzhou Haixia to the north of the Hainan Dao and to the south of Xuwen county, with a planform in the shape of a long strip in the direction of E–W (Fig.1–4).		
命名由来 Origin of Name	该海槽位于广东省湛江市徐闻县以南，因此得名。 The Trough is located to the south of Xuwen county in Zhanjiang, Guangdong province, and “Nan” means south in Chinese, so the word “Xuwennan” was used to name the Trough.		

1.7 徐闻西水道

标准名称 Standard Name	徐闻西水道 Xuwenxi Shuidao	类别 Generic Term	水道 Channel
中心点坐标 Center Coordinates	20°05.0'N, 109°39.7'E	规模（千米 × 千米） Dimension（km × km）	45 × 8
最小水深（米） Min Depth (m)	30	最大水深（米） Max Depth (m)	40
地理实体描述 Feature Description	徐闻西水道位于海南岛以北、徐闻县以西的琼州海峡西出口处，平面形态呈东—西向的长条形（图 1–4）。 Xuwenxi Shuidao is located at the west entrance of Qiongzhou Haixia to the north of Hainan Dao and to the west of Xuwen county, with a planform in the shape of a long strip in the direction of E–W (Fig.1–4).		
命名由来 Origin of Name	该水道位于广东省湛江市徐闻县以西，因此得名。 The Channel is located to the west of Xuwen county in Zhanjiang, Guangdong province, and “Xi” means west in Chinese, so the word “Xuwenxi” was used to name the Channel.		

1.8 徐闻西沙脊群

标准名称 Standard Name	徐闻西沙脊群 Xuwenxi Shajiqun	类别 Generic Term	海底沙脊群 Sand Ridges
中心点坐标 Center Coordinates	20°15.9'N, 109°35.9'E	规模（千米 × 千米） Dimension（km × km）	60 × 60
最小水深（米） Min Depth (m)	10	最大水深（米） Max Depth (m)	30
地理实体描述 Feature Description	徐闻西沙脊群位于海南岛以北、徐闻县以西的琼州海峡西出口处，由数条北西—南东向的潮流沙脊构成（图 1-4）。 Xuwenxi Shajiqun is located at the west entrance of Qiongzhou Haixia to the north of the Hainan Dao and to the west of Xuwen county and is composed of several tidal sand ridges in the direction of NW−SE (Fig.1−4).		
命名由来 Origin of Name	该沙脊群位于广东省湛江市徐闻县以西，因此得名。 These Sand Ridges are located to the west of Xuwen county in Zhanjiang, Guangdong province, and “Xi” means west in Chinese, so the word “Xuwenxi” was used to name these Sand Ridges.		

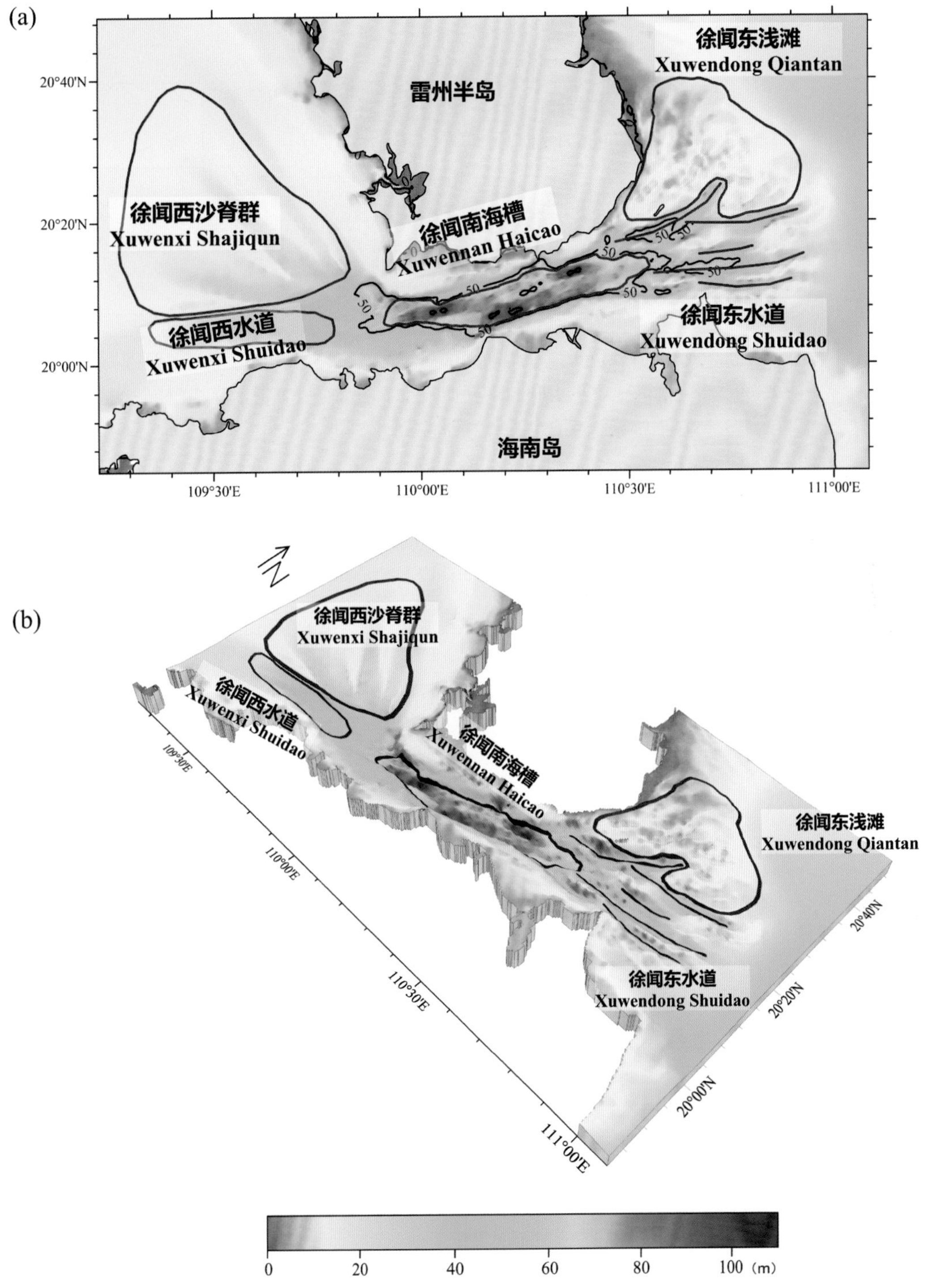

图 1-4 徐闻东浅滩、徐闻东水道、徐闻南海槽、徐闻西水道、徐闻西沙脊群

(a) 海底地形图（等深线间隔 50 米）；(b) 三维海底地形图

Fig.1-4 Xuwendong Qiantan, Xuwendong Shuidao, Xuwennan Haicao, Xuwenxi Shuidao, Xuwenxi Shajiqun

(a) Seafloor topographic map (with contour interval of 50 m); (b) 3-D seafloor topographic map

1.9 恒春海脊

标准名称 Standard Name	恒春海脊 Hengchun Haiji	类别 Generic Term	海脊 Ridge
中心点坐标 Center Coordinates	21°10.0′N, 120°50.0′E	规模（千米 × 千米） Dimension（km × km）	210 × 110
最小水深（米） Min Depth (m)	40	最大水深（米） Max Depth (m)	3700
地理实体描述 Feature Description	恒春海脊位于恒春半岛南部海域，其东侧为北吕宋海槽，西侧为马尼拉海沟，是台湾岛陆上中央山脉向南延伸至水下的部分。其宽度从 20°N 处约 70 千米，向北至恒春半岛南端约 110 千米（图 1–5）。 Hengchun Haiji is located on the sea area in the south of the Hengchun Bandao, with North Luzon Trough on the east, Manila Trench on the west, and is a submarine extension of the central ranges on the land of Taiwan Dao towards the south. Its width varies from about 70 km at 20°N to about 110 km on the south end of the Hengchun Bandao in the north (Fig.1–5).		
命名由来 Origin of Name	该海脊邻近我国台湾省恒春半岛，因此得名。 The Ridge is adjacent to Hengchun Bandao in Taiwan province, China, so the word “Hengchun” was used to name the Ridge.		

(a)

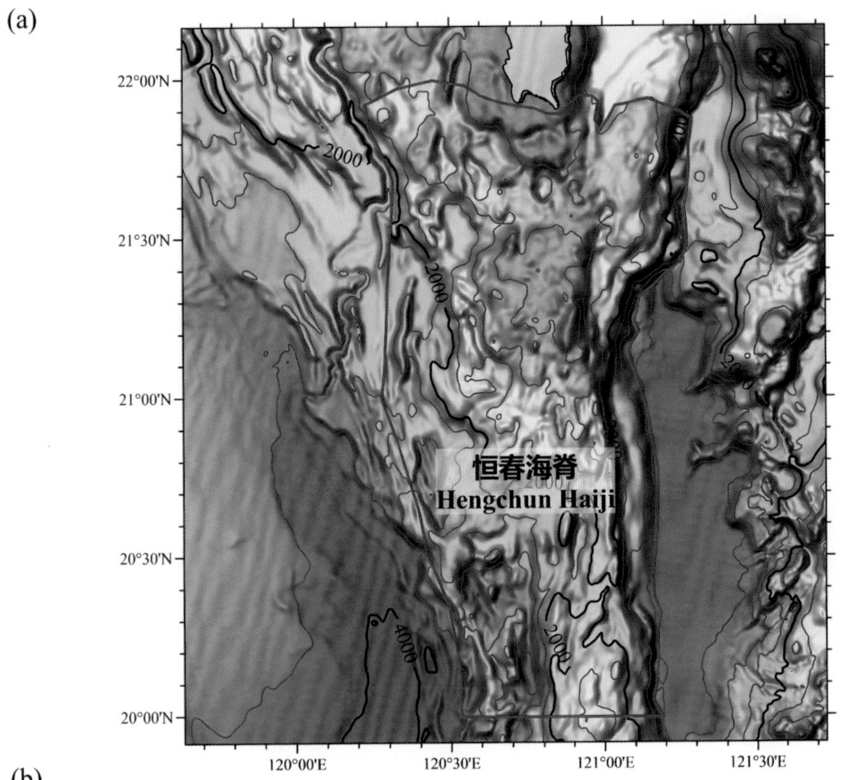

(b)

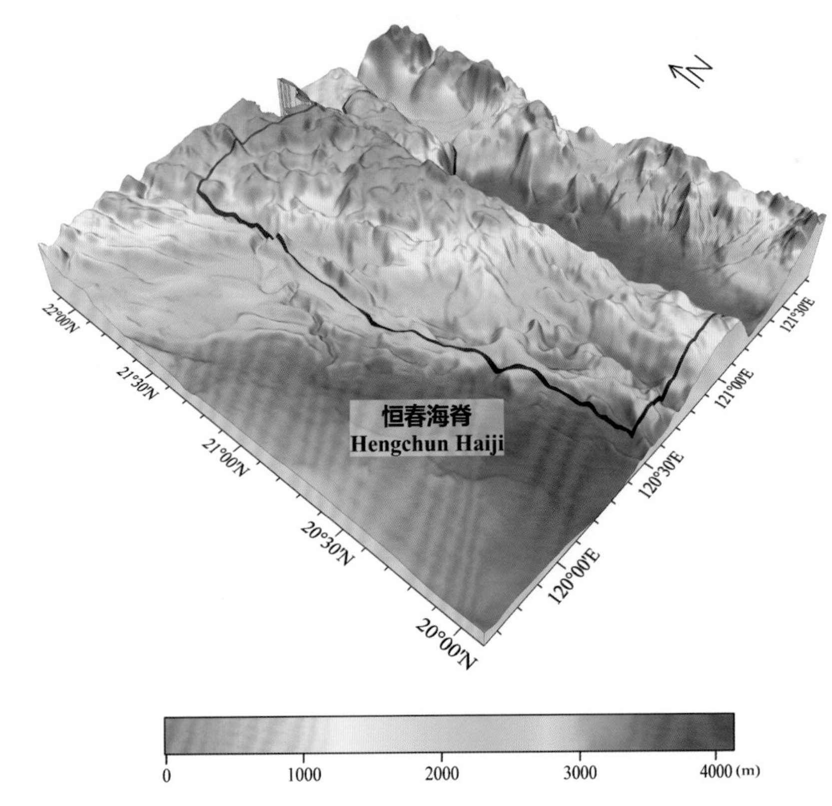

图 1-5　恒春海脊

(a) 海底地形图（等深线间隔 500 米）；(b) 三维海底地形图

Fig.1-5　Hengchun Haiji

(a) Seafloor topographic map (with contour interval of 500 m); (b) 3-D seafloor topographic map

1.10 澎湖海底峡谷群

标准名称 Standard Name	澎湖海底峡谷群 Penghu Haidixiaguqun	类别 Generic Term	海底峡谷群 Canyons
中心点坐标 Center Coordinates	22°00.0′N, 119°30.0′E	规模（千米 × 千米） Dimension（km × km）	130 × 140
最小水深（米） Min Depth (m)	300	最大水深（米） Max Depth (m)	3500
地理实体描述 Feature Description	澎湖海底峡谷群位于南海东北部的上—中陆坡处，北邻澎湖列岛和台湾浅滩，南邻台西南盆地，由数十条北西—北西西向的海底峡谷组成（图 1–6）。 Penghu Haidixiaguqun are located at the upper-middle continental slope in the northeast part of the South China Sea, with their north adjacent to Penghu Liedao and Taiwan Qiantan, and their south adjacent to Taixinan Basin, and are composed of dozens of canyons in the direction of NW–NWW (Fig.1–6).		
命名由来 Origin of Name	该海底峡谷群邻近我国台湾省澎湖列岛，因此得名。 These Canyons are adjacent to Penghu Liedao in Taiwan province, China, so the word “Penghu” was used to name these Canyons.		

1.11 台湾海底峡谷

标准名称 Standard Name	台湾海底峡谷 Taiwan Haidixiagu	类别 Generic Term	海底峡谷 Canyon
中心点坐标 Center Coordinates	21°38.5′N, 118°42.3′E	规模（千米 × 千米） Dimension（km × km）	50 × 210
最小水深（米） Min Depth (m)	200	最大水深（米） Max Depth (m)	3500
地理实体描述 Feature Description	台湾海底峡谷发源于南海东北部的陆架边缘处，其头部由两条分支峡谷沿陆坡呈北西—南东向发育，在 21°43′N, 118°38′E 处汇合形成一条主干峡谷并继续向北西—南东向延伸发育，峡谷总长约 210 千米，峡谷头部和尾部高差达 3000 米（图 1–6）。 Taiwan Haidixiagu is originated from the edge of the continental shelf in northeast of the South China Sea. Two canyons developed along the continental slope in the direction NW–SE at its head, and converge into a main canyon at 21°43′N, 118°38′E, which extends and develops in the direction of NW–SE. The total length of the Canyon is about 210 km and the height difference of the head and the tail of the Canyon is up to 3000 m (Fig.1–6).		
命名由来 Origin of Name	该海底峡谷邻近我国台湾岛，因此得名。 The Canyons is adjacent to Taiwan Dao, so the word “Taiwan” was used to name the Canyon.		

1.12 浦元海山

标准名称 Standard Name	浦元海山 Puyuan Haishan	类别 Generic Term	海山 Seamount
中心点坐标 Center Coordinates	21°07.7′N, 119°13.0′E	规模（千米 × 千米） Dimension（km × km）	10 × 10
最小水深（米） Min Depth (m)	1750	最大水深（米） Max Depth (m)	2850
地理实体描述 Feature Description	浦元海山位于东沙斜坡的东北部，其北侧紧邻台湾海底峡谷，平面形态呈圆形（图 1-6）。 Puyuan Haishan is located in the northeast of Dongsha Xiepo and is adjacent to Taiwan Haidixiagu on the north, with a circular planform (Fig.1−6).		
命名由来 Origin of Name	以中国古代科学家的名字进行地名的团组化命名。该海山以三国时期蜀国著名工匠浦元命名，纪念他在冶炼方面的卓越成就。 A group naming of undersea features after the names of famous ancient Chinese scientists. The Seamount is named after Pu Yuan (?−300 A.D.), a famous craftsman in Shu Kingdom during the Three Kingdoms Period (220−280 A.D.), to commemorate his outstanding achievements in smelting.		

(a)

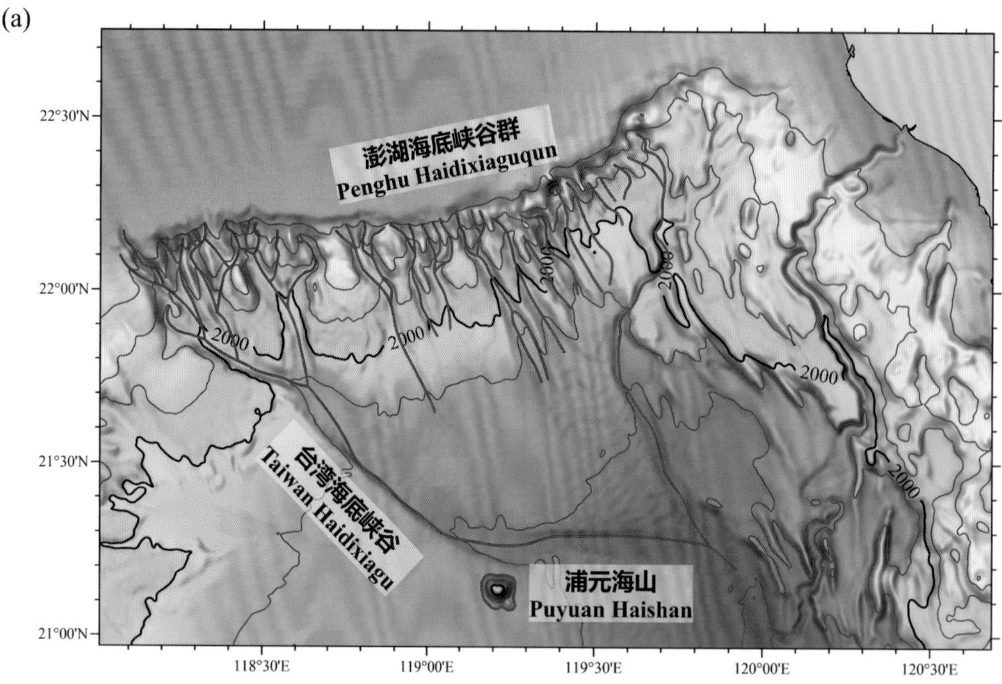

(b)

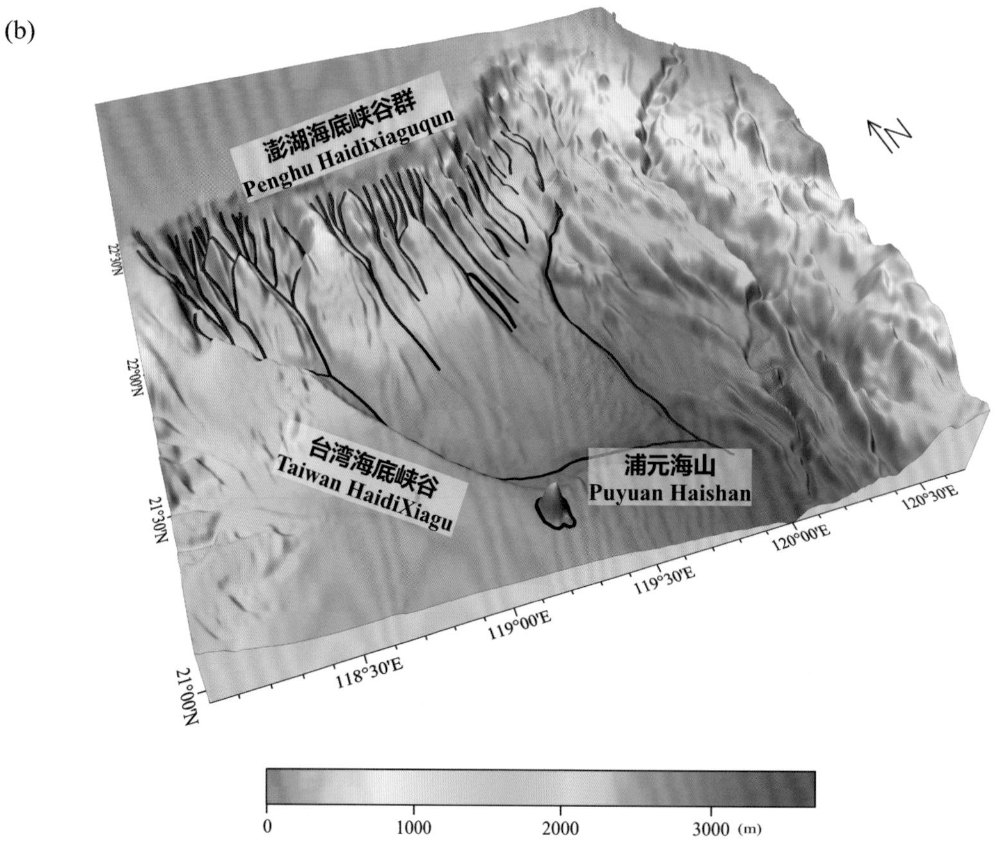

图 1-6　澎湖海底峡谷群、台湾海底峡谷、浦元海山

(a) 海底地形图（等深线间隔 500 米）；(b) 三维海底地形图

Fig.1-6　Penghu Haidixiaguqun, Taiwan Haidixiagu, Puyuan Haishan

(a) Seafloor topographic map (with contour interval of 500 m)；(b) 3-D seafloor topographic map

1.13 笔架海底峡谷群

标准名称 Standard Name	笔架海底峡谷群 Bijia Haidixiaguqun	类别 Generic Term	海底峡谷群 Canyons
中心点坐标 Center Coordinates	21°17.7'N, 117°56.7'E	规模（千米 × 千米） Dimension（km × km）	120 × 100
最小水深（米） Min Depth (m)	450	最大水深（米） Max Depth (m)	2500
地理实体描述 Feature Description	笔架海底峡谷群发育于南海北部陆坡，由数十条北西—南东向的海底峡谷组成，是典型的密集型峡谷群。峡谷群的头部平均水深约 700 米，尾部平均水深约 2500 米（图 1-7）。 Bijia Haidixiaguqun are located in Nanhaibeibu Lupo and are composed of dozens of canyons in the direction of NW-SE, with average depth of about 700 m and 2500 m at the head and at the tail respectively (Fig.1-7).		
命名由来 Origin of Name	该海底峡谷群发育在笔架海丘周边，因此得名。 These Canyons are adjacent to the Bijia Haiqiu, so the word “Bijia” was used to name these Canyons.		

(a)

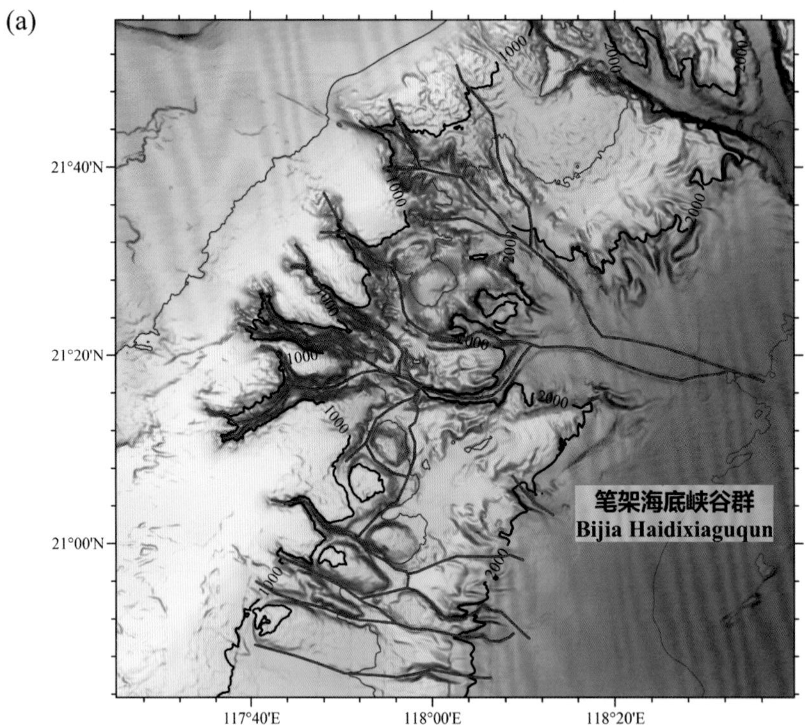

(b)

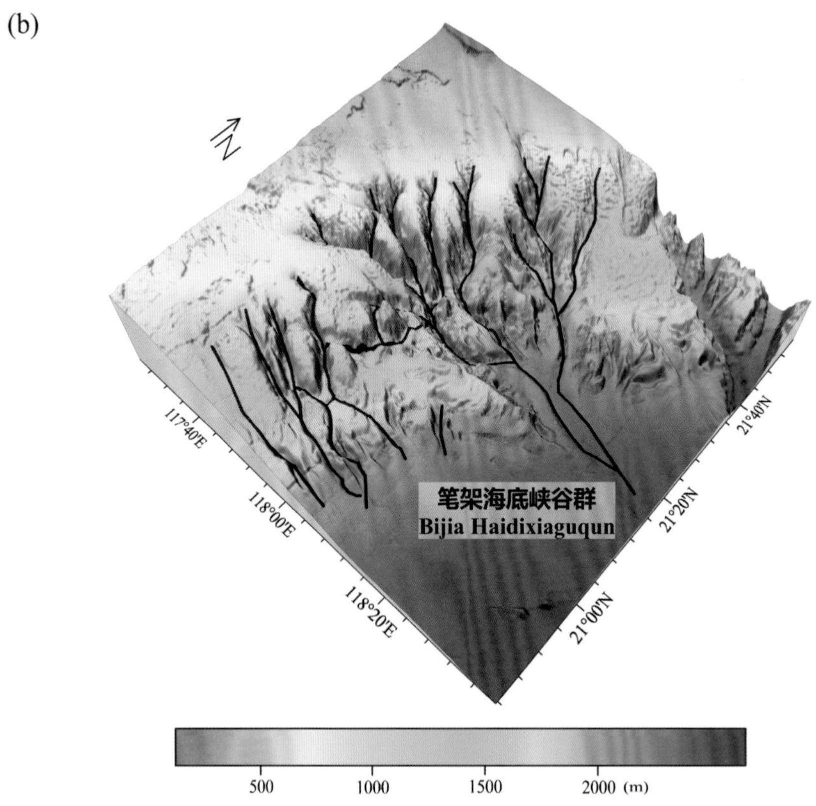

图 1-7　笔架海底峡谷群

(a) 海底地形图（等深线间隔 500 米）；(b) 三维海底地形图

Fig.1-7　Bijia Haidixiaguqun

(a) Seafloor topographic map (with contour interval of 500 m)；(b) 3-D seafloor topographic map

1.14 东沙斜坡

标准名称 Standard Name	东沙斜坡 Dongsha Xiepo	类别 Generic Term	海底斜坡 Slope
中心点坐标 Center Coordinates	21°00.0′N, 118°30.0′E	规模（千米 × 千米） Dimension（km × km）	250 × 300
最小水深（米） Min Depth (m)	400	最大水深（米） Max Depth (m)	3700
地理实体描述 Feature Description	东沙斜坡自东沙海台向东南方向延伸，西接尖峰斜坡，东侧以台湾海底峡谷为界，其北东—南西向宽 250 千米，北西—南东向长约 300 千米，整体呈北西—南东向延伸的堆积地形。斜坡上发育众多的海山、海丘和海底峡谷等（图 1–8）。 Dongsha Xiepo extends towards southeast from Dongsha Haitai and connects to Jianfeng Xiepo in the west. Its east side borders on Taiwan Haidixiagu, with a width of 250 km in the direction of NE–SW, and a length of about 300 km in the direction of NW–SE. Generally, it presents an accumulative relief extending in the direction of NW–SE with multiple seamounts, hills and canyons distributed on it (Fig.1–8).		
命名由来 Origin of Name	该斜坡邻近我国东沙群岛，因此得名。 The Slope is adjacent to Dongsha Qundao of China, so the word “Dongsha” was used to name the Slope.		

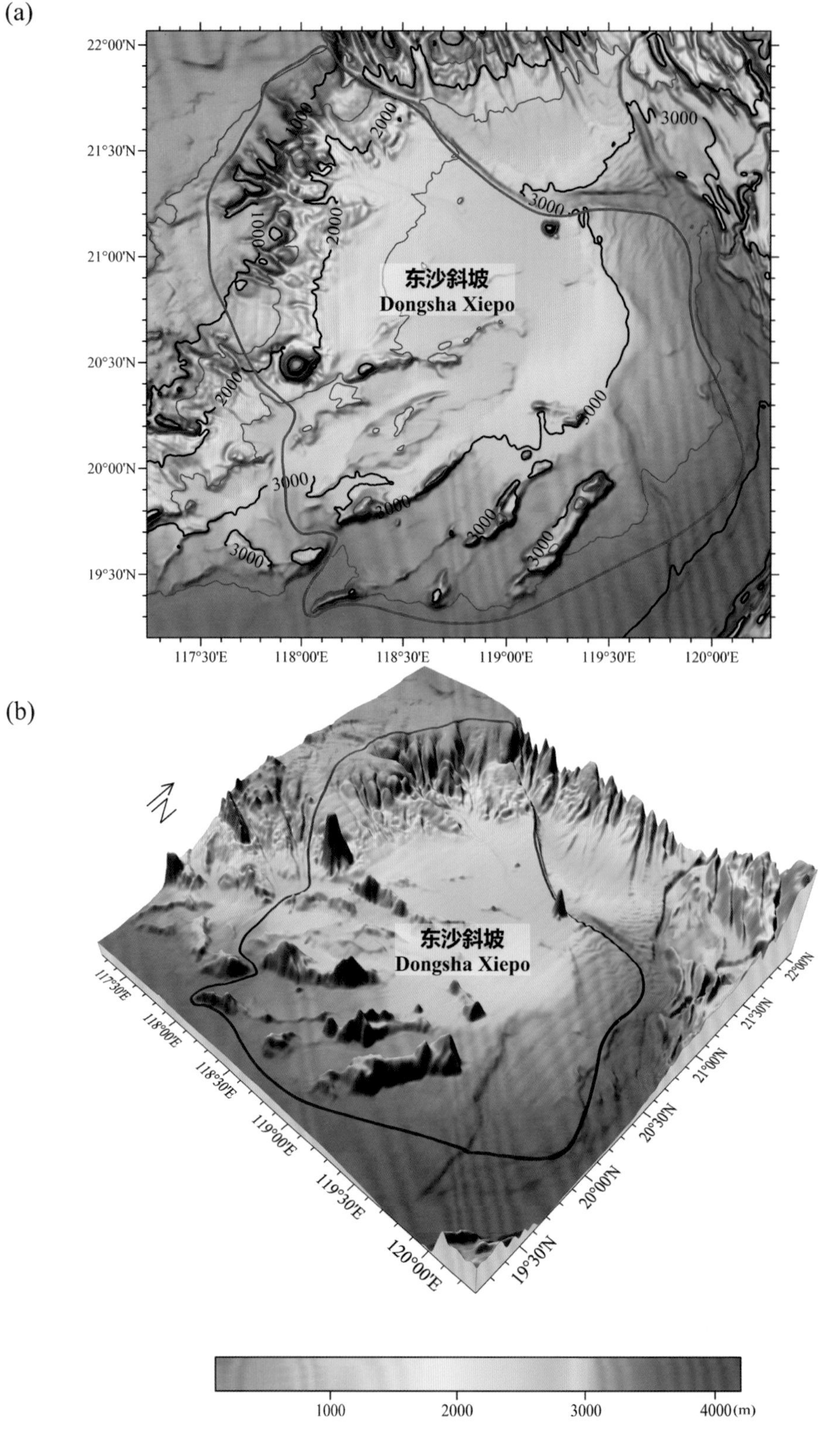

图 1-8　东沙斜坡

(a) 海底地形图（等深线间隔 500 米）；(b) 三维海底地形图

Fig.1-8　Dongsha Xiepo

(a) Seafloor topographic map (with contour interval of 500 m)；(b) 3-D seafloor topographic map

1.15 潮汕海底崖

标准名称 Standard Name	潮汕海底崖 Chaoshan Haidiya	类别 Generic Term	海底崖 Escarpment
中心点坐标 Center Coordinates	21°55.5′N, 117°34.8′E	规模（千米 × 千米） Dimension（km × km）	80 × 2
最小水深（米） Min Depth (m)	120	最大水深（米） Max Depth (m)	130
地理实体描述 Feature Description	潮汕海底崖发育在南海北部陆架边缘处，平面形态呈北东东—南西西向的曲线（图 1–9）。 Chaoshan Haidiya is developed on the edge of Nanhaibeibu Lujia with a curve planform in the direction of NEE—SWW (Fig.1–9).		
命名由来 Origin of Name	该海底崖位于我国广东省潮汕港外海，因此得名。 The Escarpment is located in the sea area off Chaoshan Gang in Guangdong province, China, so the word “Chaoshan” was used to name the Escarpment.		

1.16 东沙北斜坡

标准名称 Standard Name	东沙北斜坡 Dongshabei Xiepo	类别 Generic Term	海底斜坡 Slope
中心点坐标 Center Coordinates	21°25.6′N, 116°41.6′E	规模（千米 × 千米） Dimension（km × km）	120 × 60
最小水深（米） Min Depth (m)	200	最大水深（米） Max Depth (m)	380
地理实体描述 Feature Description	东沙北斜坡发育在南海北部陆坡上部，笔架海底峡谷群西侧，水深自西北向东南方向逐渐变深（图 1–9）。 Dongshabei Xiepo is located in the upper part of Nanhaibeibu Lupo and is on the west of the Bijia Haidixiaguqun, with depth deepening gradually from northwest to southeast (Fig.1–9).		
命名由来 Origin of Name	该斜坡位于我国东沙群岛以北海域，因此得名。 The Slope is located in the sea area to the north of Dongsha Qundao, and “Bei” means north in Chinese, so the word “Dongshabei” was used to name the Slope.		

1.17 东沙北海底崖

标准名称 Standard Name	东沙北海底崖 Dongshabei Haidiya	类别 Generic Term	海底崖 Escarpment
中心点坐标 Center Coordinates	21°25.2'N, 117°03.7'E	规模（千米 × 千米） Dimension（km × km）	110 × 5
最小水深（米） Min Depth (m)	300	最大水深（米） Max Depth (m)	450
地理实体描述 Feature Description	东沙北海底崖发育在南海北部陆坡上部，平面形态呈北东—南西向的折线（图 1-9）。 Dongshabei Haidiya is located in the upper part of Nanhaibeibu Lupo with a planform in the shape of a polyline in the direction of NE-SW (Fig.1-9).		
命名由来 Origin of Name	该海底崖位于我国东沙群岛以北海域，因此得名。 The Escarpment is located in the sea area to the north of Dongsha Qundao, and “Bei” means north in Chinese, so the word “Dongshabei” was used to name the Escarpment.		

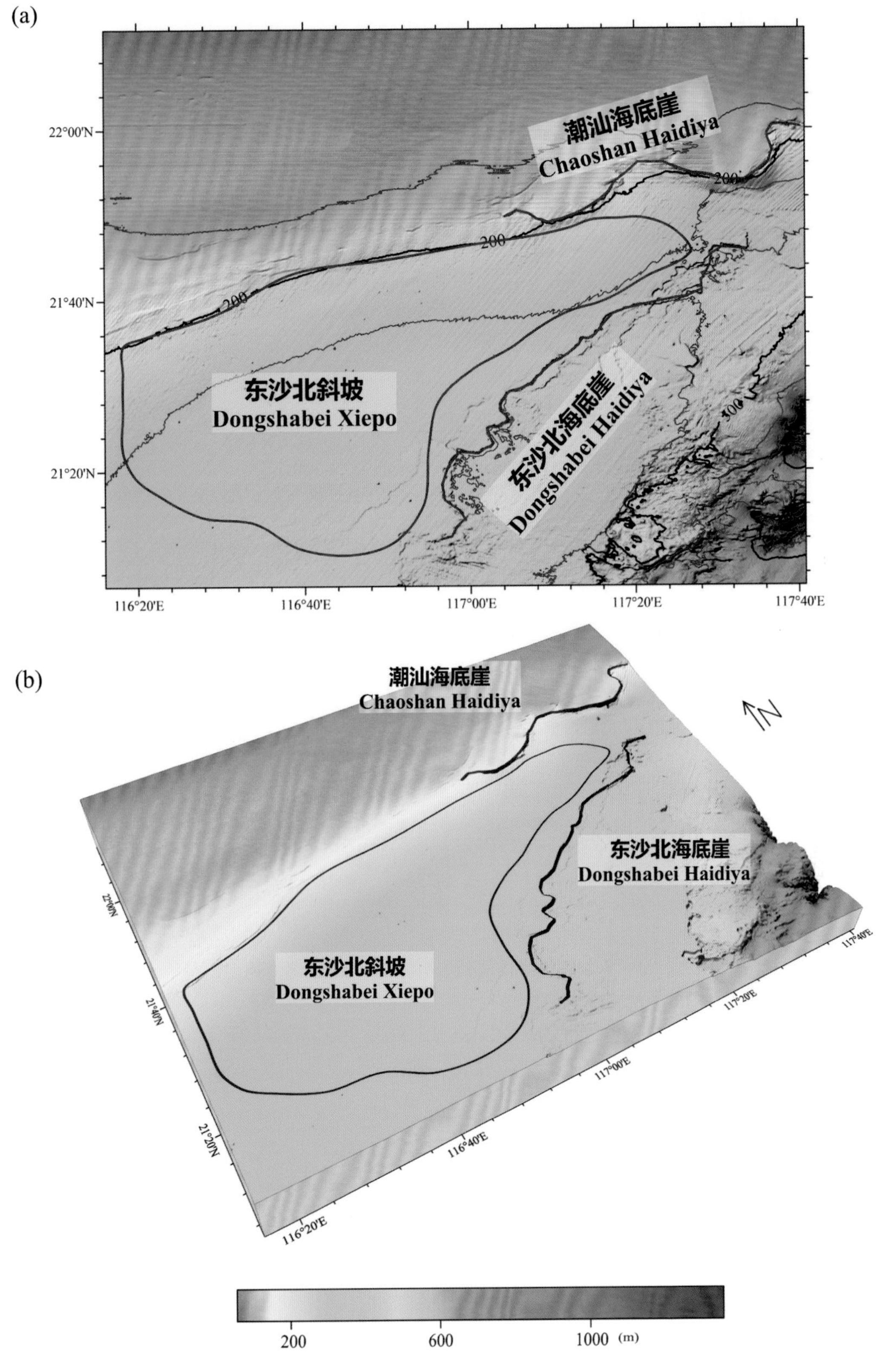

图 1-9 潮汕海底崖、东沙北斜坡、东沙北海底崖

(a) 海底地形图（等深线间隔 100 米）；(b) 三维海底地形图

Fig.1-9 Chaoshan Haidiya, Dongshabei Xiepo, Dongshabei Haidiya

(a) Seafloor topographic map (with contour interval of 100 m); (b) 3-D seafloor topographic map

1.18 北卫海丘

标准名称 Standard Name	北卫海丘 Beiwei Haiqiu	类别 Generic Term	海丘 Hill
中心点坐标 Center Coordinates	21°19.3'N, 115°51.3'E	规模（千米 × 千米） Dimension（km × km）	4 × 4
最小水深（米） Min Depth (m)	100	最大水深（米） Max Depth (m)	180
地理实体描述 Feature Description	北卫海丘位于北卫滩西北侧海域，平面形态近似圆形（图 1–10）。 Beiwei Haiqiu is located in the sea area on the northwest side of Beiwei Tan with a nearly circular planform (Fig.1–10).		
命名由来 Origin of Name	该海丘位于北卫滩周边海域，因此得名。 The Hill is located in the sea area near the Beiwei Tan, so the word "Beiwei" was used to name the Hill.		

1.19 北卫西台地

标准名称 Standard Name	北卫西台地 Beiweixi Taidi	类别 Generic Term	海台 Plateau
中心点坐标 Center Coordinates	21°08.9'N, 115°44.4'E	规模（千米 × 千米） Dimension（km × km）	10 × 3
最小水深（米） Min Depth (m)	100	最大水深（米） Max Depth (m)	140
地理实体描述 Feature Description	北卫西台地位于北卫滩西侧海域，其南部紧邻南卫西台地，平面形态呈东—西走向的椭圆形（图 1–10）。 Beiweixi Taidi is located in the sea area to the west of Beiwei Tan and its south is adjacent to Nanweixi Taidi, with an oval-shaped planform in the direction of E–W (Fig.1–10).		
命名由来 Origin of Name	该台地位于北卫滩以西海域，因此得名。 The Plateau is located in the sea area to the west of Beiwei Tan, and "Xi" means west in Chinese, so the word "Beiweixi" was used to name the Plateau.		

1.20 南卫西台地

标准名称 Standard Name	南卫西台地 Nanweixi Taidi	类别 Generic Term	台地 Plateau
中心点坐标 Center Coordinates	21°06.3'N, 115°41.4'E	规模（千米 × 千米） Dimension（km × km）	10 × 5
最小水深（米） Min Depth (m)	90	最大水深（米） Max Depth (m)	140
地理实体描述 Feature Description	南卫西台地位于南卫滩西北侧海域，其北部紧邻北卫西台地，平面形态呈北西—南东向的椭圆形（图 1–10）。 Nanweixi Taidi is located in the sea area to the northwest of Nanwei Tan and its north is adjacent to Beiweixi Taidi, with an oval-shaped planform in the direction of NW–SE (Fig.1–10).		
命名由来 Origin of Name	该台地位于南卫滩以西海域，因此得名。 The Plateau is located in the sea area to the west Nanwei Tan, and “Xi” means west in Chinese, so the word “Nanweixi” was used to name the Plateau.		

1.21 东沙海底崖

标准名称 Standard Name	东沙海底崖 Dongsha Haidiya	类别 Generic Term	海底崖 Escarpment
中心点坐标 Center Coordinates	21°11.4'N, 115°47.6'E	规模（千米 × 千米） Dimension（km × km）	90 × 1
最小水深（米） Min Depth (m)	140	最大水深（米） Max Depth (m)	180
地理实体描述 Feature Description	东沙海底崖位于东沙群岛西北海域，平面形态呈北东—南西向的曲线（图 1–10）。 Dongsha Haidiya is located in the northwest sea area of Dongsha Qundao, with a curve planform in the direction of NE–SW (Fig.1–10).		
命名由来 Origin of Name	该海底崖位于东沙群岛附近海域，因此得名。 The Escarpment is located in the sea area near Dongsha Qundao, so the word “Dongsha” was used to name the Escarpment.		

(a)

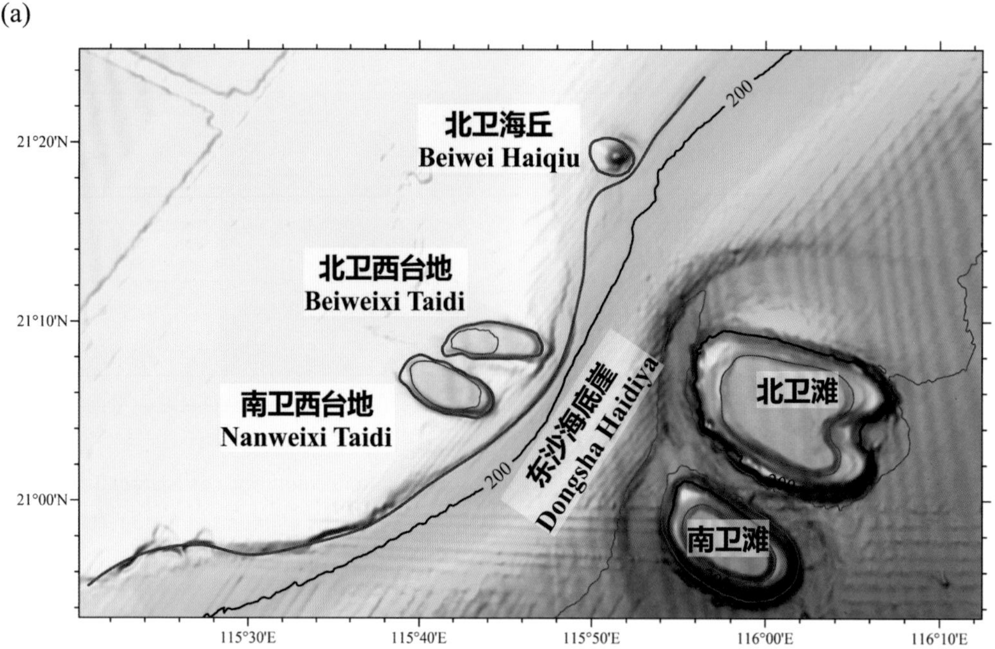

(b)

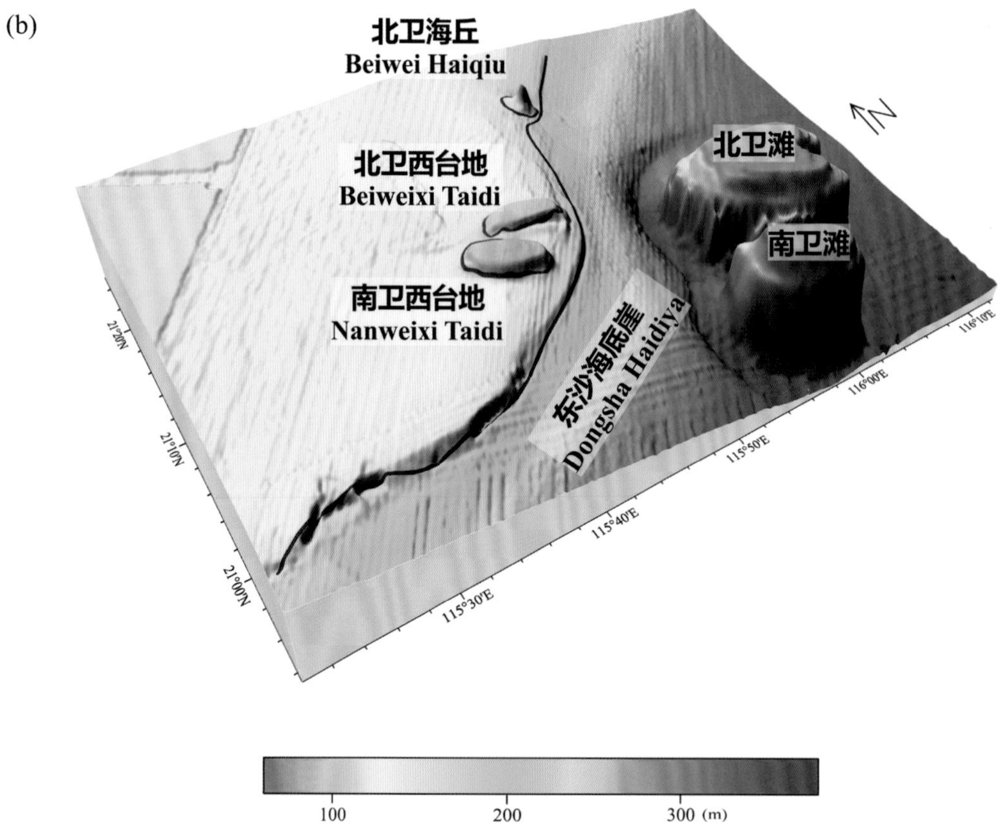

图 1-10　北卫海丘、北卫西台地、南卫西台地、东沙海底崖

(a) 海底地形图（等深线间隔 100 米）；(b) 三维海底地形图

Fig.1-10　Beiwei Haiqiu, Beiweixi Taidi, Nanweixi Taidi, Dongsha Haidiya

(a) Seafloor topographic map (with contour interval of 100 m); (b) 3-D seafloor topographic map

1.22 尖峰斜坡

标准名称 Standard Name	尖峰斜坡 Jianfeng Xiepo	类别 Generic Term	海底斜坡 Slope
中心点坐标 Center Coordinates	19°40.0′N, 117°00.0′E	规模（千米 × 千米） Dimension（km × km）	260 × 200
最小水深（米） Min Depth (m)	400	最大水深（米） Max Depth (m)	3850
地理实体描述 Feature Description	尖峰斜坡为多级台阶式下降地形，其西侧以珠江海谷相隔与一统斜坡相邻，东侧与东沙斜坡相邻，北侧与东沙海台相邻，整体水深自北向南逐渐增大，并最终过渡到南海海盆；斜坡上发育众多的海山、海底峡谷等（图 1-11）。 Jianfeng Xiepo is a multi-step descending terrain. It is adjacent to Yitong Xiepo across Zhujiang Haigu on the west, is adjacent to Dongsha Xiepo on the east and to Dongsha Haitai on the north. Generally, the depth increases gradually from north to south until it finally transitions to Nanhai Haipen. A multitude of seamounts and canyons have been developed on the Slope (Fig.1-11).		
命名由来 Origin of Name	该斜坡上发育尖峰圆丘，故以“尖峰”命名该斜坡。 Jianfeng Yuanqiu is developed on this Slope, so the word “Jianfeng” was used to name the Slope.		

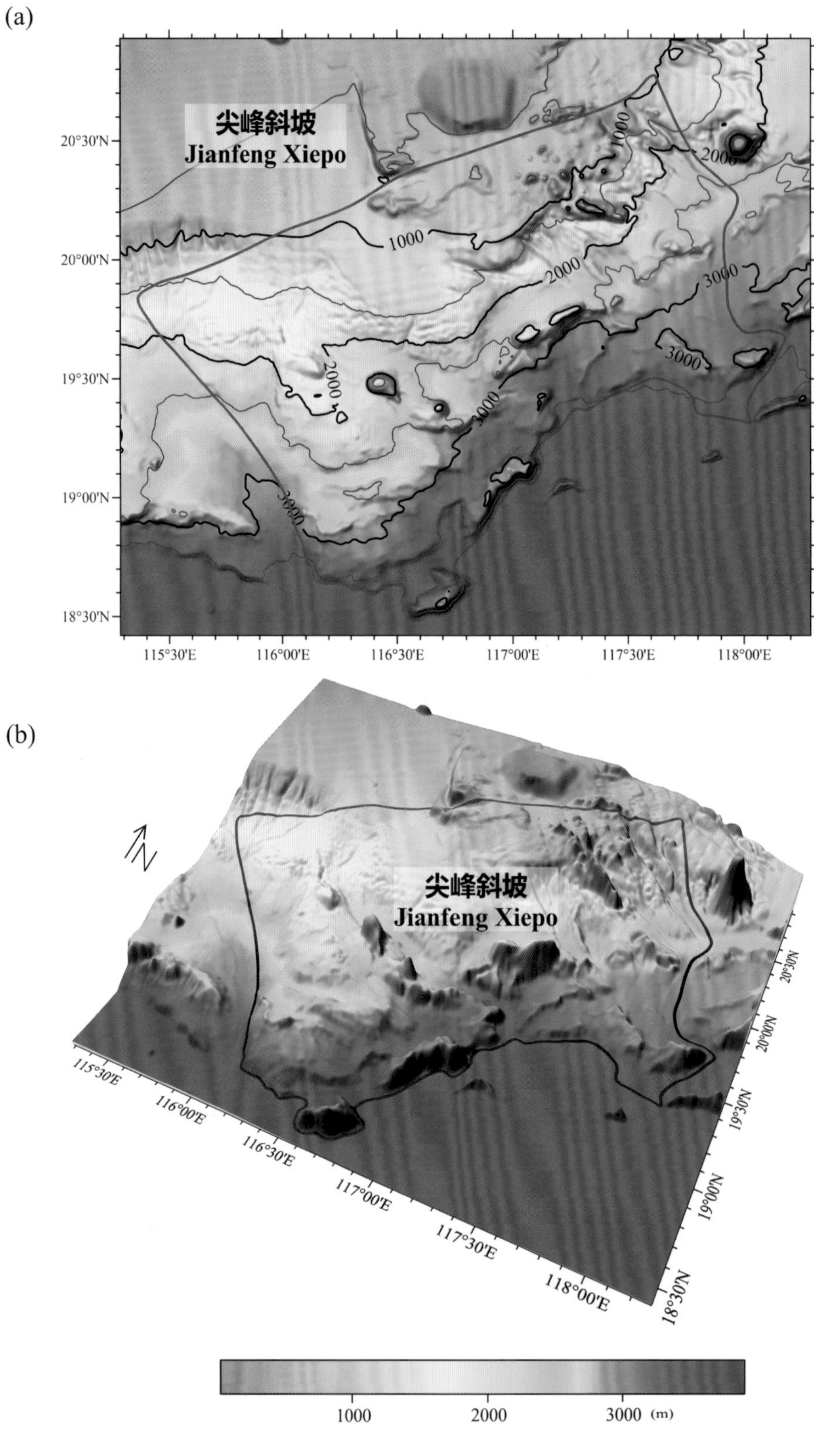

图 1-11　尖峰斜坡

(a) 海底地形图（等深线间隔 500 米）；(b) 三维海底地形图

Fig.1-11　Jianfeng Xiepo

(a) Seafloor topographic map (with contour interval of 500 m)；(b) 3-D seafloor topographic map

1.23 西樵海丘

标准名称 Standard Name	西樵海丘 Xiqiao Haiqiu	类别 Generic Term	海丘 Hill
中心点坐标 Center Coordinates	20°21.5'N, 117°13.0'E	规模（千米 × 千米） Dimension（km × km）	5 × 6
最小水深（米） Min Depth (m)	330	最大水深（米） Max Depth (m)	950
地理实体描述 Feature Description	西樵海丘位于东沙群岛东南侧海域，平面形态呈圆形（图 1–12）。 Xiqiao Haiqiu is located in the sea area on the southeast of Dongsha Qundao with a circular planform (Fig.1–12).		
命名由来 Origin of Name	该海丘位于我国广东省外海，故以广东省名山进行地名的团组化命名。“西樵山”为广东省名山之一。 The Hill is located to the sea area off Guangdong province, so it is named after a group of famous mountains in Guangdong province. “Xiqiao Shan” is one of the famous mountains in Guangdong province, so the word “Xiqiao” was used to name the Hill.		

1.24 丹霞海丘

标准名称 Standard Name	丹霞海丘 Danxia Haiqiu	类别 Generic Term	海丘 Hill
中心点坐标 Center Coordinates	20°18.4'N, 117°10.8'E	规模（千米 × 千米） Dimension（km × km）	10 × 7
最小水深（米） Min Depth (m)	430	最大水深（米） Max Depth (m)	850
地理实体描述 Feature Description	丹霞海丘位于东沙群岛东南侧海域，平面形态呈北西—南东向的椭圆形（图 1–12）。 Danxia Haiqiu is located in the sea area on the southeast of Dongsha Qundao, with an oval-shaped planform in the direction of NW–SE (Fig.1–12).		
命名由来 Origin of Name	该海丘位于我国广东省外海，故以广东省名山进行地名的团组化命名。“丹霞山”为广东省名山之一。 The Hill is located to the sea area off Guangdong province, so it is named after a group of famous mountains in Guangdong province. “Danxia Shan” is one of the famous mountains in Guangdong province, so the word “Danxia” was used to name the Hill.		

1.25 罗浮海丘

标准名称 Standard Name	罗浮海丘 Luofu Haiqiu	类别 Generic Term	海丘 Hill
中心点坐标 Center Coordinates	20°13.5'N, 117°18.7'E	规模（千米 × 千米） Dimension（km × km）	17 × 4
最小水深（米） Min Depth (m)	490	最大水深（米） Max Depth (m)	1400
地理实体描述 Feature Description	罗浮海丘位于东沙群岛东南侧海域，平面形态呈北西—南东向的长条形（图 1−12）。 Luofu Haiqiu is located in the sea area on the southeast of Dongsha Qundao, with a planform in the shape of a long strip in the direction of NW−SE (Fig.1−12).		
命名由来 Origin of Name	该海丘位于我国广东省外海，故以广东省名山进行地名的团组化命名。“罗浮山”为广东省名山之一。 The Hill is located to the sea area off Guangdong province, so it is named after a group of famous mountains in Guangdong province. “Luofu Shan” is one of the famous mountains in Guangdong province, so the word “Luofu” was used to name the Hill.		

(a)

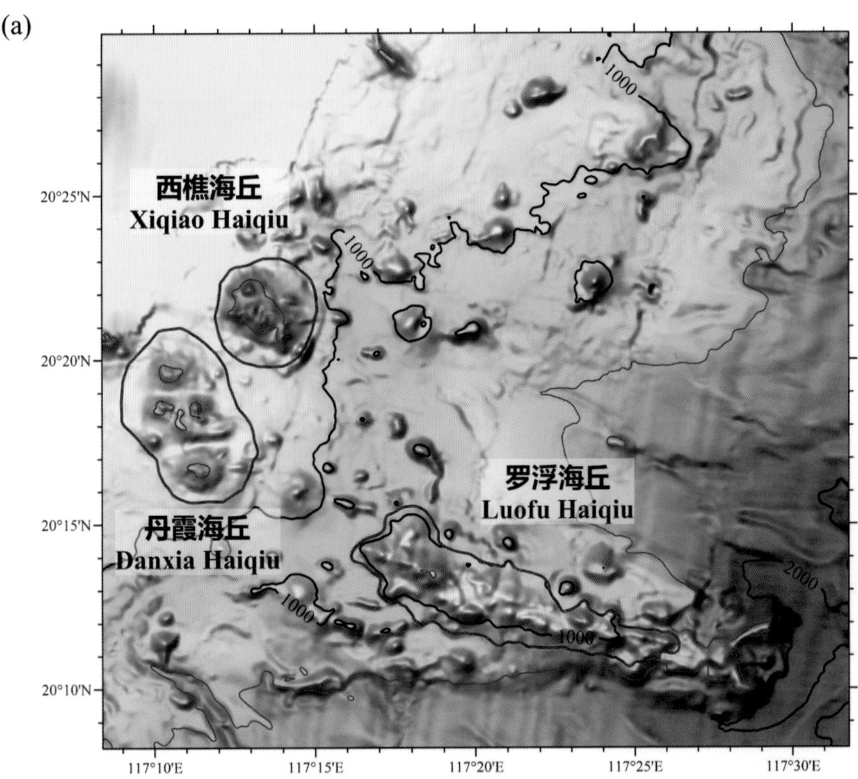

(b)

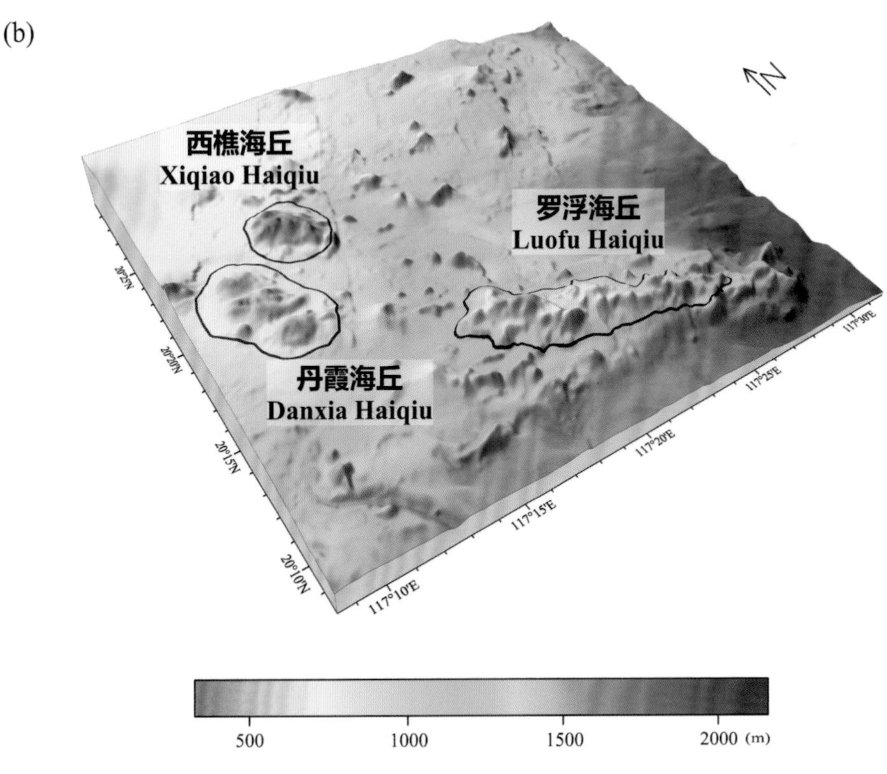

图 1-12　西樵海丘、丹霞海丘、罗浮海丘

(a) 海底地形图（等深线间隔 500 米）；(b) 三维海底地形图

Fig.1-12　Xiqiao Haiqiu, Danxia Haiqiu, Luofu Haiqiu

(a) Seafloor topographic map (with contour interval of 500 m); (b) 3-D seafloor topographic map

1.26 北坡海山

标准名称 Standard Name	北坡海山 Beipo Haishan	类别 Generic Term	海山 Seamount
中心点坐标 Center Coordinates	20°31.1′N, 117°59.0′E	规模（千米 × 千米） Dimension（km × km）	20 × 16
最小水深（米） Min Depth (m)	380	最大水深（米） Max Depth (m)	2150
地理实体描述 Feature Description	北坡海山发育在东沙斜坡上，其顶部非常平坦，边坡陡峭，平面形态呈葫芦形，其东北侧延伸出一小山脊（图 1−13）。 Beipo Haishan is located in Dongsha Xiepo, with a flat top, a steep side slope and a gourd-shaped planform. A small ridge is extending from its northeast side (Fig.1−13).		
命名由来 Origin of Name	1986 年，国务院、外交部和中国地名委员会批准了前地矿部第二海洋地质调查大队（现广州海洋地质调查局）对南海 22 个海底地理实体进行的命名，“北坡海山”是这 22 个海底地理实体名称之一。 In 1986, the State Council, the Ministry of Foreign Affairs and Chinese Toponymy Committee approved 22 undersea feature names in Nanhai named by former Second Marine Geological Survey Brigade of the Ministry of Geology and Mineral Resources (present Guangzhou Marine Geological Survey of China Geological Survey). Beipo Haishan is one of the 22 undersea feature names.		

1.27 北坡南海底峡谷群

标准名称 Standard Name	北坡南海底峡谷群 Beiponan Haidixiaguqun	类别 Generic Term	海底峡谷群 Canyons
中心点坐标 Center Coordinates	20°22.1'N, 117°46.2'E	规模（千米 × 千米） Dimension（km × km）	60 × 40
最小水深（米） Min Depth (m)	950	最大水深（米） Max Depth (m)	2600
地理实体描述 Feature Description	北坡南海底峡谷群发育在尖峰斜坡上，由数条北西—南东向的海底峡谷组成，其头部水深约 950 米，尾部水深约 2600 米（图 1−13）。 Beiponan Haidixiaguqun are located in Jianfeng Xiepo and are composed of several canyons in the direction of NW−SE, with depth of about 950 m and 2600 m at the head and at the tail respectively (Fig.1−13).		
命名由来 Origin of Name	该海底峡谷群发育在北坡海山以南海域，因此得名。 These Canyons are developed in the sea area to the south of Beipo Haishan, and “Nan” means south in Chinese, so the word “Beiponan” was used to name these Canyons.		

1.28 北坡南海丘

标准名称 Standard Name	北坡南海丘 Beiponan Haiqiu	类别 Generic Term	海丘 Hill
中心点坐标 Center Coordinates	20°03.6'N, 117°40.6'E	规模（千米 × 千米） Dimension（km × km）	28 × 13
最小水深（米） Min Depth (m)	1930	最大水深（米） Max Depth (m)	2900
地理实体描述 Feature Description	北坡南海丘发育在尖峰斜坡上，平面形态呈东—西向的椭圆形（图 1–13）。 Beiponan Haiqiu is located in Jianfeng Xiepo with an oval-shaped planform in the direction of E–W (Fig.1–13).		
命名由来 Origin of Name	该海丘是发育在北坡海山以南海域，因此得名。 The Hill is developed in the sea area to the south of Beipo Haishan, and “Nan” means south in Chinese, so the word “Beiponan” was used to name the Hill.		

(a)

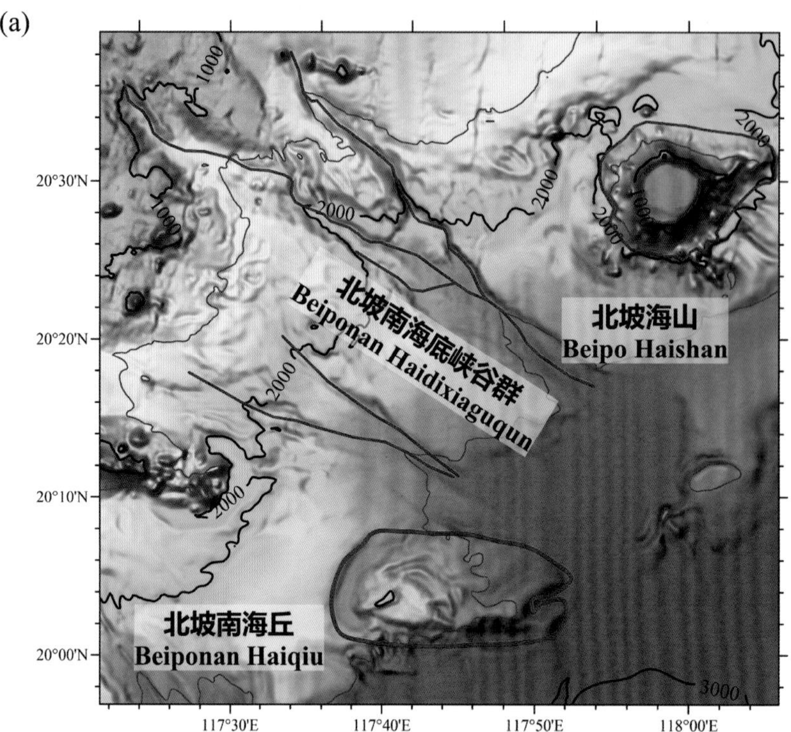

(b)

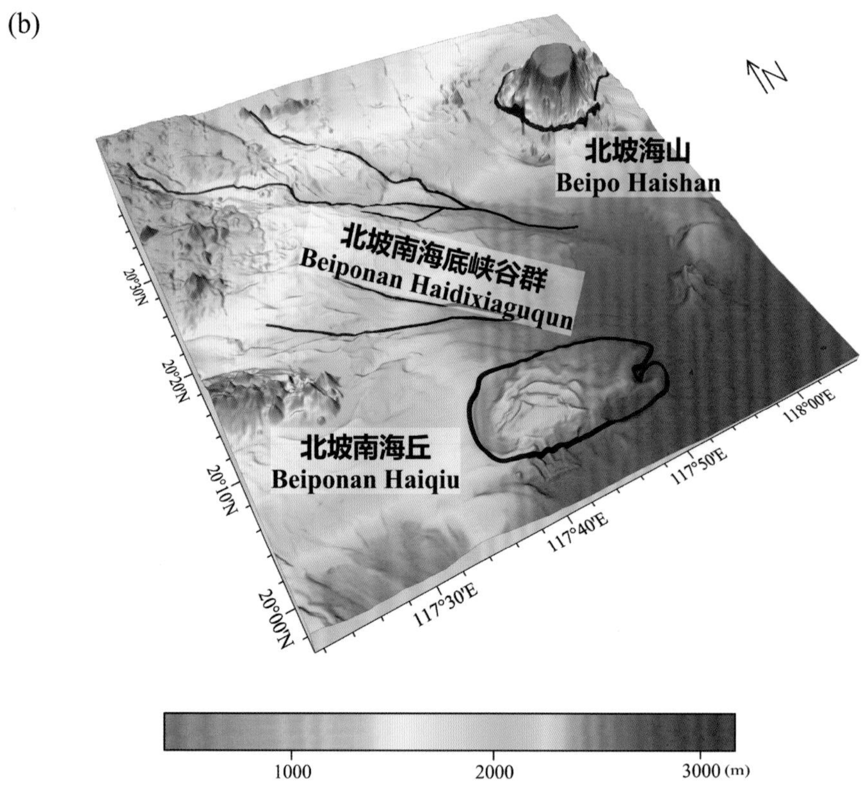

图 1-13　北坡海山、北坡南海底峡谷群、北坡南海丘

(a) 海底地形图（等深线间隔 500 米）；(b) 三维海底地形图

Fig.1-13　Beipo Haishan, Beiponan Haidixiaguqun, Beiponan Haiqiu

(a) Seafloor topographic map (with contour interval of 500 m)；(b) 3-D seafloor topographic map

1.29 李春北海底峡谷群

标准名称 Standard Name	李春北海底峡谷群 Lichunbei Haidixiaguqun	类别 Generic Term	海底峡谷群 Canyons
中心点坐标 Center Coordinates	19°53.7'N, 117°27.7'E	规模（千米 × 千米） Dimension（km × km）	40 × 25
最小水深（米） Min Depth (m)	1400	最大水深（米） Max Depth (m)	2900
地理实体描述 Feature Description	李春北海底峡谷群发育在尖峰斜坡上，由数条北西—南东向的海底峡谷组成，其头部平均水深约 130 米，尾部平均水深约 2900 米（图 1–14）。 Lichunbei Haidixiaguqun are located in Jianfeng Xiepo and are composed of several canyons in the direction of NW–SE, with average depth of about 130 m and about 2900 m at the head and at the tail respectively (Fig.1–14).		
命名由来 Origin of Name	该海底峡谷群发育在李春海山以北海域，因此得名。 These Canyons are developed in the sea area to the north of Lichun Haishan, and “Bei” means north in Chinese, so the word “Lichunbei” was used to name these Canyons.		

1.30 李春海山

标准名称 Standard Name	李春海山 Lichun Haishan	类别 Generic Term	海山 Seamount
中心点坐标 Center Coordinates	19°47.7′N, 117°16.3′E	规模（千米 × 千米） Dimension（km × km）	48 × 19
最小水深（米） Min Depth (m)	1720	最大水深（米） Max Depth (m)	3100
地理实体描述 Feature Description	李春海山发育在尖峰斜坡上，平面形态呈北东—南西向的长条形（图 1–14）。 Lichun Haishan is located in Jianfeng Xiepo, with a planform in the shape of a long strip in the direction of NE–SW (Fig.1–14).		
命名由来 Origin of Name	以中国古代科学家的名字进行地名的团组化命名。该海山以隋朝造桥工匠李春命名，纪念他在桥梁工程方面的卓越成就。 A group naming of undersea features after the names of famous ancient Chinese scientists. The Seamount is named after Li Chun, a bridge builder in Sui Dynasty (581–618 A.D.), to commemorate his outstanding achievements in bridge engineering.		

1.31 李春西海底峡谷

标准名称 Standard Name	李春西海底峡谷 Lichunxi Haidixiagu	类别 Generic Term	海底峡谷 Canyon
中心点坐标 Center Coordinates	19°40.5'N, 116°56.1'E	规模（千米 × 千米） Dimension（km × km）	60 × 11
最小水深（米） Min Depth (m)	2300	最大水深（米） Max Depth (m)	2850
地理实体描述 Feature Description	李春西海底峡谷发育在尖峰斜坡上，整体平面形态呈北东—南西向延伸发育（图 1-14）。 Lichunxi Haidixiagu is located in Jianfeng Xiepo, and stretches from northeast to southwest (Fig.1-14).		
命名由来 Origin of Name	该海底峡谷发育在李春海山以西海域，因此得名。 The Canyon is developed in the sea area to the west of Lichun Haishan, and “Xi” means west in Chinese, so the word “Lichunxi” was used to name the Canyon.		

1.32 东沙南海底峡谷

标准名称 Standard Name	东沙南海底峡谷 Dongshanan Haidixiagu	类别 Generic Term	海底峡谷 Canyon
中心点坐标 Center Coordinates	19°58.1'N, 116°49.9'E	规模（千米 × 千米） Dimension（km × km）	45 × 10
最小水深（米） Min Depth (m)	1230	最大水深（米） Max Depth (m)	2150
地理实体描述 Feature Description	东沙南海底峡谷发育在尖峰斜坡上，整体平面形态呈北西—南东向延伸发育（图 1-14）。 Dongshanan Haidixiagu is located in Jianfeng Xiepo and stretches from northwest to southeast (Fig.1-14).		
命名由来 Origin of Name	该海底峡谷发育在东沙群岛以南海域，因此得名。 The Canyon is developed in the sea area to the south of Dongsha Qundao, and “Nan” means south in Chinese, so the word “Dongshanan” was used to name the Canyon.		

(a)

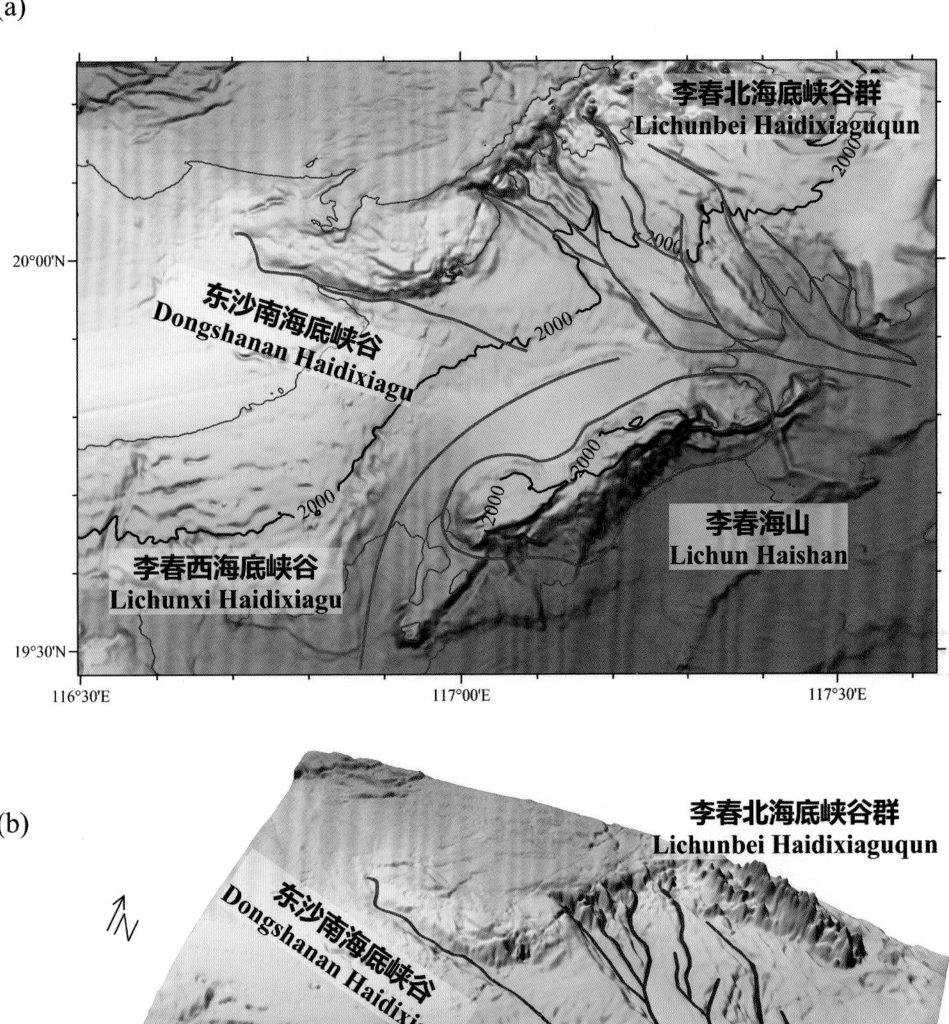

(b)

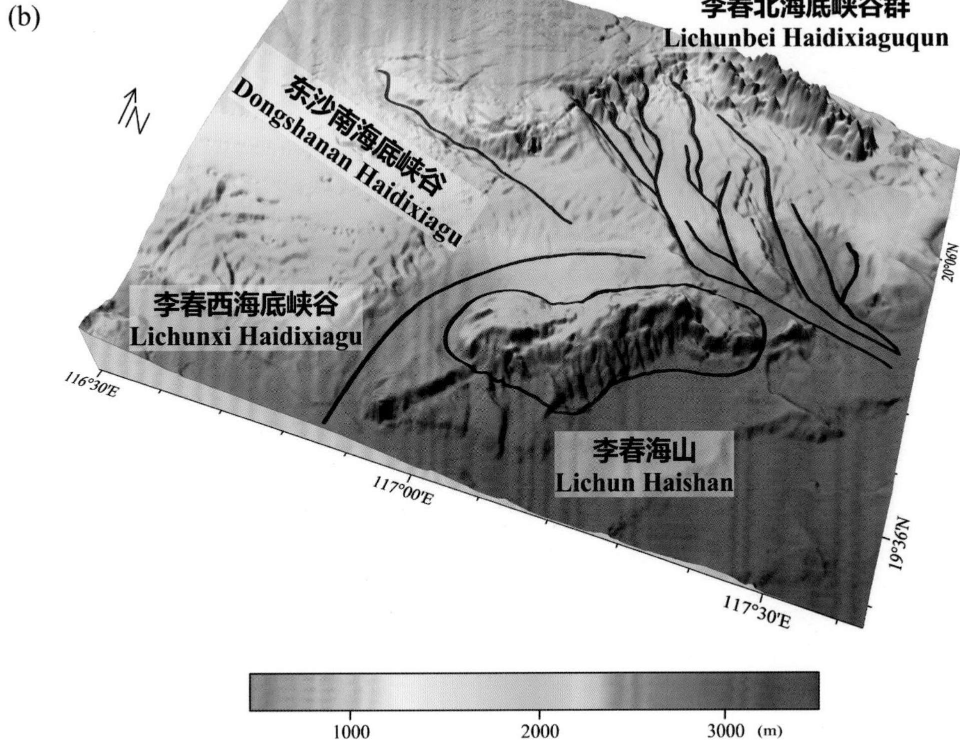

图 1-14　李春北海底峡谷群、李春海山、李春西海底峡谷、东沙南海底峡谷

(a) 海底地形图（等深线间隔 500 米）；(b) 三维海底地形图

Fig.1-14　Lichunbei Haidixiaguqun, Lichun Haishan, Lichunxi Haidixiagu, Dongshanan Haidixiagu

(a) Seafloor topographic map (with contour interval of 500 m); (b) 3-D seafloor topographic map

1.33 沈括海山

标准名称 Standard Name	沈括海山 Shenkuo Haishan	类别 Generic Term	海山 Seamount
中心点坐标 Center Coordinates	19°23.5′N, 116°41.8′E	规模（千米 × 千米） Dimension（km × km）	60 × 15
最小水深（米） Min Depth (m)	1800	最大水深（米） Max Depth (m)	3200
地理实体描述 Feature Description	沈括海山发育在尖峰斜坡上，其西北侧与尖峰圆丘相邻，平面形态近似东—西向展布的长条形，顶部发育多座峰（图 1-15）。 Shenkuo Haishan is located in Jianfeng Xiepo and is adjacent to Jianfeng Yuanqiu on the northwest, with a planform in the shape of a long strip nearly in the direction of E−W. Several peaks have developed on the top of the Seamount (Fig.1−15).		
命名由来 Origin of Name	以中国古代科学家的名字进行地名的团组化命名。该海山以宋代著名科学家沈括命名，纪念其对古代科学技术的卓越贡献。 A group naming of undersea features after the names of famous ancient Chinese scientists. The Seamount is named after Shen Kuo (1031−1095 A.D.), a famous scientist in the Song Dynasty (960−1279 A.D.), to commemorate his outstanding contributions to the ancient scientific technology.		

1.34 尖峰圆丘

标准名称 Standard Name	尖峰圆丘 Jianfeng Yuanqiu	类别 Generic Term	圆丘 Knoll
中心点坐标 Center Coordinates	19°29.6′N, 116°24.7′E	规模（千米 × 千米） Dimension（km × km）	16 × 14
最小水深（米） Min Depth (m)	1391	最大水深（米） Max Depth (m)	2346
地理实体描述 Feature Description	尖峰圆丘发育在尖峰斜坡上，其东南侧与沈括海山相邻，平面形态呈不规则多边形。圆丘的西部和南部地形较陡，东部和北部地形较缓（图 1-15）。 Jianfeng Yuanqiu is located in Jianfeng Xiepo and is adjacent to Shenkuo Haishan on the southeast, with a planform in the shape of an irregular polygon. It is steeper in the west and south and is gentler in the east and north (Fig.1−15).		
命名由来 Origin of Name	1986 年，国务院、外交部和中国地名委员会批准了前地矿部第二海洋地质调查大队（现广州海洋地质调查局）对南海 22 个海底地理实体进行的命名，“尖峰海山”是这 22 个海底地理实体名称之一。2017 年，SCUFN 第 30 次会议上认定该地理实体为圆丘，故更名为“尖峰圆丘”。 In 1986, the State Council, the Ministry of Foreign Affairs and Chinese Toponymy Committee approved 22 undersea feature names in Nanhai named by former Second Marine Geological Survey Brigade of the Ministry of Geology and Mineral Resources (present Guangzhou Marine Geological Survey of China Geological Survey). Jianfeng Haishan is one of the 22 undersea feature names. In 2017, at the 30th SCUFN meeting , the generic term of the feature was identified as a knoll (Yuanqiu in Chinese), so it was renamed as “Jianfeng Yuanqiu”.		

1.35 尖峰北海底峡谷

标准名称 Standard Name	尖峰北海底峡谷 Jianfengbei Haidixiagu	类别 Generic Term	海底峡谷 Canyon
中心点坐标 Center Coordinates	19°33.2′N, 116°32.1′E	规模（千米 × 千米） Dimension（km × km）	120 × 10
最小水深（米） Min Depth (m)	1750	最大水深（米） Max Depth (m)	3300
地理实体描述 Feature Description	尖峰北海底峡谷发育在尖峰斜坡上，其南侧与尖峰圆丘和沈括海山相邻，整体平面形态呈北西—南东走向（图 1–15）。 Jianfengbei Haidixiagu is located in Jianfeng Xiepo and its south side is adjacent to Jianfeng Yuanqiu and Shenkuo Haishan. It generally stretches from northwest to southeast (Fig.1–15).		
命名由来 Origin of Name	该海底峡谷是发育于尖峰斜坡北部海域，因此得名。 The feature develops in the north sea area of Jianfeng Xiepo, and “Bei” means north in Chinese, so the word “Jianfengbei” was used to name the Canyon.		

(a)

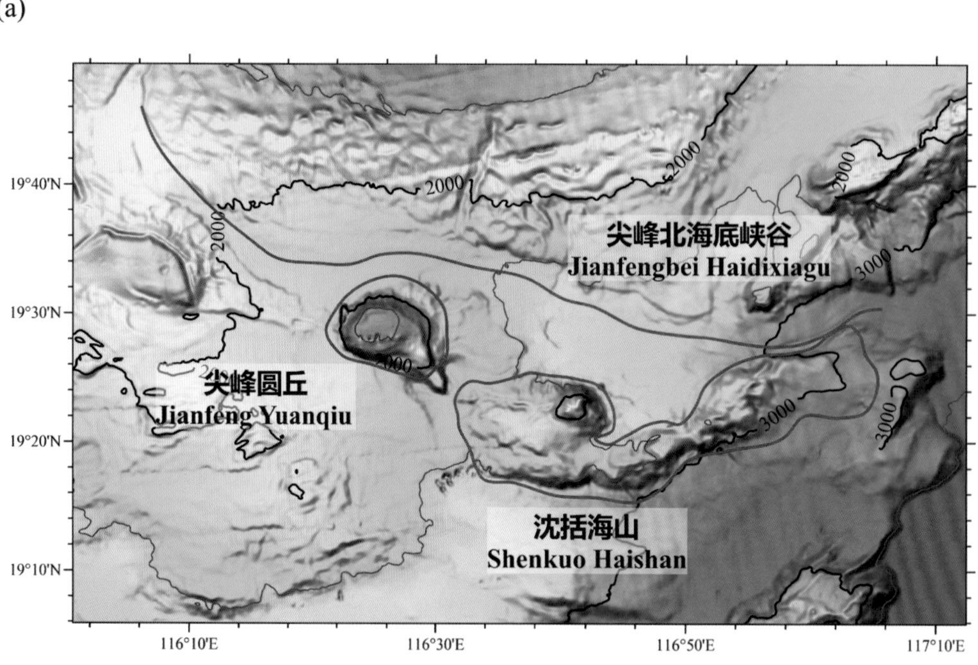

(b)

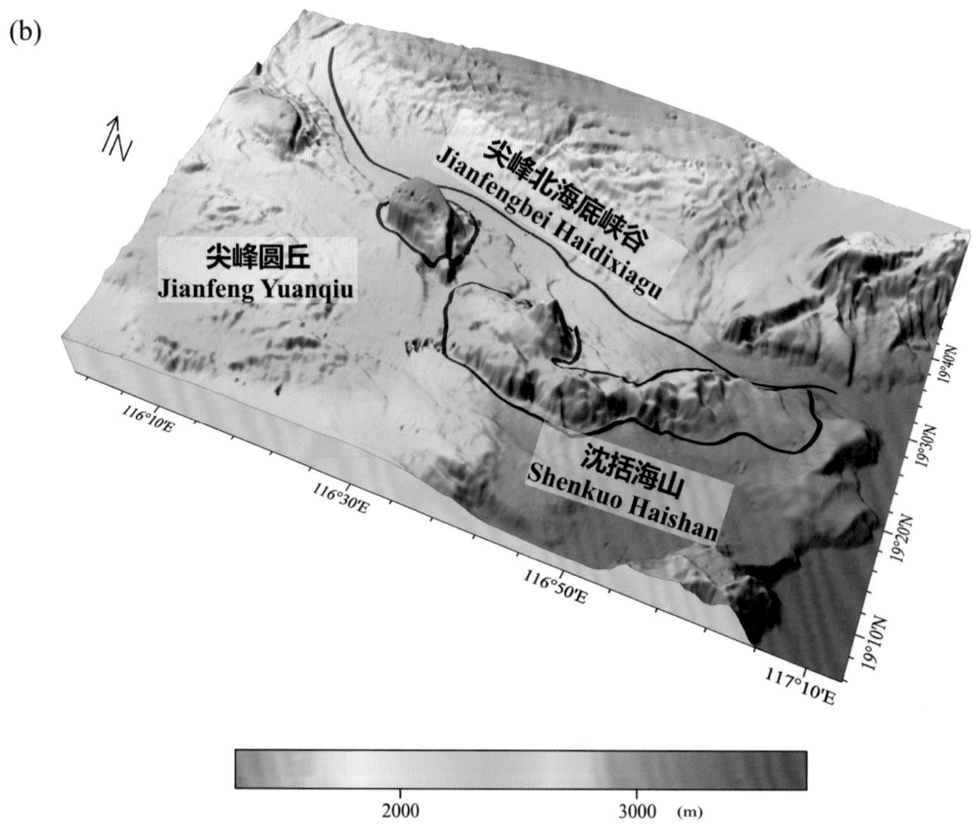

图 1-15　沈括海山、尖峰圆丘、尖峰北海底峡谷

(a) 海底地形图（等深线间隔 500 米）；(b) 三维海底地形图

Fig.1-15　Shenkuo Haishan, Jianfeng Yuanqiu, Jianfengbei Haidixiagu

(a) Seafloor topographic map (with contour interval of 500 m); (b) 3-D seafloor topographic map

1.36 石申海山

标准名称 Standard Name	石申海山 Shishen Haishan	类别 Generic Term	海山 Seamount
中心点坐标 Center Coordinates	18°32.9′N, 116°41.9′E	规模（千米 × 千米） Dimension（km × km）	32 × 20
最小水深（米） Min Depth (m)	2726	最大水深（米） Max Depth (m)	3862
地理实体描述 Feature Description	石申海山发育在尖峰斜坡南端，平面形态呈北东—南西向展布。海山的北坡较缓，南坡较陡（图 1−16）。 Shishen Haishan is located in the south end of Jianfeng Xiepo and stretches in the direction of NE−SW. Its north slope is relatively gentler and its south slope is relatively steeper (Fig.1−16).		
命名由来 Origin of Name	以中国古代科学家的名字进行地名的团组化命名。该海山以战国时期著名科学家石申命名，纪念其在天文学上的重要贡献。 A group naming of undersea features after the names of famous ancient Chinese scientists. The Seamount is named after Shi Shen, a famous scientist in the Warring State period (453 B.C.−221 B.C.), to commemorate his important contribution to the astronomy.		

1.37 杜诗海山

标准名称 Standard Name	杜诗海山 Dushi Haishan	类别 Generic Term	海山 Seamount
中心点坐标 Center Coordinates	19°06.8′N, 117°00.1′E	规模（千米 × 千米） Dimension（km × km）	50 × 15
最小水深（米） Min Depth (m)	2750	最大水深（米） Max Depth (m)	3700
地理实体描述 Feature Description	杜诗海山发育在尖峰斜坡南端，平面形态呈北东—南西向展布，海山的西北坡较缓，东南坡较陡（图 1−16）。 Dushi Haishan is located in the south end of Jianfeng Xiepo and stretches in the direction of NE−SW. Its northwest slope is relatively gentler and its southwest slope is relatively steeper (Fig.1−16).		
命名由来 Origin of Name	以中国古代科学家的名字进行地名的团组化命名。该海山以汉朝著名科学家杜诗命名，纪念其在水利学方面的贡献。 A group naming of undersea features after the names of famous ancient Chinese scientists. The Seamount is named after Du Shi (?−38 A.D.), a famous scientist in the Han Dynasty (202 B.C.−220 A.D.), to commemorate his contribution to the field water conservancy. He invented the hydraulic bellows machine, which can make the air feed into iron smelting furnace to cast agriculture tools, thus significantly increasing the production efficiency. This is one of the most important inventions in the history of world mechanical engineering.		

1.38 秦九韶海丘

标准名称 Standard Name	秦九韶海丘 Qinjiushao Haiqiu	类别 Generic Term	海丘 Hill
中心点坐标 Center Coordinates	19°03.5′N, 117°13.5′E	规模（千米 × 千米） Dimension（km × km）	30 × 10
最小水深（米） Min Depth (m)	3200	最大水深（米） Max Depth (m)	3740
地理实体描述 Feature Description	秦九韶海丘位于杜诗海山东南侧，平面形态呈东—西走向且向北凸出的弯月形（图 1-16）。 Qinjiushao Haiqiu is located on the southeast of Dushi Haishan, with a planform in the shape of a crescent protruding northward in the direction of E-W (Fig.1-16).		
命名由来 Origin of Name	以中国古代科学家的名字进行地名的团组化命名。该海丘以南宋朝著名数学家秦九韶命名，纪念其对世界数学的重要贡献。 A group naming of undersea features after the names of famous ancient Chinese scientists. The Hill is named after Qin Jiushao (1208-1268 A.D.), a famous mathematician in the Southern Song Dynasty (1127-1279 A.D.), to commemorate his important contribution to world's mathematics.		

(a)

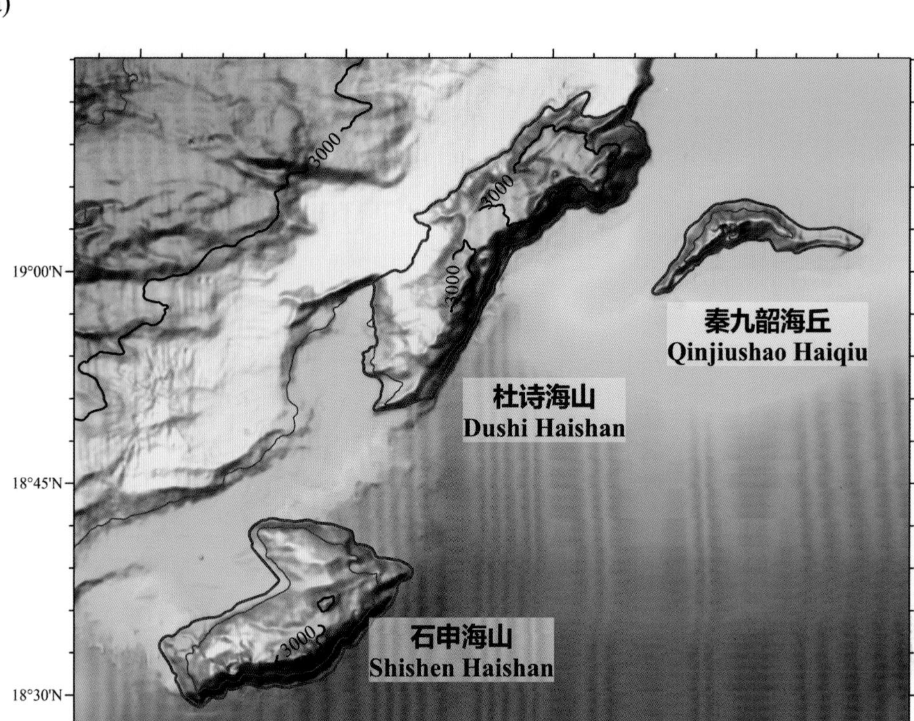

(b)

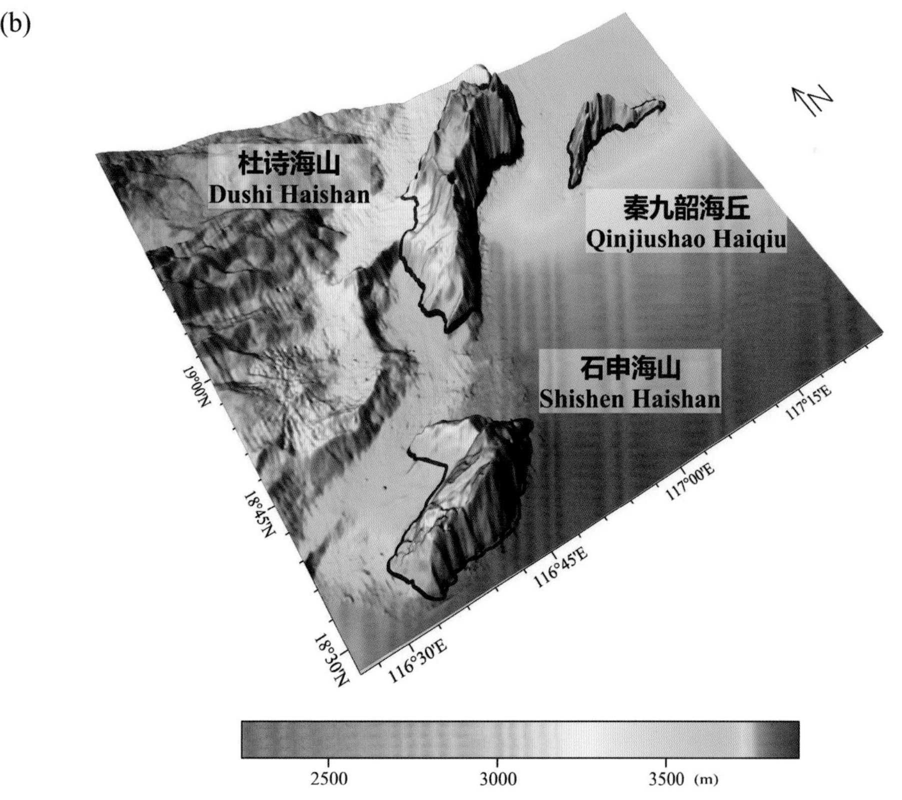

图 1-16　石申海山、杜诗海山、秦九韶海丘

(a) 海底地形图（等深线间隔 500 米）；(b) 三维海底地形图

Fig.1-16　Shishen Haishan, Dushi Haishan, Qinjiushao Haiqiu

(a) Seafloor topographic map (with contour interval of 500 m); (b) 3-D seafloor topographic map

1.39 徐光启海丘

标准名称 Standard Name	徐光启海丘 Xuguangqi Haiqiu	类别 Generic Term	海丘 Hill
中心点坐标 Center Coordinates	19°10.8′N, 117°52.2′E	规模（千米 × 千米） Dimension（km × km）	12 × 5
最小水深（米） Min Depth (m)	3100	最大水深（米） Max Depth (m)	3700
地理实体描述 Feature Description	徐光启海丘位于宋应星海丘的南侧，平面形态近似北东东—南西西向的椭圆形（图 1-17）。 Xuguangqi Haiqiu is located on the south of Songyingxing Haiqiu, with an oval-shaped planform in the direction of NEE–SWW (Fig.1–17).		
命名由来 Origin of Name	以中国古代科学家的名字进行地名的团组化命名。该海丘以明朝著名科学家徐光启命名，纪念其对中西文化交流的贡献。 A group naming of undersea features after the names of famous ancient Chinese scientists. The Hill is named after Xu Guangqi (1562–1633 A.D.), a famous scientist in the Ming Dynasty (1368–1644 A.D.), to commemorate his contribution to the exchange of Chinese and western cultures.		

1.40 宋应星海丘

标准名称 Standard Name	宋应星海丘 Songyingxing Haiqiu	类别 Generic Term	海丘 Hill
中心点坐标 Center Coordinates	19°35.0′N, 118°03.3′E	规模（千米 × 千米） Dimension（km × km）	65 × 28
最小水深（米） Min Depth (m)	2800	最大水深（米） Max Depth (m)	3600
地理实体描述 Feature Description	宋应星海丘发育在尖峰斜坡的东南部，平面形态近似东—西走向且向南凸出的弯月形，其顶部发育多座峰（图 1-17）。 Songyingxing Haiqiu is located in the southeast of Jianfeng Xiepo, with a planform in the shape of a crescent protruding southward nearly in the direction of E–W. Several peaks have developed on its top (Fig.1–17).		
命名由来 Origin of Name	以中国古代科学家的名字进行地名的团组化命名。该海丘以明朝著名科学家宋应星命名，纪念其对工艺学的贡献。 A group naming of undersea features after the names of famous ancient Chinese scientists. The Hill is named after Song Yingxing (1587 A.D.–?), a famous scientist in the Ming Dynasty (1368–1644 A.D.), to commemorate his contribution to the technical science.		

1.41 裴秀海丘

标准名称 Standard Name	裴秀海丘 Peixiu Haiqiu	类别 Generic Term	海丘 Hill
中心点坐标 Center Coordinates	19°24.7′N, 118°14.3′E	规模（千米 × 千米） Dimension（km × km）	33 × 11
最小水深（米） Min Depth (m)	2850	最大水深（米） Max Depth (m)	3670
地理实体描述 Feature Description	裴秀海丘发育在东沙斜坡西南部，南临南海海盆，平面形态呈北东—南西向的长条形（图 1–17）。 Peixiu Haiqiu is located in the southwest of Dongsha Xiepo and is adjacent to Nanhai Haipen on the south, with a planform in the shape of a long strip in the direction of NE–SW (Fig.1–17).		
命名由来 Origin of Name	以中国古代科学家的名字进行地名的团组化命名。该海丘以魏晋时期著名科学家裴秀命名，纪念其制图学方面的贡献。 A group naming of undersea features after the names of famous ancient Chinese scientists. The Hill is named after Pei Xiu (224–271 A.D.), a famous scientist in Wei and Jin Dynasty (220–420 A.D.), to commemorate his contribution to the cartography.		

(a)

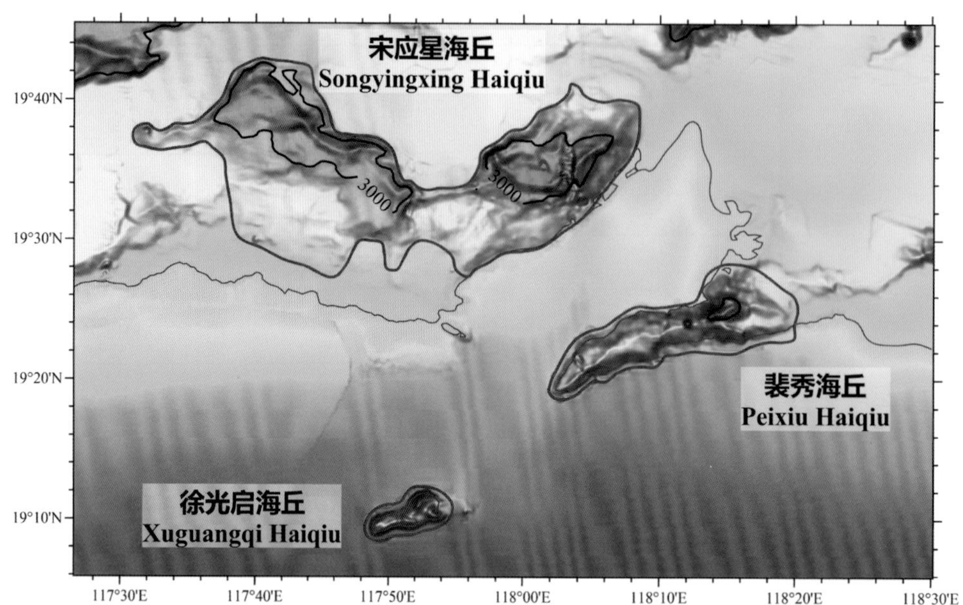

(b)

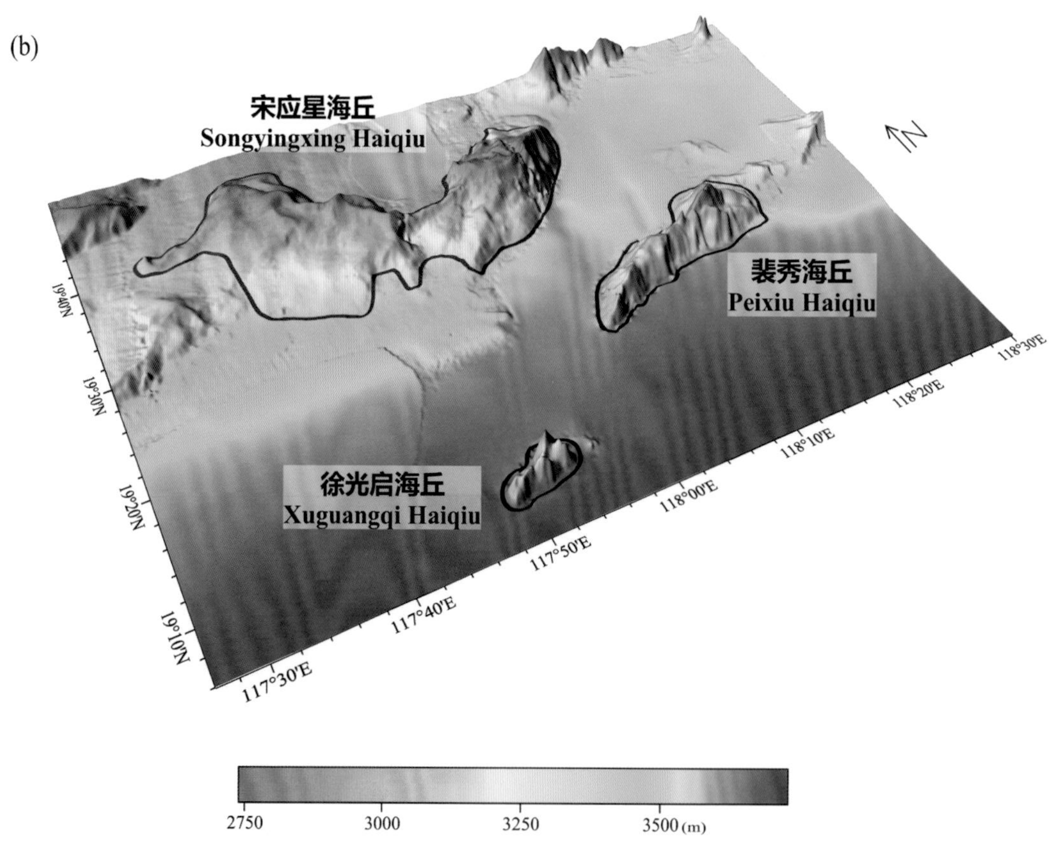

图 1-17　徐光启海丘、宋应星海丘、裴秀海丘

(a) 海底地形图（等深线间隔 500 米）；(b) 三维海底地形图

Fig.1-17　Xuguangqi Haiqiu , Songyingxing Haiqiu, Peixiu Haiqiu

(a) Seafloor topographic map (with contour interval of 500 m)；(b) 3-D seafloor topographic map

1.42 郭守敬海丘

标准名称 Standard Name	郭守敬海丘 Guoshoujing Haiqiu	类别 Generic Term	海丘 Hill
中心点坐标 Center Coordinates	19°22.6′N, 118°39.7′E	规模（千米 × 千米）Dimension（km × km）	22 × 20
最小水深（米）Min Depth (m)	2850	最大水深（米）Max Depth (m)	3580
地理实体描述 Feature Description	郭守敬海丘发育在东沙斜坡南部，平面形态呈北北东—南南西向展布的椭圆形，其顶部发育多座峰（图 1–18）。 Guoshoujing Haiqiu is located in the south of Dongsha Xiepo, with an oval-shaped planform in the direction of NNE–SSW. Several peaks have developed on its top (Fig.1–18).		
命名由来 Origin of Name	以中国古代科学家的名字进行地名的团组化命名。该海丘以元朝著名科学家郭守敬命名，纪念其在天文历法方面的杰出贡献。 A group naming of undersea features after the names of famous ancient Chinese scientists. The Hill is named after Guo Shoujing (1231–1316 A.D.), a famous scientist in the Yuan Dynasty (1271–1368 A.D.), to commemorate his outstanding contributions to the field of astronomical calender.		

1.43 蔡伦海山

标准名称 Standard Name	蔡伦海山 Cailun Haishan	类别 Generic Term	海山 Seamount
中心点坐标 Center Coordinates	19°45.4′N, 118°58.3′E	规模（千米 × 千米）Dimension（km × km）	45 × 13
最小水深（米）Min Depth (m)	2100	最大水深（米）Max Depth (m)	3380
地理实体描述 Feature Description	蔡伦海山发育在东沙斜坡南部，平面形态呈北东—南西向展布的长条形，其顶部发育多座峰（图 1–18）。 Cailun Haishan is located in the south of Dongsha Xiepo, with a planform in the shape of a long strip in the direction of NE–SW. Multiple peaks have developed on its top (Fig.1–18).		
命名由来 Origin of Name	以中国古代科学家的名字进行地名的团组化命名。该海山以汉朝发明造纸术的著名科学家蔡伦命名，纪念其对世界文明进步做出的杰出贡献。 A group naming of undersea features after the names of famous ancient Chinese scientists. The Seamount is named after Cai Lun (61–121 A.D.), a famous scientist who invented paper-making technology in the Han Dynasty (202 B.C.–220 A.D.), to commemorate his outstanding contributions to the progress of world civilization.		

(a)

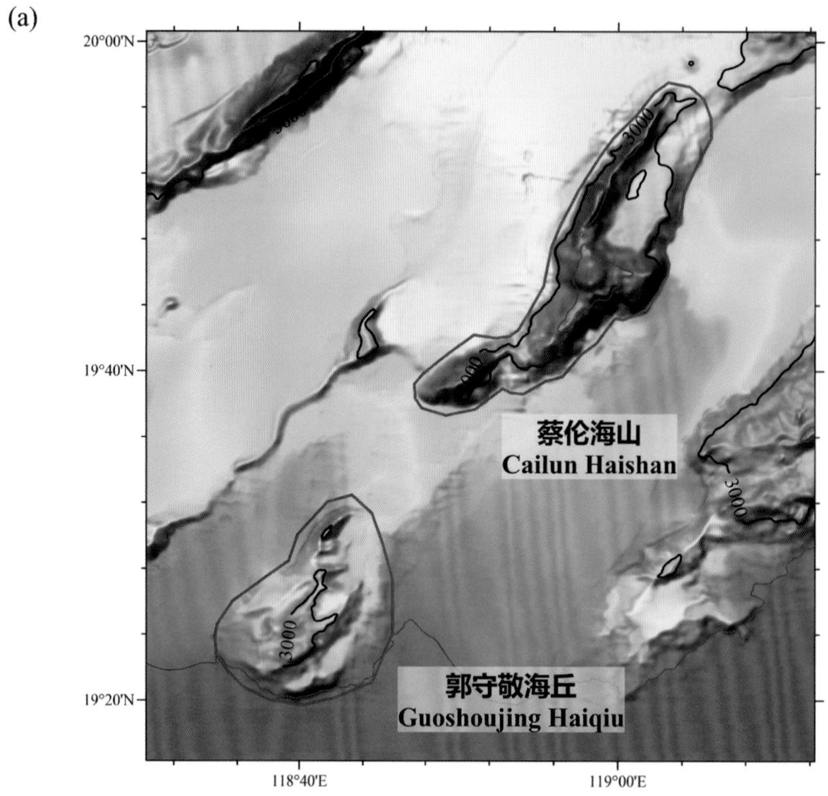

(b)

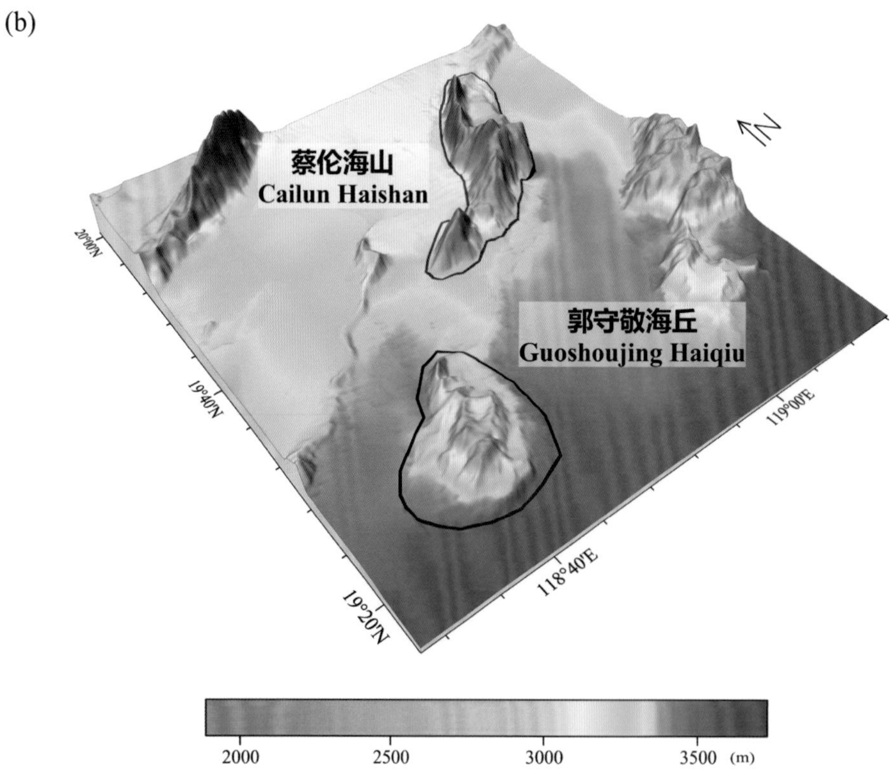

图 1-18 郭守敬海丘、蔡伦海山

(a) 海底地形图（等深线间隔 500 米）；(b) 三维海底地形图

Fig.1-18 Guoshoujing Haiqiu, Cailun Haishan

(a) Seafloor topographic map (with contour interval of 500 m); (b) 3-D seafloor topographic map

1.44 笔架海丘

标准名称 Standard Name	笔架海丘 Bijia Haiqiu	类别 Generic Term	海丘 Hill
中心点坐标 Center Coordinates	20°14.9′N, 118°30.0′E	规模（千米 × 千米） Dimension（km × km）	90 × 30
最小水深（米） Min Depth (m)	2340	最大水深（米） Max Depth (m)	3050
地理实体描述 Feature Description	笔架海丘发育于东沙斜坡南部、丁缓海山西北侧，平面形态呈北东—南西向的扇形。海丘的西北坡较陡，东南坡较缓，顶部发育多座峰（图 1−19）。 Bijia Haiqiu is located in the south of Dongsha Xiepo and on the northwest side of Dinghuan Haishan, with a fan-shaped planform in the direction of NE−SW. Its northwest slope is relatively steeper while its southeast slope is relatively gentler. Several peaks have developed on its top (Fig.1−19).		
命名由来 Origin of Name	1986 年，国务院、外交部和中国地名委员会批准了前地矿部第二海洋地质调查大队（现广州海洋地质调查局）对南海 22 个海底地理实体进行的命名，"笔架海山"是这 22 个海底地理实体名称之一。但最新的多波束海底地形调查测量显示其高差小于 1000 米，故更名为"笔架海丘"。 In 1986, the State Council, the Ministry of Foreign Affairs and Chinese Toponymy Committee approved 22 undersea feature names in Nanhai named by former Second Marine Geological Survey Brigade of the Ministry of Geology and Mineral Resources (present Guangzhou Marine Geological Survey of China Geological Survey). Bijia Haishan is one of the 22 undersea feature names. However, the latest multi-beam bathymetric survey shows that the depth difference is less than 1000 m, so it is renamed as Bijia Haiqiu.		

1.45 丁缓海山

标准名称 Standard Name	丁缓海山 Dinghuan Haishan	类别 Generic Term	海山 Seamount
中心点坐标 Center Coordinates	19°58.0′N, 118°40.4′E	规模（千米 × 千米） Dimension（km × km）	80 × 18
最小水深（米） Min Depth (m)	1900	最大水深（米） Max Depth (m)	3400
地理实体描述 Feature Description	丁缓海山发育于东沙斜坡南部、笔架海丘东南侧，平面形态呈北东—南西向的长条形，其顶部发育多座峰（图 1−19）。 Dinghuan Haishan is located in the south of Dongsha Xiepo and is on the southeast side of Bijia Haiqiu, with a planform in the shape of a long strip in the direction of NE−SW. Several peaks have developed on its top (Fig.1−19).		
命名由来 Origin of Name	以中国古代科学家的名字进行地名的团组化命名。该海山以汉朝著名工匠丁缓命名，纪念其在手工艺方面的贡献。 A group naming of undersea features after the names of famous ancient Chinese scientists. The Seamount is named after Ding Huan, a famous craftsman in the Han Dynasty (202 B.C.−220 A.D.), to commemorate his contribution to the handcraft.		

(a)

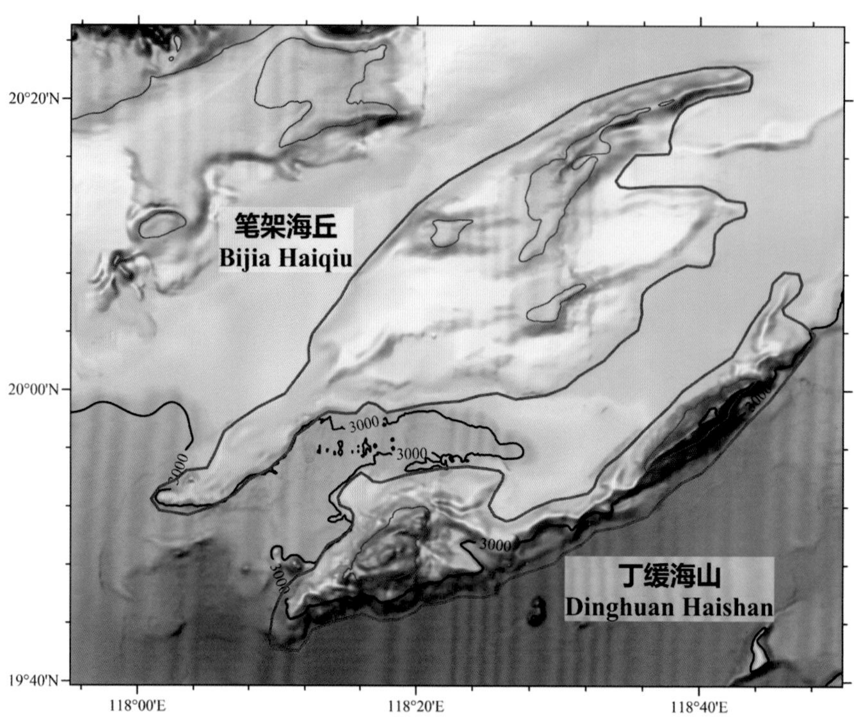

(b)

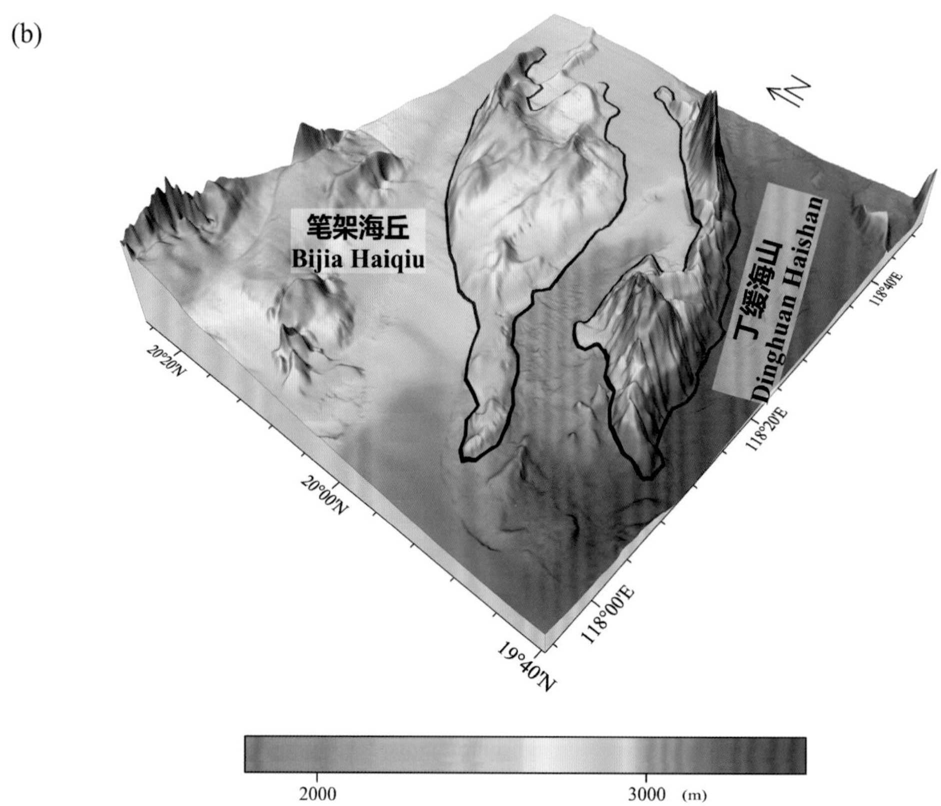

图 1-19 笔架海丘、丁缓海山

(a) 海底地形图（等深线间隔 500 米）；(b) 三维海底地形图

Fig.1-19 Bijia Haiqiu, Dinghuan Haishan

(a) Seafloor topographic map (with contour interval of 500 m)；(b) 3-D seafloor topographic map

1.46 马钧海丘

标准名称 Standard Name	马钧海丘 Majun Haiqiu	类别 Generic Term	海丘 Hill
中心点坐标 Center Coordinates	20°14.9′N, 119°21.3′E	规模（千米 × 千米） Dimension（km × km）	30 × 11
最小水深（米） Min Depth (m)	2350	最大水深（米） Max Depth (m)	3120
地理实体描述 Feature Description	马钧海丘发育于东沙斜坡东南部，平面形态呈北西西—南东东向的长条形。海丘的西侧和东侧各发育一座峰（图 1–20）。 Majun Haiqiu is located in the southwest of Dongsha Xiepo, with a planform in the shape of a long strip in the direction of NWW–SEE. One peak has developed on its west side and east side respectively (Fig.1–20).		
命名由来 Origin of Name	以中国古代科学家的名字进行地名的团组化命名。该海丘以三国时期著名发明家马钧命名，纪念其在机械发明方面的贡献。 A group naming of undersea features after the names of famous ancient Chinese scientists. The Hill is named after Ma Jun, a famous inventor in the Three Kingdoms period (220–280 A.D.), to commemorate his contributions to the field of machinery invention.		

1.47 墨子海山

标准名称 Standard Name	墨子海山 Mozi Haishan	类别 Generic Term	海山 Seamount
中心点坐标 Center Coordinates	19°56.5′N, 119°26.3′E	规模（千米 × 千米） Dimension（km × km）	90 × 18
最小水深（米） Min Depth (m)	2000	最大水深（米） Max Depth (m)	3500
地理实体描述 Feature Description	墨子海山发育于东沙斜坡东南部，平面形态呈北东—南西向的长条形，其顶部发育多座峰（图 1–20）。 Mozi Haishan is located in the southwest of Dongsha Xiepo, with a planform in the shape of a long strip in the direction of NE–SW. Multiple peaks have developed on its top (Fig.1–20).		
命名由来 Origin of Name	以中国古代科学家的名字进行地名的团组化命名。该海山以战国时期著名哲学家和科学家墨子命名。 A group naming of undersea features after the names of famous ancient Chinese scientists. The Seamount is named after Mozi, a famous philosopher and scientist in the Warring States period (475 B.C.–221 B.C.).		

1.48 刘焯海谷

标准名称 Standard Name	刘焯海谷 Liuzhuo Haigu	类别 Generic Term	海谷 Valley
中心点坐标 Center Coordinates	19°52.8′N, 119°43.8′E	规模（千米 × 千米） Dimension（km × km）	110 × 4
最小水深（米） Min Depth (m)	3200	最大水深（米） Max Depth (m)	4200
地理实体描述 Feature Description	刘焯海谷发育于东沙斜坡东南部，为斜坡下端一条北北西—南南东向延伸至深海平原中的谷地，其头部水深约 3200 米，尾部水深约 4200 米（图 1–20）。 Liuzhuo Haigu is located in the southeast of Dongsha Xiepo and is a valley at the bottom of the slope that extends to the abyssal plain in the direction of NNW–SSE, with depth of about 3200 m and about 4200 m at the head and at the tail respectively (Fig.1–20).		
命名由来 Origin of Name	以中国古代科学家的名字进行地名的团组化命名。该海谷以隋朝著名天文学家刘焯命名，纪念他在天文学方面的贡献。 A group naming of undersea features after the names of famous ancient Chinese scientists. The Valley is named after Liu Zhuo (544–610 A.D.), a famous astronomer in the Sui Dynasty (581–618 A.D.), to commemorate his contribution to the astronomy.		

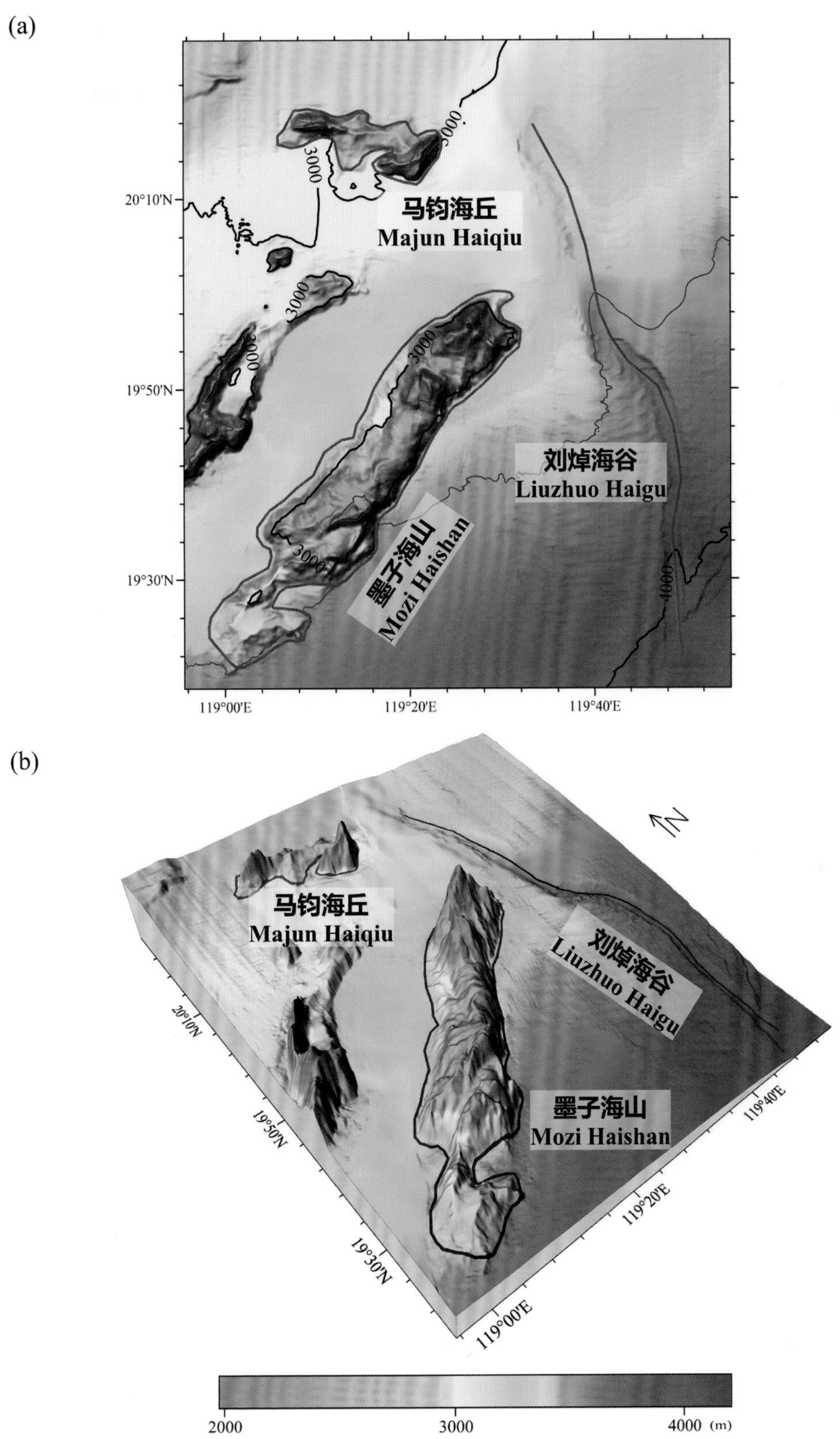

图 1-20 马钧海丘、墨子海山、刘焯海谷

(a) 海底地形图（等深线间隔 500 米）；(b) 三维海底地形图

Fig.1-20 Majun Haiqiu, Mozi Haishan, Liuzhuo Haigu

(a) Seafloor topographic map (with contour interval of 500 m)；(b) 3-D seafloor topographic map

1.49 神狐海底峡谷特征区

标准名称 Standard Name	神狐海底峡谷特征区 Shenhu Haidixiagutezhengqu	类别 Generic Term	海底峡谷特征区 Canyon Province
中心点坐标 Center Coordinates	19°54.0′N, 115°10.0′E	规模（千米 × 千米） Dimension（km × km）	127 × 36
最小水深（米） Min Depth (m)	296	最大水深（米） Max Depth (m)	1714
地理实体描述 Feature Description	神狐海底峡谷特征区发源于南海北部陆架坡脚线附近，由 17 条北北西—南南东向发育的海底峡谷组成，其中最大的峡谷长 38 千米，宽 8 千米。这些峡谷的头部起始于 450 米水深处，尾部终止于 1500 米水深处，长度 30 ~ 60 千米，是典型的陆坡限制型海底峡谷特征区（图 1-21）。 Shenhu Haidixiagutezhengqu is originated from somewhere near the foot line of slope of Nanhaibeibu Lujia and is composed of 17 canyons developed in the direction of NNW–SSE, while the largest one being 38 km long and 8 km wide. The depth is 450 m and 1500 m at the head and at the tail of the Canyons respectively, with lengths of between 30~60 km (Fig.1–21).		
命名由来 Origin of Name	该海底峡谷特征区邻近南海神狐暗沙，因此得名。 The Canyon Province is adjacent to Shenhu Ansha (Shoal), so the word “Shenhu” was used to name the Canyon Province.		

1.50 珠江海谷

标准名称 Standard Name	珠江海谷 Zhujiang Haigu	类别 Generic Term	海谷 Valley
中心点坐标 Center Coordinates	19°30.0′N, 115°20.0′E	规模（千米 × 千米） Dimension（km × km）	250 × 30
最小水深（米） Min Depth (m)	450	最大水深（米） Max Depth (m)	3600
地理实体描述 Feature Description	珠江海谷位于一统斜坡、神狐海底峡谷特征区和尖峰斜坡之间，平面形态呈横“S”型。海谷的头部可上溯至南海北部陆架，尾部与南海海盆相连接，是一条深切南海北部陆坡的凹陷海谷（图 1-21）。 Zhujiang Haigu is located among Yitong Xiepo, Shenhu Haidixiagutezhengqu and Jianfeng Xiepo, with a S-shaped planform. Its head may be traced to Nanhaibeibu Lujia and its tail is connected to Nanhai Haipen. It’s a valley that deeply cut through Nanhaibeibu Lupo (Fig.1–21).		
命名由来 Origin of Name	该海谷发育在珠江口外海，因此得名。 This Valley developed at the sea area off Zhujiang Estuary, so the word “Zhujiang” was used to name the Valley.		

1.51 毕昇海丘

标准名称 Standard Name	毕昇海丘 Bisheng Haiqiu	类别 Generic Term	海丘 Hill
中心点坐标 Center Coordinates	18°55.7′N, 115°21.5′E	规模（千米 × 千米） Dimension（km × km）	80 × 10
最小水深（米） Min Depth (m)	2350	最大水深（米） Max Depth (m)	3000
地理实体描述 Feature Description	毕昇海丘位于珠江海谷南部，整体平面形态呈东—西向的长条形（图 1-21）。 Bisheng Haiqiu is located on the south of Zhujiang Haigu, with an overall planform in the shape of a long strip in the direction of E−W (Fig.1−21).		
命名由来 Origin of Name	以中国古代科学家的名字进行地名的团组化命名。该海丘以宋朝著名科学家毕昇命名，纪念其对活字印刷术技术的贡献。 A group naming of undersea features after the names of famous ancient Chinese scientists. The Hill is named after Bi Sheng (972−1051 A.D.), a famous scientist in the Song Dynasty (960−1279 A.D.), to commemorate his contribution to the typography technology.		

(a)

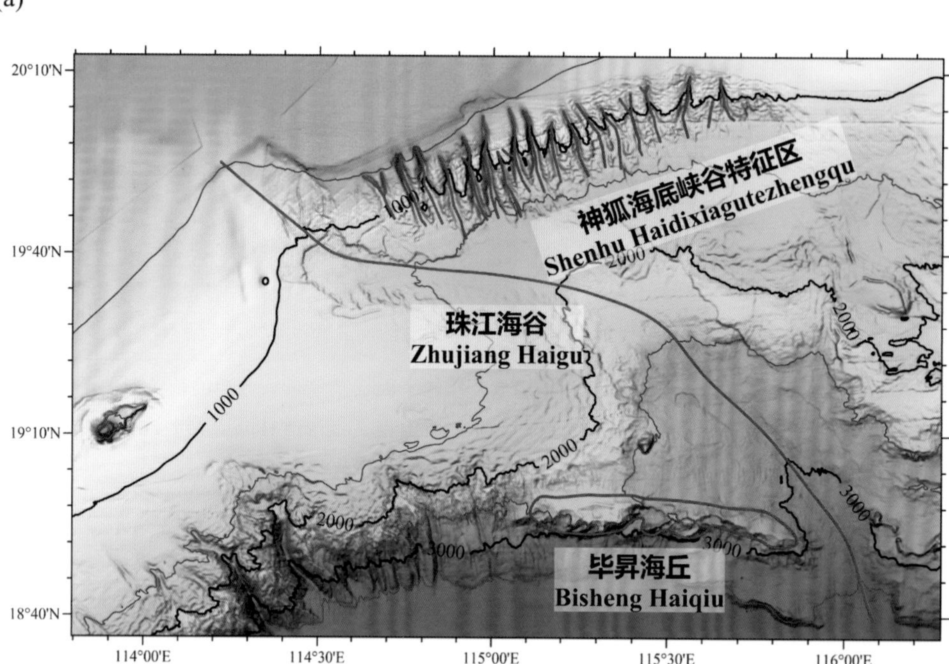

(b)

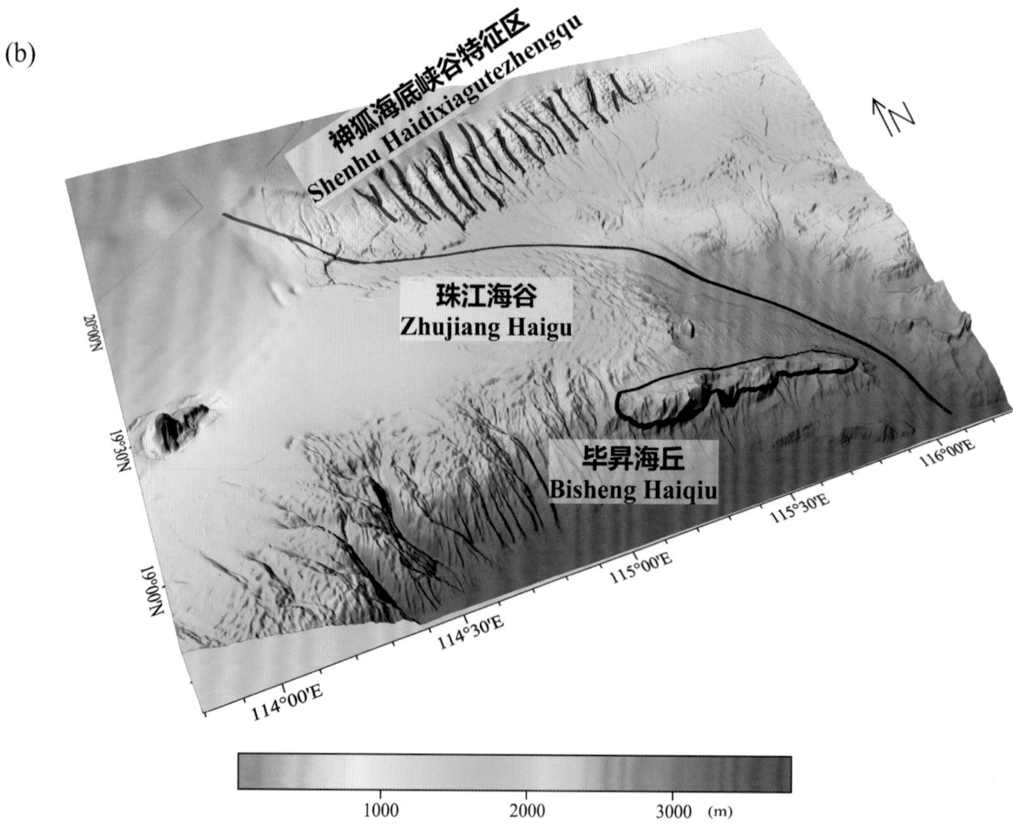

图 1-21　神狐海底峡谷特征区、珠江海谷、毕昇海丘

(a) 海底地形图（等深线间隔 500 米）；(b) 三维海底地形图

Fig.1-21　Shenhu Haidixiagutezhengqu, Zhujiang Haigu, Bisheng Haiqiu

(a) Seafloor topographic map (with contour interval of 500 m)；(b) 3-D seafloor topographic map

1.52 一统斜坡

标准名称 Standard Name	一统斜坡 Yitong Xiepo	类别 Generic Term	海底斜坡 Slope
中心点坐标 Center Coordinates	19°00.0′N, 113°30.0′E	规模（千米 × 千米） Dimension（km × km）	410 × 140
最小水深（米） Min Depth (m)	60	最大水深（米） Max Depth (m)	2040
地理实体描述 Feature Description	一统斜坡是发育在南海北部陆坡中部的一个较大范围的地貌单元，水深自西北向东南方向逐渐加大。斜坡的地形相对平坦，与南海其他陆坡区陡峭的地形特征差异甚大（图 1−22）。 Yitong Xiepo is a large-extent geomorphic unit in the middle of Nanhaibeibu Lupo, with the depth increasing gradually in the direction of NW−SE. Its topography is relative flat, and is distinctively different from the topographic features of other steep continental slopes in the South China Sea (Fig.1−22).		
命名由来 Origin of Name	该斜坡之上发育一统暗沙，故以“一统”命名该斜坡。 As Yitong Ansha has developed on the Slope, so the word “Yitong” was used to name the Slope.		

(a)

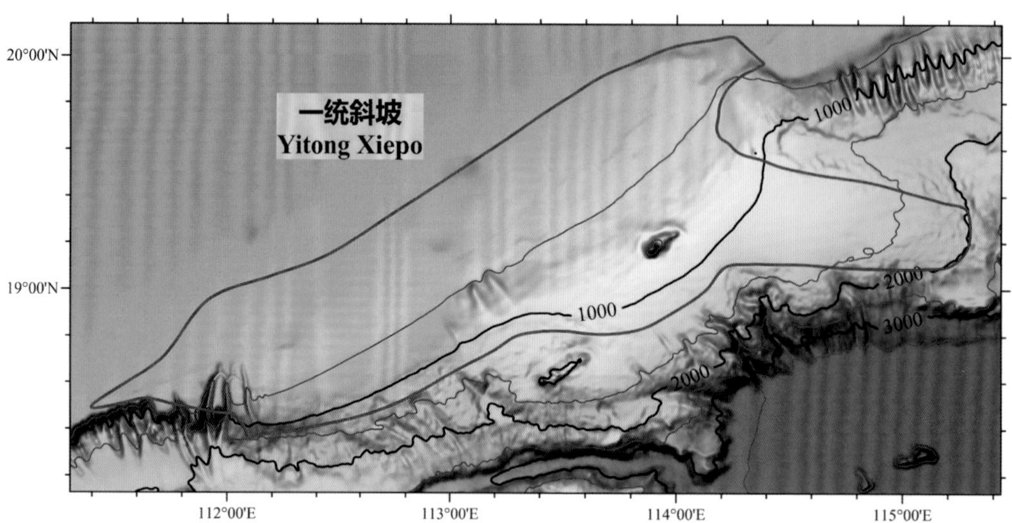

(b)

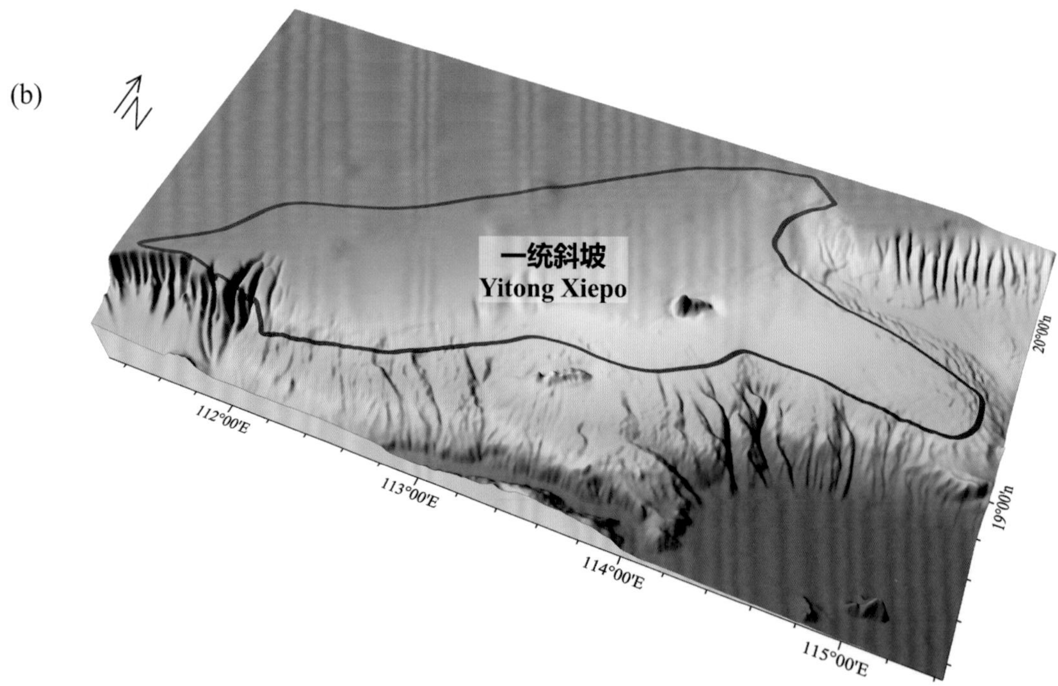

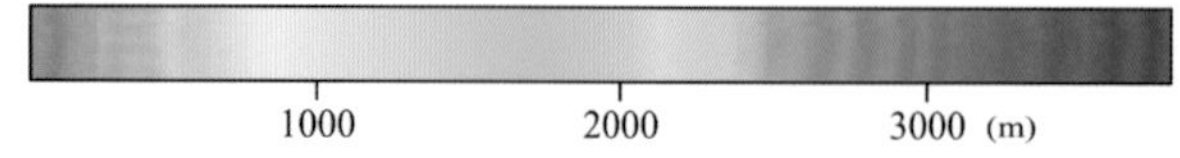

图 1-22　一统斜坡

(a) 海底地形图（等深线间隔 500 米）；(b) 三维海底地形图

Fig.1-22　Yitong Xiepo

(a) Seafloor topographic map (with contour interval of 500 m); (b) 3-D seafloor topographic map

1.53 一统海底峡谷群

标准名称 Standard Name	一统海底峡谷群 Yitong Haidixiaguqun	类别 Generic Term	海底峡谷群 Canyons
中心点坐标 Center Coordinates	18°50.0′N, 114°30.0′E	规模（千米 × 千米） Dimension（km×km）	140 × 62
最小水深（米） Min Depth (m)	1100	最大水深（米） Max Depth (m)	3611
地理实体描述 Feature Description	一统海底峡谷群发育在一统斜坡南部，由多条北北西—南南东向和北西—南东向发育的海底峡谷组成，大体上平行排列（图 1−23）。 Yitong Haidixiaguqun are located on the south of Yitong Xiepo and are composed of several canyons that have developed in the directions of NNW−SSE and NW−SE and are roughly parallel in arrangement (Fig.1−23).		
命名由来 Origin of Name	该海底峡谷群邻近一统暗沙，因此得名。 These Canyons are adjacent to Yitong Ansha, so the word “Yitong” was used to name these Canyons.		

1.54 张衡海丘

标准名称 Standard Name	张衡海丘 Zhangheng Haiqiu	类别 Generic Term	海丘 Hill
中心点坐标 Center Coordinates	18°36.4′N, 113°24.4′E	规模（千米 × 千米） Dimension（km×km）	30 × 10
最小水深（米） Min Depth (m)	480	最大水深（米） Max Depth (m)	1300
地理实体描述 Feature Description	张衡海丘发育在一统斜坡南部，整体平面形态呈北东—南西向的椭圆形（图 1−23）。 Zhangheng Haiqiu is located in the south of Yitong Xiepo, with an overall planform in the shape of an oval in the direction of NE−SW (Fig.1−23).		
命名由来 Origin of Name	以中国古代科学家的名字进行地名的团组化命名。该海丘以汉朝著名科学家张衡命名，纪念其在天文学、机械技术、地震学方面的贡献。 A group naming of undersea features after the names of famous ancient Chinese scientists. The Hill is named after Zhang Heng (78−139 A.D.), a famous scientist in the Han Dynasty (202 B.C.−220 A.D.), to commemorate his contribution to the astronomy, mechanical technology and seismology.		

(a)

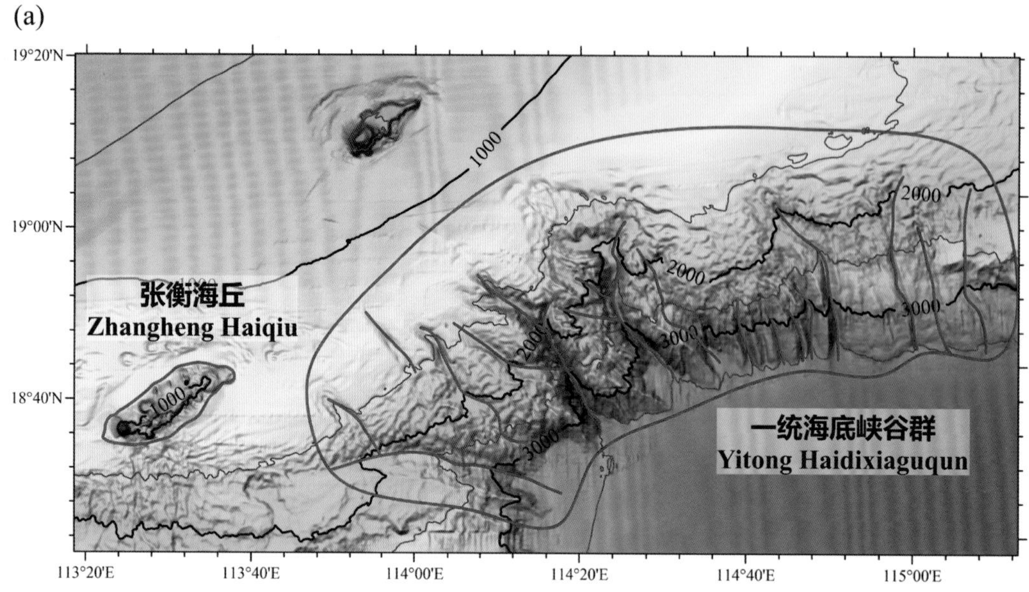

(b)

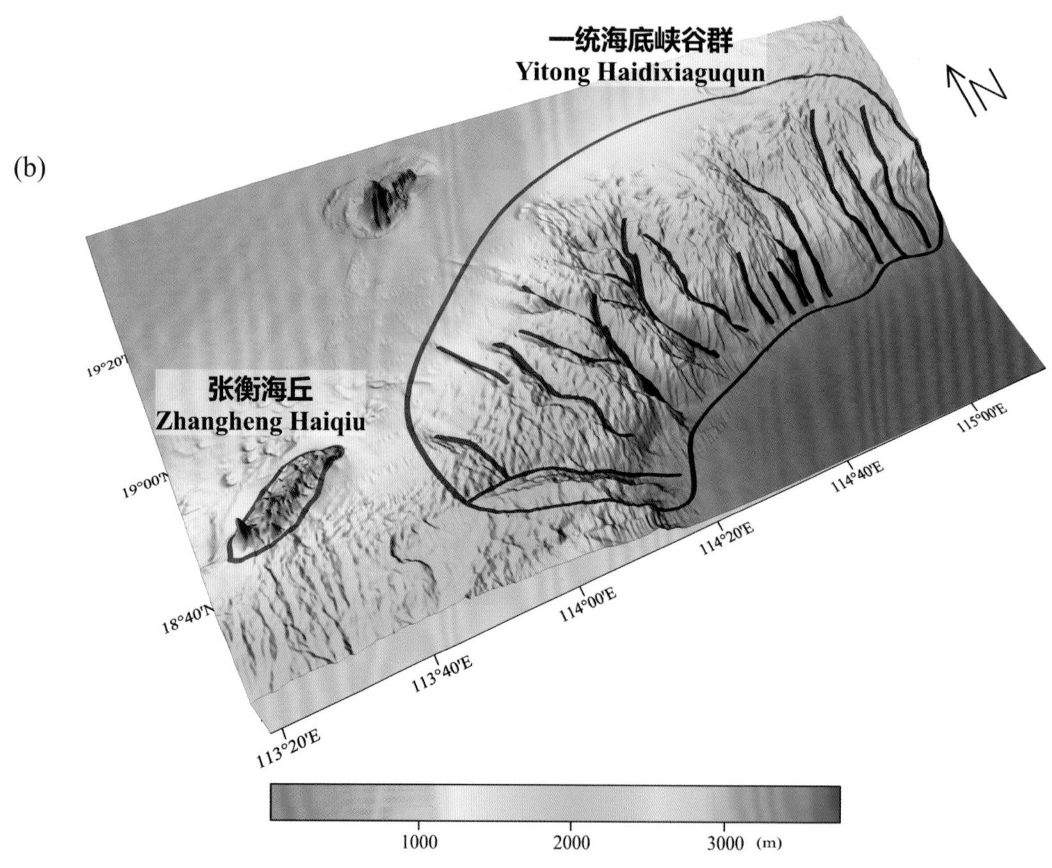

图 1-23 一统海底峡谷群、张衡海丘

(a) 海底地形图（等深线间隔 500 米）；(b) 三维海底地形图

Fig.1-23 Yitong Haidixiaguqun, Zhangheng Haiqiu

(a) Seafloor topographic map (with contour interval of 500 m); (b) 3-D seafloor topographic map

1.55 万泉海底峡谷

标准名称 Standard Name	万泉海底峡谷 Wanquan Haidixiagu	类别 Generic Term	海底峡谷 Canyon
中心点坐标 Center Coordinates	18°28.9′N, 113°12.0′E	规模（千米 × 千米） Dimension（km × km）	50 × 3
最小水深（米） Min Depth (m)	1400	最大水深（米） Max Depth (m)	3000
地理实体描述 Feature Description	万泉海底峡谷发育在一统斜坡南部，其头部水深约 1400 米，尾部水深约 3000 米，总体形态呈北北西—南南东向延伸至西沙海槽（图 1–24）。 Wanquan Haidixiagu is located in the south of Yitong Xiepo, with depth of about 1400 m and 3000 m at the head and at the tail respectively, and generally stretches to Xisha Haicao in the direction of NNW–SSE (Fig.1–24).		
命名由来 Origin of Name	该海底峡谷邻近我国海南岛，故以海南岛上的主要河流名进行地名的团组化命名。“万泉河”为海南岛上的主要河流之一，故取“万泉”两字命名该海底峡谷。 The Canyon is adjacent to Hainan Dao in China, so it is named after a group of main rivers in Hainan Dao. “Wanquan He” is one of the main rivers in Hainan Dao, so the word “Wanquan” was used to name the Canyon.		

1.56 昌化海底峡谷

标准名称 Standard Name	昌化海底峡谷 Changhua Haidixiagu	类别 Generic Term	海底峡谷 Canyon
中心点坐标 Center Coordinates	18°23.4′N, 112°59.3′E	规模（千米 × 千米） Dimension（km × km）	44 × 2
最小水深（米） Min Depth (m)	1600	最大水深（米） Max Depth (m)	2900
地理实体描述 Feature Description	昌化海底峡谷发育在一统斜坡南部，其头部水深约 1600 米，尾部水深约 2900 米，总体形态呈北北西—南南东向延伸至西沙海槽（图 1–24）。 Changhua Haidixiagu is located in the south of Yitong Xiepo, with depth of about 1600 m and about 2900 m at the head and at the tail respectively, and generally extends to Xisha Haicao in the direction of NNW–SSE (Fig.1–24).		
命名由来 Origin of Name	该海底峡谷邻近我国海南岛，故以海南岛上的主要河流名进行地名的团组化命名。“昌化河”为海南岛上的主要河流之一，故取“昌化”两字命名该海底峡谷。 The Canyon is adjacent to Hainan Dao in China, so it is named after a group of main rivers in Hainan Dao. “Changhua He” is one of the main rivers in Hainan Dao, so the word “Changhua” was used to name the Canyon.		

1.57 南渡海底峡谷

标准名称 Standard Name	南渡海底峡谷 Nandu Haidixiagu	类别 Generic Term	海底峡谷 Canyon
中心点坐标 Center Coordinates	18°21.9′N, 112°50.1′E	规模（千米 × 千米） Dimension（km × km）	40 × 2
最小水深（米） Min Depth (m)	1600	最大水深（米） Max Depth (m)	2800
地理实体描述 Feature Description	南渡海底峡谷发育在一统斜坡南部，峡谷头部水深约1600米，尾部水深约2800米，总体形态呈北北西—南南东向延伸至西沙海槽（图1–24）。 Nandu Haidixiagu is located in the south of Yitong Xiepo, with depth of about 1600 m and about 2900 m at the head and at the tail respectively, and generally extends to Xisha Haicao in the direction of NNW–SSE (Fig.1–24).		
命名由来 Origin of Name	该海底峡谷邻近我国海南岛，故以海南岛上的主要河流名进行地名的团组化命名。“南渡河”为海南岛上的主要河流之一，故取“南渡”两字命名该海底峡谷。 The Canyon is adjacent to Hainan Dao in China, so it is named after a group of main rivers in Hainan Dao. “Nandu He” is one of the main rivers in Hainan Dao, so the word “Nandu” was used to name the Canyon.		

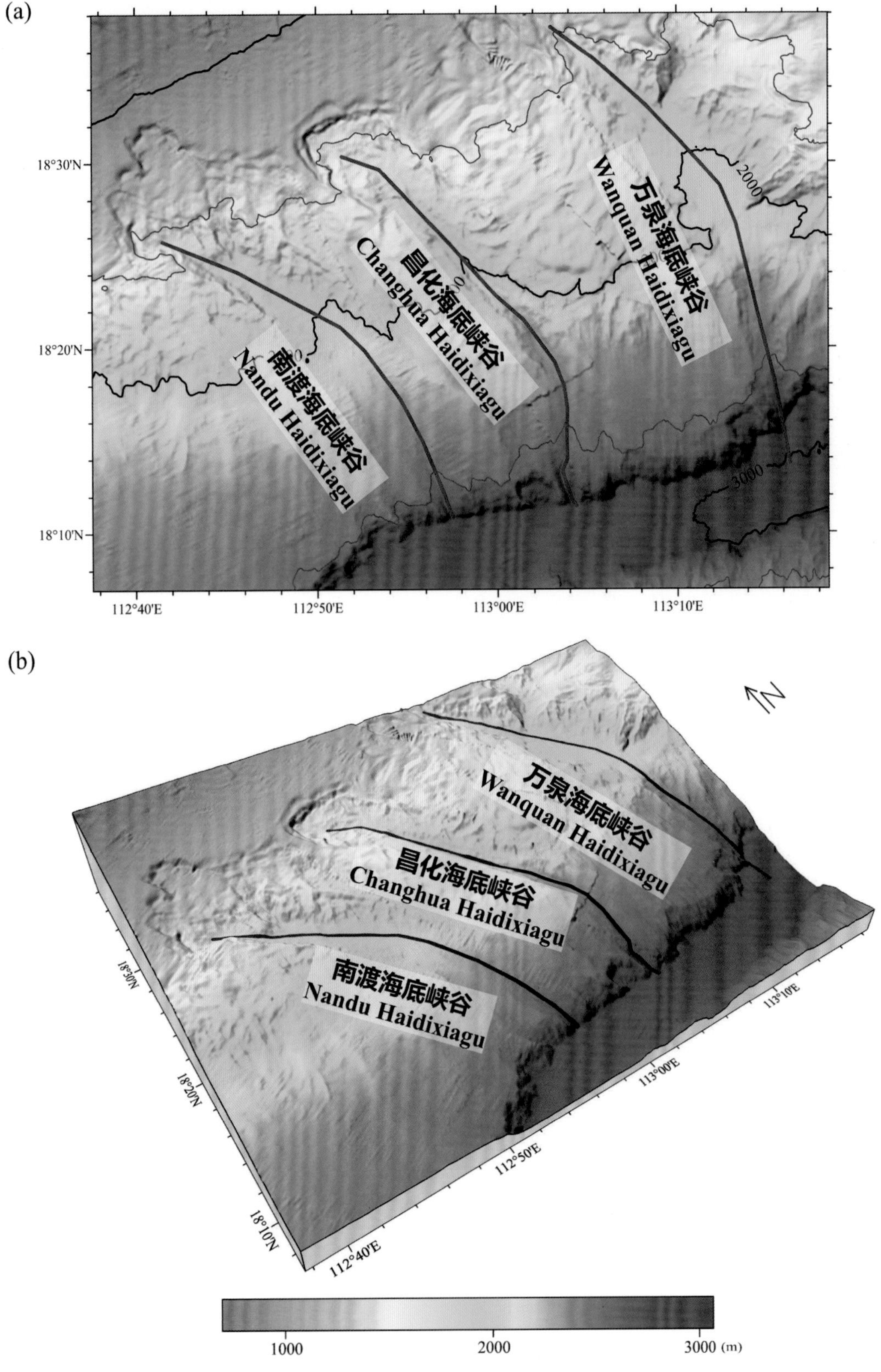

图 1-24　万泉海底峡谷、昌化海底峡谷、南渡海底峡谷

(a) 海底地形图（等深线间隔 500 米）；(b) 三维海底地形图

Fig.1-24　Wanquan Haidixiagu, Changhua Haidixiagu, Nandu Haidixiagu

(a) Seafloor topographic map (with contour interval of 500 m); (b) 3-D seafloor topographic map

1.58 西沙北海底峡谷群

标准名称 Standard Name	西沙北海底峡谷群 Xishabei Haidixiaguqun	类别 Generic Term	海底峡谷群 Canyons
中心点坐标 Center Coordinates	18°20.0′N, 111°20.0′E	规模（千米 × 千米） Dimension（km × km）	220 × 65
最小水深（米） Min Depth (m)	300	最大水深（米） Max Depth (m)	2400
地理实体描述 Feature Description	西沙北海底峡谷群发育在西沙海槽的西北槽坡，由近 30 条南—北和北西—南东向的海底峡谷构成。这些峡谷垂直切割槽坡，自西南向东北其长度和规模逐渐增大（图 1–25）。 Xishabei Haidixiaguqun are developed on the northwest slope of Xisha Haicao and are composed of approximate 30 canyons in the directions of S–N and NW–SE. These Canyons cut through the trough slope vertically, with their lengths and scales increasing gradually from southwest to northeast (Fig.1–25).		
命名由来 Origin of Name	该峡谷群位于西沙海槽北部海域，因此得名。 These Canyons are located in the north sea area of Xisha Haicao, and "Bei" means north in Chinese, so the word "Xishabei" was used to name these Canyons.		

1.59 万户海山群

标准名称 Standard Name	万户海山群 Wanhu Haishanqun	类别 Generic Term	海山群 Seamounts
中心点坐标 Center Coordinates	18°06.5′N, 111°45.4′E	规模（千米 × 千米） Dimension（km × km）	15 × 9
最小水深（米） Min Depth (m)	1215	最大水深（米） Max Depth (m)	2174
地理实体描述 Feature Description	万户海山群发育在西沙北海底峡谷群南部、西沙海槽内，平面形态呈东—西向的椭圆形。海山的顶部发育两座峰（图 1–25）。 Wanhu Haishanqun is located in Xisha Haicao, south of Xishabei Haidixiaguqun, with an oval-shaped planform in the direction of E–W. Two peaks have developed on its top (Fig.1–25).		
命名由来 Origin of Name	以中国古代科学家的名字进行地名的团组化命名。该海山群以明朝著名科学家万户（公元 17 世纪—生卒年不详）命名，事实上，万户是他的官职头衔。他的名字是陶成道。 A group naming of undersea features after the names of famous ancient Chinese scientists. The Seamounts is named after Wanhu (17th century A.D.–years of birth and death unknown), a Chinese scientist during the Ming Dynasty (1368–1644 A.D.). In fact, Wanhu was the title of his official position. His name was Tao Chengdao.		

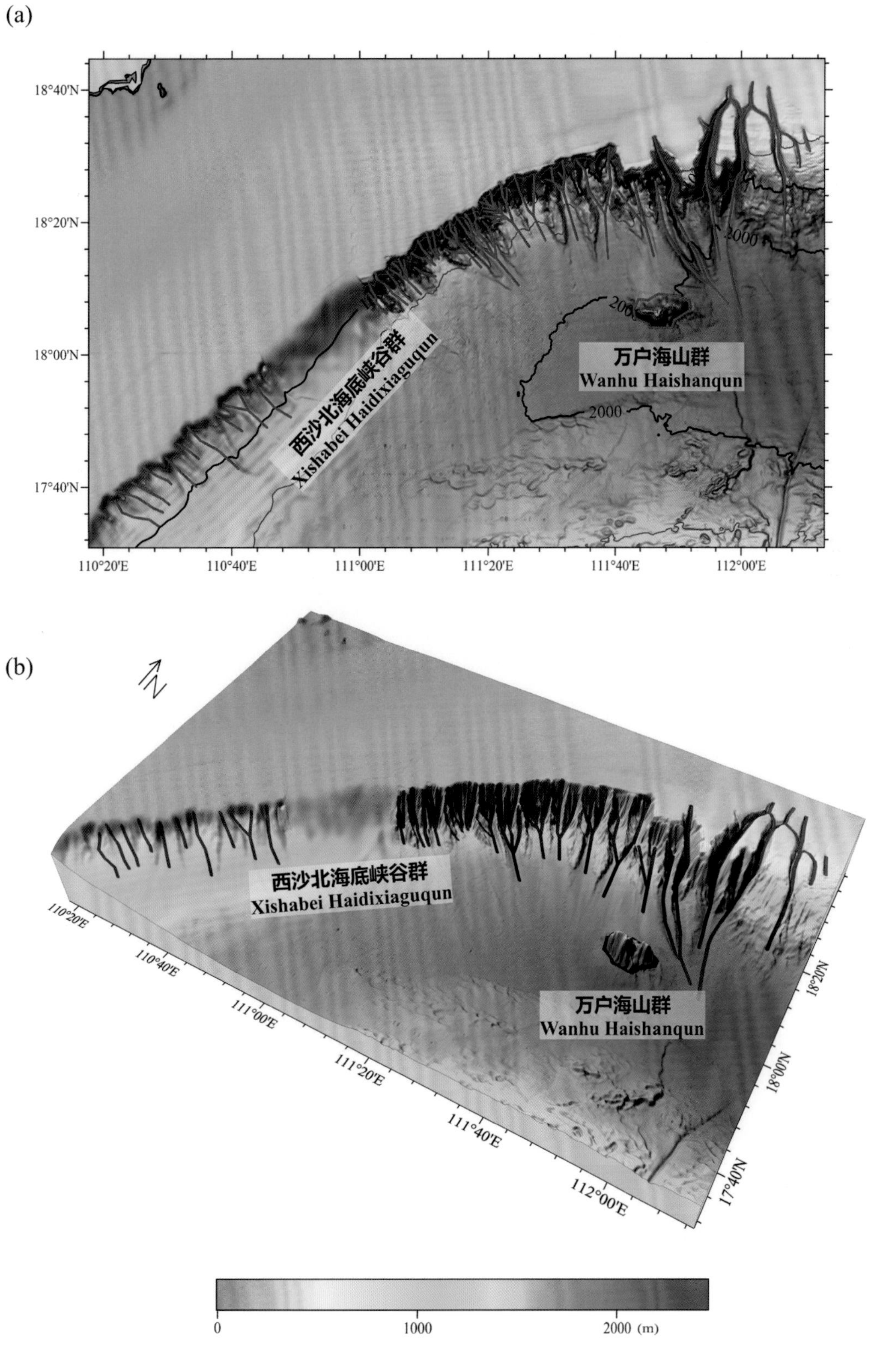

图 1-25 西沙北海底峡谷群、万户海山群

(a) 海底地形图（等深线间隔 500 米）；(b) 三维海底地形图

Fig.1-25 Xishabei Haidixiaguqun, Wanhu Haishanqun

(a) Seafloor topographic map (with contour interval of 500 m); (b) 3-D seafloor topographic map

南海北部海域海底地理实体图集
Atlas of Undersea Features of Northern
South China Sea

2

西沙群岛周边海区海底地理实体

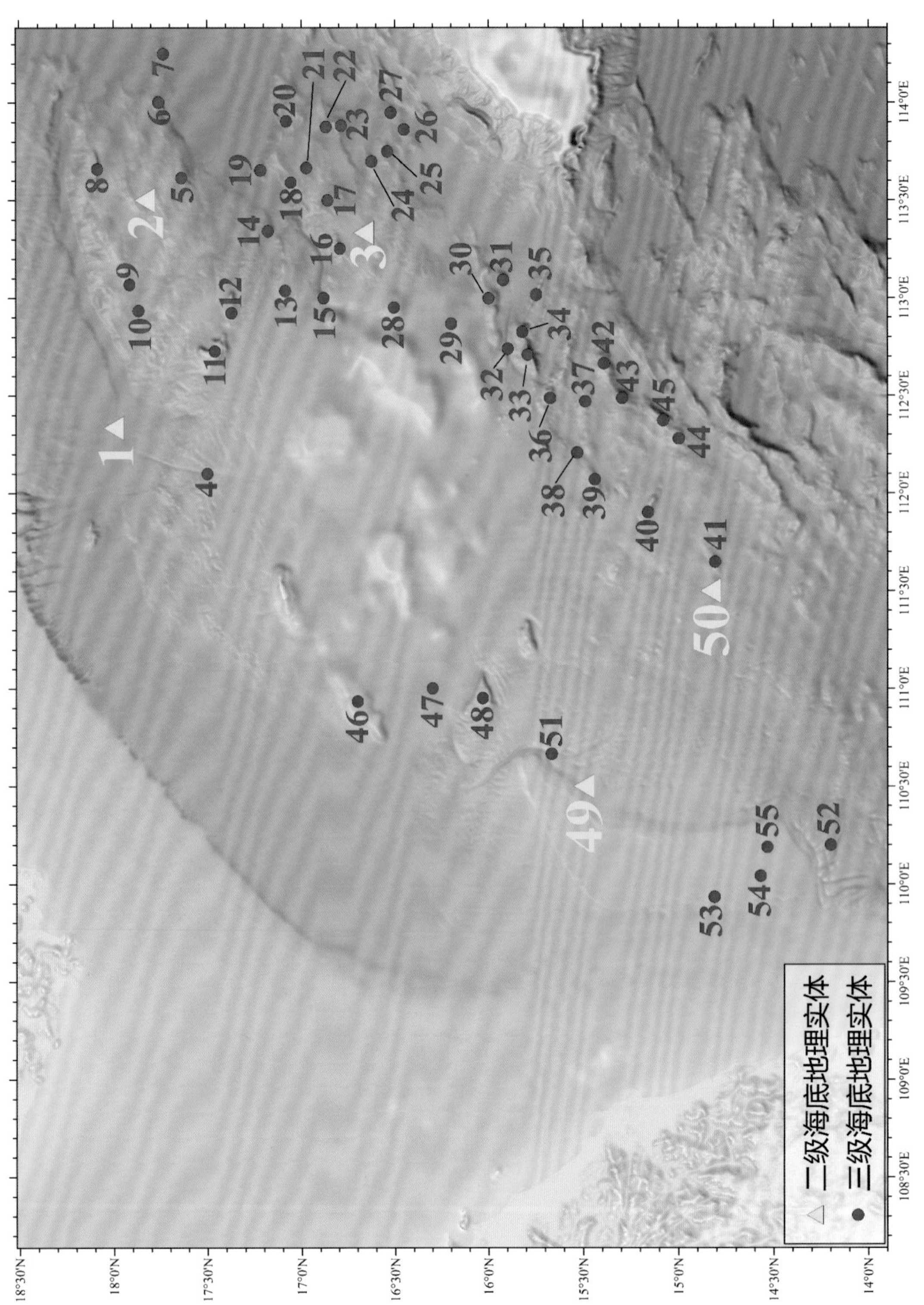

图 2-1　西沙群岛周边海区海底地理实体中心点位置示意图，序号含义见表 2-1

Fig.2-1　Location of center coordinates of undersea features in the sea area off Xisha Qundao, with the meanings of the serial numbers shown in Tab. 2-1

表 2-1　西沙群岛周边海区海底地理实体列表

Tab.2-1　List of undersea features in the sea area off Xisha Qundao

序号 No.	标准名称 Standard Name	汉语拼音 Chinese Phonetic Alphabet	类别 Generic Term	中心点坐标 Center Coordinates		实体等级 Order
				纬度 Latitude	经度 Longitude	
1	西沙海槽 Xisha Haicao	Xīshā Hǎicáo	海槽 Trough	18°00.0'N	112°20.0'E	2
2	永乐海隆 Yongle Hailong	Yǒnglè Hǎilóng	海隆 Rise	17°50.0'N	113°30.0'E	2
3	西沙海隆 Xisha Hailong	Xīshā Hǎilóng	海隆 Rise	16°40.0'N	113°20.0'E	2
4	永兴海谷 Yongxing Haigu	Yǒngxīng Hǎigǔ	海谷 Valley	17°30.0'N	112°06.0'E	3
5	银砾海丘 Yinli Haiqiu	Yínlì Hǎiqiū	海丘 Hill	17°38.3'N	113°36.9'E	3
6	红草海丘 Hongcao Haiqiu	Hóngcǎo Hǎiqiū	海丘 Hill	17°45.6'N	114°00.1'E	3
7	红草东海丘 Hongcaodong Haiqiu	Hóngcǎodōng Hǎiqiū	海丘 Hill	17°44.1'N	114°15.0'E	3
8	高尖海山 Gaojian Haishan	Gāojiān Hǎishān	海山 Seamount	18°05.2'N	113°39.6'E	3
9	珊瑚海山 Shanhu Haishan	Shānhú Hǎishān	海山 Seamount	17°54.9'N	113°04.1'E	3
10	赵述海谷 Zhaoshu Haigu	Zhàoshù Hǎigǔ	海谷 Valley	17°52.0'N	112°56.0'E	3
11	赵述海台 Zhaoshu Haitai	Zhàoshù Hǎitái	海台 Plateau	17°27.6'N	112°43.6'E	3
12	赵述海山 Zhaoshu Haishan	Zhàoshù Hǎishān	海山 Seamount	17°22.3'N	112°55.3'E	3
13	银屿海台 Yinyu Haitai	Yínyǔ Hǎitái	海台 Plateau	17°05.3'N	113°02.2'E	3
14	银屿北海丘 Yinyubei Haiqiu	Yínyǔběi Hǎiqiū	海丘 Hill	17°10.7'N	113°20.5'E	3
15	永乐海底峡谷 Yongle Haidixiagu	Yǒnglè Hǎidǐxiágǔ	海底峡谷 Canyon	16°53.0'N	113°00.0'E	3
16	西渡海山 Xidu Haishan	Xīdù Hǎishān	海山 Seamount	16°47.7'N	113°15.1'E	3
17	西新海丘 Xixin Haiqiu	Xīxīn Hǎiqiū	海丘 Hill	16°51.7'N	113°29.9'E	3

续表

序号 No.	标准名称 Standard Name	汉语拼音 Chinese Phonetic Alphabet	类别 Generic Term	中心点坐标 Center Coordinates 纬度 Latitude	 经度 Longitude	实体等级 Order
18	华光海山 Huaguang Haishan	Huáguāng Hǎishān	海山 Seamount	17°03.4'N	113°35.4'E	3
19	北边廊海丘 Beibianlang Haiqiu	Běibiānláng Hǎiqiū	海丘 Hill	17°13.1'N	113°39.2'E	3
20	盘石海山 Panshi Haishan	Pánshí Hǎishān	海山 Seamount	17°04.9'N	113°54.3'E	3
21	滨湄海丘 Binmei Haiqiu	Bīnméi Hǎiqiū	海丘 Hill	16°58.5'N	113°40.0'E	3
22	嵩焘海丘 Songtao Haiqiu	Sōngtāo Hǎiqiū	海丘 Hill	16°52.2'N	113°52.5'E	3
23	东新海丘 Dongxin Haiqiu	Dōngxīn Hǎiqiū	海丘 Hill	16°47.3'N	113°52.9'E	3
24	鸭公海丘 Yagong Haiqiu	Yāgōng Hǎiqiū	海丘 Hill	16°37.5'N	113°41.9'E	3
25	咸舍海丘 Xianshe Haiqiu	Xiánshě Hǎiqiū	海丘 Hill	16°32.5'N	113°45.0'E	3
26	全富海丘 Quanfu Haiqiu	Quánfù Hǎiqiū	海丘 Hill	16°27.2'N	113°51.7'E	3
27	七连海丘 Qilian Haiqiu	Qīlián Hǎiqiū	海丘 Hill	16°31.4'N	113°56.9'E	3
28	湛涵海台 Zhanhan Haitai	Zhànhán Hǎitái	海台 Plateau	16°30.4'N	112°57.0'E	3
29	浪花海谷 Langhua Haigu	Lànghuā Hǎigǔ	海谷 Valley	16°12.0'N	112°52.0'E	3
30	浪花海山 Langhua Haishan	Lànghuā Hǎishān	海山 Seamount	16°00.0'N	112°59.9'E	3
31	宣德海丘 Xuande Haiqiu	Xuāndé Hǎiqiū	海丘 Hill	15°55.4'N	113°05.5'E	3
32	玉琢海底峡谷 Yuzhuo Haidixiagu	Yùzhuó Hǎidǐxiágǔ	海底峡谷 Canyon	15°53.9'N	112°44.3'E	3
33	银砾海山 Yinli Haishan	Yínlì Hǎishān	海山 Seamount	15°47.5'N	112°42.5'E	3
34	筐仔海山 Kuangzai Haishan	Kuāngzǎi Hǎishān	海山 Seamount	15°49.2'N	112°49.5'E	3
35	金银海脊 Jinyin Haiji	Jīnyín Hǎijǐ	海脊 Ridge	15°44.7'N	113°01.0'E	3
36	广金海脊 Guangjin Haiji	Guǎngjīn Hǎijǐ	海脊 Ridge	15°40.5'N	112°29.2'E	3

续表

序号 No.	标准名称 Standard Name	汉语拼音 Chinese Phonetic Alphabet	类别 Generic Term	中心点坐标 Center Coordinates 纬度 Latitude	经度 Longitude	实体等级 Order
37	湛涵海脊 Zhanhan Haiji	Zhànhán Hǎijǐ	海脊 Ridge	15°29.5'N	112°28.2'E	3
38	湛涵北海丘 Zhanhanbei Haiqiu	Zhànhánběi Hǎiqiū	海丘 Hill	15°32.0'N	112°12.4'E	3
39	羚羊海山 Lingyang Haishan	Língyáng Hǎishān	海山 Seamount	15°26.5'N	112°04.2'E	3
40	羚羊南海山 Lingyangnan Haishan	Língyángnán Hǎishān	海山 Seamount	15°09.6'N	111°54.2'E	3
41	杨柳西海丘 Yangliuxi Haiqiu	Yángliǔxī Hǎiqiū	海丘 Hill	14°48.5'N	111°38.9'E	3
42	晋卿海山 Jinqing Haishan	Jìnqīng Hǎishān	海山 Seamount	15°23.6'N	112°39.8'E	3
43	晋卿西海山 Jinqingxi Haishan	Jìnqīngxī Hǎishān	海山 Seamount	15°17.8'N	112°29.2'E	3
44	琛航海山 Chenhang Haishan	Chēnháng Hǎishān	海山 Seamount	15°00.0'N	112°16.7'E	3
45	琛航东海山 Chenhangdong Haishan	Chēnhángdōng Hǎishān	海山 Seamount	15°04.9'N	112°22.4'E	3
46	甘泉海台 Ganquan Haitai	Gānquán Hǎitái	海台 Plateau	16°42.0'N	110°56.0'E	3
47	甘泉海谷 Ganquan Haigu	Gānquán Hǎigǔ	海谷 Valley	16°18.0'N	111°00.0'E	3
48	中建北海台 Zhongjianbei Haitai	Zhōngjiànběi Hǎitái	海台 Plateau	16°01.9'N	110°57.0'E	3
49	中建阶地 Zhongjian Jiedi	Zhōngjiàn Jiēdì	海底阶地 Terrace	15°30.0'N	110°30.0'E	2
50	中建南斜坡 Zhongjiannan Xiepo	Zhōngjiànnán Xiépō	海底斜坡 Slope	14°50.0'N	111°30.0'E	2
51	中建西海底峡谷群 Zhongjianxi Haidixiaguqun	Zhōngjiànxī Hǎidǐxiágǔqún	海底峡谷群 Canyons	15°40.0'N	110°40.0'E	3
52	中建南海底峡谷群 Zhongjiannan Haidixiaguqun	Zhōngjiànnán Hǎidǐxiágǔqún	海底峡谷群 Canyons	14°12.0'N	110°12.0'E	3
53	金吒海丘 Jinzha Haiqiu	Jīnzhā Hǎiqiū	海丘 Hill	14°48.7'N	109°56.1'E	3
54	木吒海丘 Muzha Haiqiu	Mùzhā Hǎiqiū	海丘 Hill	14°34.2'N	110°02.5'E	3
55	哪吒海丘 Nezha Haiqiu	Nézhā Hǎiqiū	海丘 Hill	14°32.1'N	110°11.3'E	3

2.1 西沙海槽

标准名称 Standard Name	西沙海槽 Xisha Haicao	类别 Generic Term	海槽 Trough
中心点坐标 Center Coordinates	18°00.0′N, 112°20.0′E	规模（千米 × 千米） Dimension（km × km）	400 × 50
最小水深（米） Min Depth (m)	1350	最大水深（米） Max Depth (m)	3400
地理实体描述 Feature Description	西沙海槽位于南海北部陆坡的西北部，西沙群岛以北，海南岛东南面，其西缘与中建阶地相接，东端与南海海盆相连。海槽平面形态似弓形，呈近东—西向展布，其两侧槽坡地形陡峭，底部平原地形平缓，总长度约 420 千米（图 2−2）。 Xisha Haicao is located in the northwest of Nanhaibeibu Lupo, to the north of Xisha Qundao and to the southeast of Hainan Dao. It connects to Zhongjian Jiedi on the west and to Nanhai Haipen on the east. The planform of the Trough is bow-shaped in the direction of E−W, with steep slopes on the both sides and gentle plain at the bottom, with a total length of 420 km. (Fig.2−2)		
命名由来 Origin of Name	1986 年，国务院、外交部和中国地名委员会批准了前地矿部第二海洋地质调查大队（现广州海洋地质调查局）对南海 22 个海底地理实体进行的命名，“西沙海槽”是这 22 个海底地理实体名称之一。 In 1986, the State Council, the Ministry of Foreign Affairs and Chinese Toponymy Committee approved 22 undersea feature names in Nanhai named by former Second Marine Geological Survey Brigade of the Ministry of Geology and Mineral Resources (present Guangzhou Marine Geological Survey of China Geological Survey). Xisha Haicao is one of the 22 undersea feature names.		

2.2 永乐海隆

标准名称 Standard Name	永乐海隆 Yongle Hailong	类别 Generic Term	海隆 Rise
中心点坐标 Center Coordinates	17°50.0′N, 113°30.0′E	规模（千米 × 千米） Dimension（km × km）	220 × 130
最小水深（米） Min Depth (m)	310	最大水深（米） Max Depth (m)	3580
地理实体描述 Feature Description	永乐海隆位于西沙海槽以南，其南侧以永乐海底峡谷为界与西沙海隆相邻，平面形态大致呈四方形展布，水深自西南向东北方向逐渐增大，最终与南海海盆相接（图 2−2）。 Yongle Hailong is located to the south of Xisha Haicao and is adjacent to Xisha Hailong on the south with Yongle Haidixiagu as the boundary. The planform is roughly a square in shape and the depth gradually increases from southwest to northeast, until it finally connects to Nanhai Haipen (Fig.2−2).		
命名由来 Origin of Name	该海隆位于我国西沙群岛附近，故以西沙群岛中的岛屿、暗礁、暗滩、环礁、水道、沙洲名进行地名的团组化命名。“永乐群岛”为西沙群岛的一组岛屿，取“永乐”两字命名该海隆。 The Rise is located near the Xisha Qundao in China, so it is named after a group of islands, reefs, shoals, atolls, channels and sandbars in Xisha Qundao. “Yongle Qundao” is a group of the islands belonging to Xisha Qundao, so the word “Yongle” was used to name the Rise.		

2.3 西沙海隆

标准名称 Standard Name	西沙海隆 Xisha Hailong	类别 Generic Term	海隆 Rise
中心点坐标 Center Coordinates	16°40.0′N, 113°20.0′E	规模（千米 × 千米） Dimension（km × km）	190 × 90
最小水深（米） Min Depth (m)	460	最大水深（米） Max Depth (m)	3480
地理实体描述 Feature Description	西沙海隆位于西沙群岛西部，永乐海隆与中沙海槽之间，平面形态大致呈北东—南西向的条带状，水深自西南向东北方向逐渐加大，最终与南海海盆相接（图 2–2）。 Xisha Hailong is located between Yongle Hailong and Zhongsha Haicao in the west of Xisha Qundao, with a roughly strip-shaped planform in the direction of NE–SW. The depth increases gradually from southwest to northeast, until it finally connects to Nanhai Haipen (Fig.2–2).		
命名由来 Origin of Name	该海隆位于西沙群岛附近，因此得名。 The Rise is located near Xisha Qundao, so the word “Xisha” was used to name the Rise.		

2.4 永兴海谷

标准名称 Standard Name	永兴海谷 Yongxing Haigu	类别 Generic Term	海谷 Valley
中心点坐标 Center Coordinates	17°30.0′N, 112°06.0′E	规模（千米 × 千米） Dimension（km × km）	70 × 2
最小水深（米） Min Depth (m)	1130	最大水深（米） Max Depth (m)	2500
地理实体描述 Feature Description	永兴海谷发源于西沙海台，北北东—南南西向延伸发育，最终到达西沙海槽（图 2–2）。 Yongxing Haigu is originated from Xisha Haitai. It stretches in the direction of NNE–SSW and ends at Xisha Haicao (Fig.2–2).		
命名由来 Origin of Name	该海谷位于我国西沙群岛附近，故以西沙群岛中的岛屿、暗礁、暗滩、环礁、水道、沙洲名进行地名的团组化命名。“永兴岛”属于西沙群岛的一个岛屿，取“永兴”两字命名该海谷。 The Valley is located near the Xisha Qundao in China, so it is named after a group of islands, reefs, shoals, atolls, channels and sandbars in Xisha Qundao. “Yongxing Dao” is one of the islands belonging to Xisha Qundao, so the word “Yongxing” was used to name the Valley.		

(a)

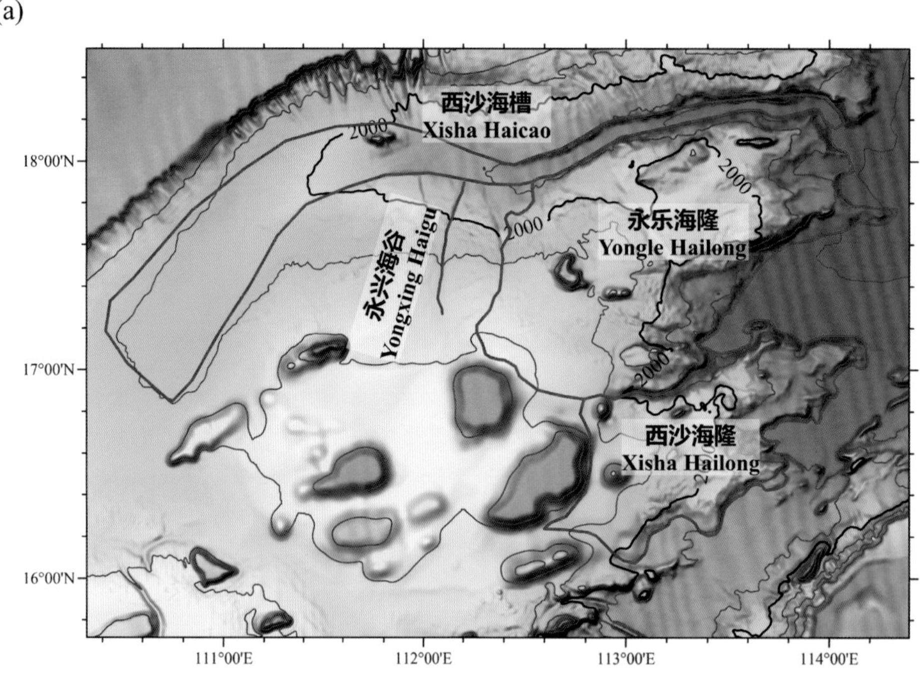

(b)

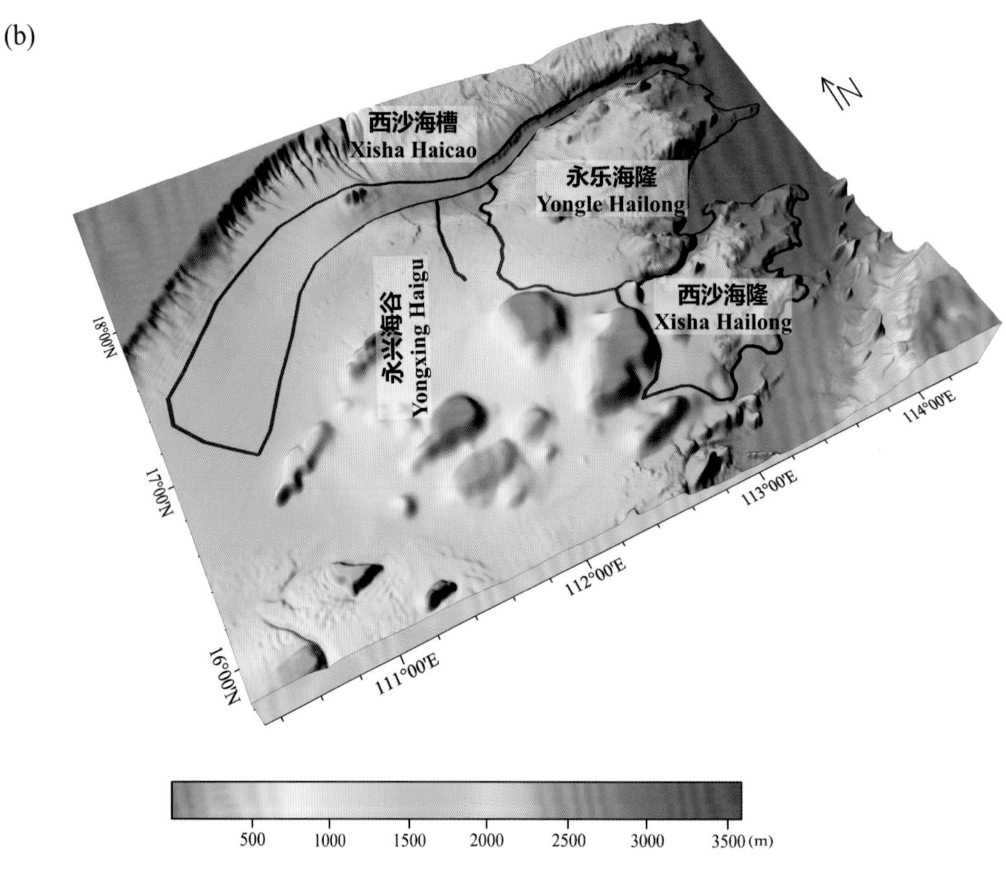

图 2-2　西沙海槽、永乐海隆、西沙海隆、永兴海谷

(a) 海底地形图（等深线间隔 500 米）；(b) 三维海底地形图

Fig.2-2　Xisha Haicao, Yongle Hailong, Xisha Hailong, Yongxing Haigu

(a) Seafloor topographic map (with contour interval of 500 m)；(b) 3-D seafloor topographic map

2.5 银砾海丘

标准名称 Standard Name	银砾海丘 Yinli Haiqiu	类别 Generic Term	海丘 Hill
中心点坐标 Center Coordinates	17°38.3′N, 113°36.9′E	规模（千米 × 千米） Dimension（km × km）	14 × 7
最小水深（米） Min Depth (m)	1780	最大水深（米） Max Depth (m)	2420
地理实体描述 Feature Description	银砾海丘发育在永乐海隆东南部，平面形态呈东—西向展布的纺锤形（图 2–3）。 Yinli Haiqiu is located in the southeast of Yongle Hailong with a spindle-shaped planform in the direction of E–W (Fig.2–3).		
命名由来 Origin of Name	该海丘位于我国西沙群岛附近，故以西沙群岛中的岛屿、暗礁、暗滩、环礁、水道、沙洲名进行地名的团组化命名。“银砾滩”属于西沙群岛的一个暗滩，取“银砾”两字命名该海丘。 The Hill is located near the Xisha Qundao in China, so it is named after a group of islands, reefs, shoals, atolls, channels and sandbars in Xisha Qundao. “Yinli Tan” is one of the shoals belonging to Xisha Qundao, so the word “Yinli” was used to name the Hill.		

2.6 红草海丘

标准名称 Standard Name	红草海丘 Hongcao Haiqiu	类别 Generic Term	海丘 Hill
中心点坐标 Center Coordinates	17°45.6′N, 114°00.1′E	规模（千米 × 千米） Dimension（km × km）	10 × 8
最小水深（米） Min Depth (m)	2660	最大水深（米） Max Depth (m)	3300
地理实体描述 Feature Description	红草海丘发育在永乐海隆东部，平面形态呈不规则多边形（图 2–3）。 Hongcao Haiqiu is located in the east of Yongle Hailong, with a planform in the shape of an irregular polygon (Fig.2–3).		
命名由来 Origin of Name	该海丘位于我国西沙群岛附近，故以西沙群岛中的岛屿、暗礁、暗滩、环礁、水道、沙洲名进行地名的团组化命名。“红草门”属于西沙群岛的一个水道，取“红草”两字命名该海丘。 The Hill is located near the Xisha Qundao in China, so it is named after a group of islands, reefs, shoals, atolls, channels and sandbars in Xisha Qundao. “Hongcao Men” is one of the channels belonging to Xisha Qundao, so the word “Hongcao” was used to name the Hill.		

2.7 红草东海丘

标准名称 Standard Name	红草东海丘 Hongcaodong Haiqiu	类别 Generic Term	海丘 Hill
中心点坐标 Center Coordinates	17°44.1′N, 114°15.0′E	规模（千米 × 千米） Dimension（km × km）	13 × 7
最小水深（米） Min Depth (m)	3180	最大水深（米） Max Depth (m)	3580
地理实体描述 Feature Description	红草东海丘发育在永乐海隆最东端，平面形态呈东—西向的不规则多边形（图 2−3）。 Hongcaodong Haiqiu is located in the easternmost end of Yongle Hailong, with a planform in the shape of an irregular polygon in the direction of E−W (Fig.2−3).		
命名由来 Origin of Name	该海丘位于红草海丘以东，因此得名。 The Hill is located to the east of Hongcao Haiqiu, and “Dong” means east in Chinese, so the word “Hongcaodong” was used to name the Hill.		

2.8 高尖海山

标准名称 Standard Name	高尖海山 Gaojian Haishan	类别 Generic Term	海山 Seamount
中心点坐标 Center Coordinates	18°05.2′N, 113°39.6′E	规模（千米 × 千米） Dimension（km × km）	30 × 20
最小水深（米） Min Depth (m)	1500	最大水深（米） Max Depth (m)	2580
地理实体描述 Feature Description	高尖海山发育在永乐海隆北部，北邻西沙海槽，平面形态呈四方形（图 2−3）。 Gaojian Haishan is located in the north of Yongle Hailong and is adjacent to Xisha Haicao on the north, with a quadrate planform (Fig.2−3).		
命名由来 Origin of Name	该海山位于我国西沙群岛附近，故以西沙群岛中的岛屿、暗礁、暗滩、环礁、水道、沙洲名进行地名的团组化命名。“高尖石”属于西沙群岛的一个岛屿，取“高尖”两字命名该海山。 The Seamount is located near the Xisha Qundao in China, so it is named after a group of islands, reefs, shoals, atolls, channels and sandbars in Xisha Qundao. “Gaojian Shi” is one of the islands belonging to Xisha Qundao, so the word “Gaojian” was used to name the Seamount.		

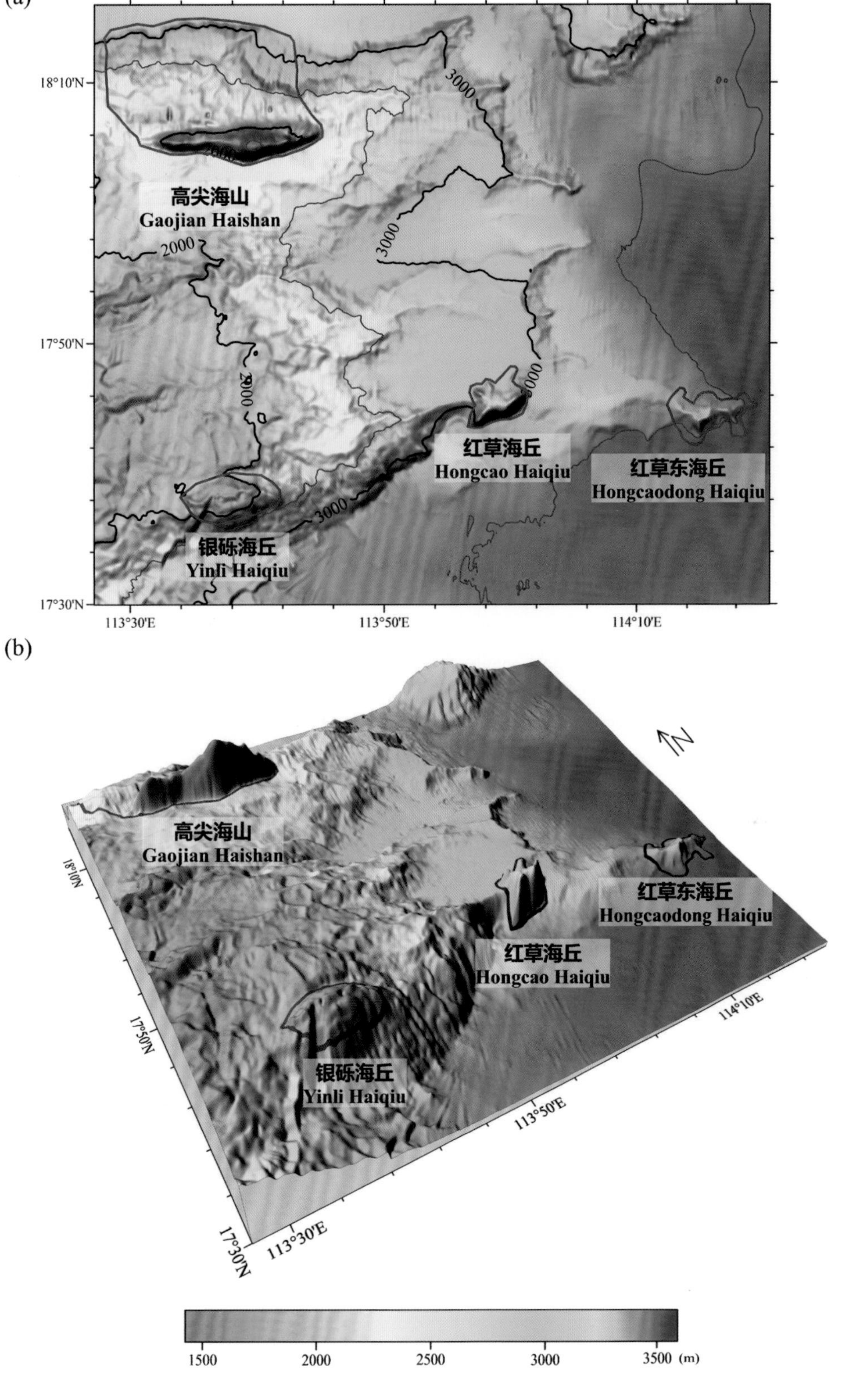

图 2-3 银砾海丘、红草海丘、红草东海丘、高尖海山

(a) 海底地形图（等深线间隔 500 米）；(b) 三维海底地形图

Fig.2-3 Yinli Haiqiu, Hongcao Haiqiu, Hongcaodong Haiqiu, Gaojian Haishan

(a) Seafloor topographic map (with contour interval of 500 m); (b) 3-D seafloor topographic map

2.9 珊瑚海山

标准名称 Standard Name	珊瑚海山 Shanhu Haishan	类别 Generic Term	海山 Seamount
中心点坐标 Center Coordinates	17°54.9′N, 113°04.1′E	规模（千米 × 千米） Dimension（km × km）	45 × 19
最小水深（米） Min Depth (m)	1450	最大水深（米） Max Depth (m)	2340
地理实体描述 Feature Description	珊瑚海山发育在永乐海隆北部，北邻西沙海槽，平面形态呈北东—南西向的长条形，其顶部发育三座峰（图 2-4）。 Shanhu Haishan is located in the north of Yongle Hailong and is adjacent to Xisha Haicao on the north, with a planform in the shape of a long strip in the direction of NE-SW, and has three peaks developed on its top (Fig.2-4).		
命名由来 Origin of Name	该海山位于我国西沙群岛附近，故以西沙群岛中的岛屿、暗礁、暗滩、环礁、水道、沙洲名进行地名的团组化命名。“珊瑚岛”属于西沙群岛的一个岛屿，取“珊瑚”两字命名该海山。 The Seamount is located near the Xisha Qundao in China, so it is named after a group of islands, reefs, shoals, atolls, channels and sandbars in Xisha Qundao. “Shanhu Dao” is one of the islands belonging to Xisha Qundao, so the word “Shanhu” was used to name the Seamount.		

2.10 赵述海谷

标准名称 Standard Name	赵述海谷 Zhaoshu Haigu	类别 Generic Term	海谷 Valley
中心点坐标 Center Coordinates	17°52.0′N, 112°56.0′E	规模（千米 × 千米） Dimension（km × km）	85 × 8
最小水深（米） Min Depth (m)	1800	最大水深（米） Max Depth (m)	2560
地理实体描述 Feature Description	赵述海谷发育在永乐海隆中部，平面形态呈北西—南东走向，海谷两头水深较深，中部水深较浅（图 2-4）。 Zhaoshu Haigu is located in the middle of Yongle Hailong and stretches in the direction NW-SE. The depth of the Valley is deeper on the both sides and is shallower in the middle (Fig.2-4).		
命名由来 Origin of Name	该海谷位于我国西沙群岛附近，故以西沙群岛中的岛屿、暗礁、暗滩、环礁、水道、沙洲名进行地名的团组化命名。“赵述岛”属于西沙群岛的一个岛屿，取“赵述”两字命名该海谷。 The Valley is located near the Xisha Qundao in China, so it is named after a group of islands, reefs, shoals, atolls, channels and sandbars in Xisha Qundao. “Zhaoshu Dao” is one of the islands belonging to Xisha Qundao, so the word “Zhaoshu” was used to name the Valley.		

2.11 赵述海台

标准名称 Standard Name	赵述海台 Zhaoshu Haitai	类别 Generic Term	海台 Plateau
中心点坐标 Center Coordinates	17°27.6′N, 112°43.6′E	规模（千米 × 千米） Dimension（km × km）	20 × 18
最小水深（米） Min Depth (m)	300	最大水深（米） Max Depth (m)	1500
地理实体描述 Feature Description	赵述海台发育在永乐海隆西南部，平面形态呈弯钩状，整体呈近北西—南东走向。海台的顶部较平坦，平均水深310米，宽5～12千米，向四周水深迅速变大（图2–4）。 Zhaoshu Haitai is located in the southwest of Yongle Hailong with a hook-shaped planform. The Plateau stretches in the direction of NW–SE and has a flat top, with an average depth of 310 m, a width of 5~12 km. The depth increases rapidly from the center to all directions (Fig.2–4).		
命名由来 Origin of Name	该海台位于赵述海谷附近，因此得名。 The Plateau is located near Zhaoshu Haigu, so the word “Zhaoshu” was used to name the Plateau.		

2.12 赵述海山

标准名称 Standard Name	赵述海山 Zhaoshu Haishan	类别 Generic Term	海山 Seamount
中心点坐标 Center Coordinates	17°22.3′N, 112°55.3′E	规模（千米 × 千米） Dimension（km × km）	18 × 10
最小水深（米） Min Depth (m)	500	最大水深（米） Max Depth (m)	1700
地理实体描述 Feature Description	赵述海山发育在永乐海隆西南部，赵述海台东侧，平面形态呈东—西向的锥形（图2–4）。 Zhaoshu Haishan is located in the southwest of Yongle Hailong and is on the east of Zhaoshu Haitai with a taper-shaped planform in the direction of E–W (Fig.2–4).		
命名由来 Origin of Name	该海山位于赵述海谷附近，因此得名。 The Seamount is located near Zhaoshu Haigu, so the word “Zhaoshu” was used to name the Seamount.		

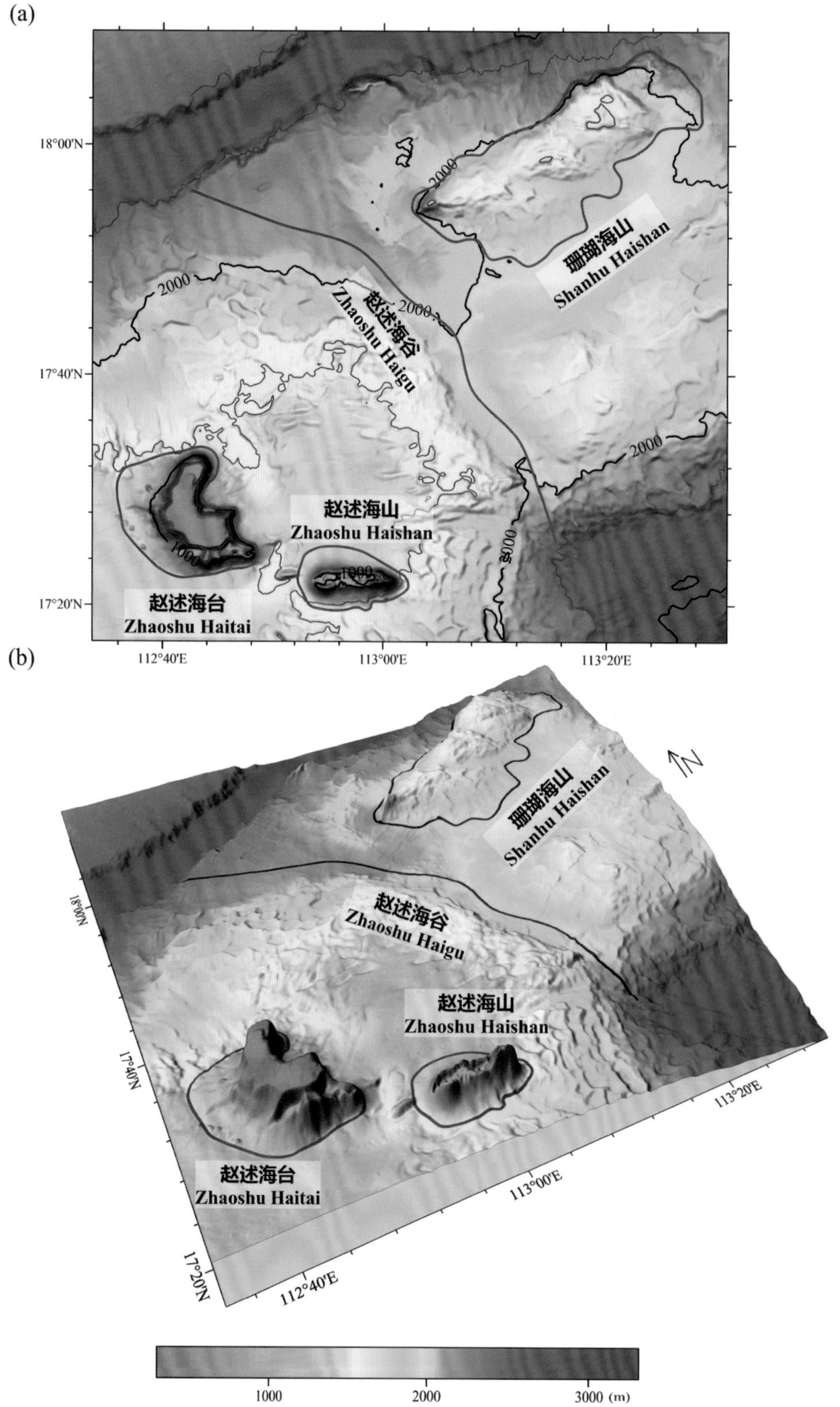

图 2-4　珊瑚海山、赵述海谷、赵述海台、赵述海山

(a) 海底地形图（等深线间隔 500 米）；(b) 三维海底地形图

Fig.2-4　Shanhu Haishan, Zhaoshu Haigu, Zhaoshu Haitai, Zhaoshu Haishan

(a) Seafloor topographic map (with contour interval of 500 m); (b) 3-D seafloor topographic map

2.13 银屿海台

标准名称 Standard Name	银屿海台 Yinyu Haitai	类别 Generic Term	海台 Plateau
中心点坐标 Center Coordinates	17°05.3'N, 113°02.2'E	规模（千米 × 千米） Dimension（km × km）	22 × 12
最小水深（米） Min Depth (m)	1300	最大水深（米） Max Depth (m)	1780
地理实体描述 Feature Description	银屿海台发育在永乐海隆南部，平面形态不规则，整体呈东—西走向（图 2-5）。 Yinyu Haitai is located in the south of Yongle Hailong with an irregular planform and generally stretches in the direction of E−W (Fig.2−5).		
命名由来 Origin of Name	该海台位于我国西沙群岛附近，故以西沙群岛中的岛屿、暗礁、暗滩、环礁、水道、沙洲名进行地名的团组化命名。“银屿门”为西沙群岛的一个水道，取“银屿”两字命名该海台。 The Plateau is located near the Xisha Qundao in China, so it is named after a group of islands, reefs, shoals, atolls, channels and sandbars in Xisha Qundao. “Yinyu Men” is one of the channels belonging to Xisha Qundao, so the word “Yinyu” was used to name the Plateau.		

2.14 银屿北海丘

标准名称 Standard Name	银屿北海丘 Yinyubei Haiqiu	类别 Generic Term	海丘 Hill
中心点坐标 Center Coordinates	17°10.7'N, 113°20.5'E	规模（千米 × 千米） Dimension（km × km）	20 × 10
最小水深（米） Min Depth (m)	2420	最大水深（米） Max Depth (m)	3200
地理实体描述 Feature Description	银屿北海丘发育在永乐海隆南部，平面形态呈北东东—南西西向的三角形（图 2-5）。 Yinyubei Haiqiu is located in the south of Yongle Hailong, with a triangle-shaped planform in the direction of NEE−SWW (Fig.2−5).		
命名由来 Origin of Name	该海丘位于银屿海台以北，因此得名。 The Hill is located to the north of Yinyu Haitai, and “Bei” means north in Chinese, so the word “Yinyubei” was used to name the Hill.		

(a)

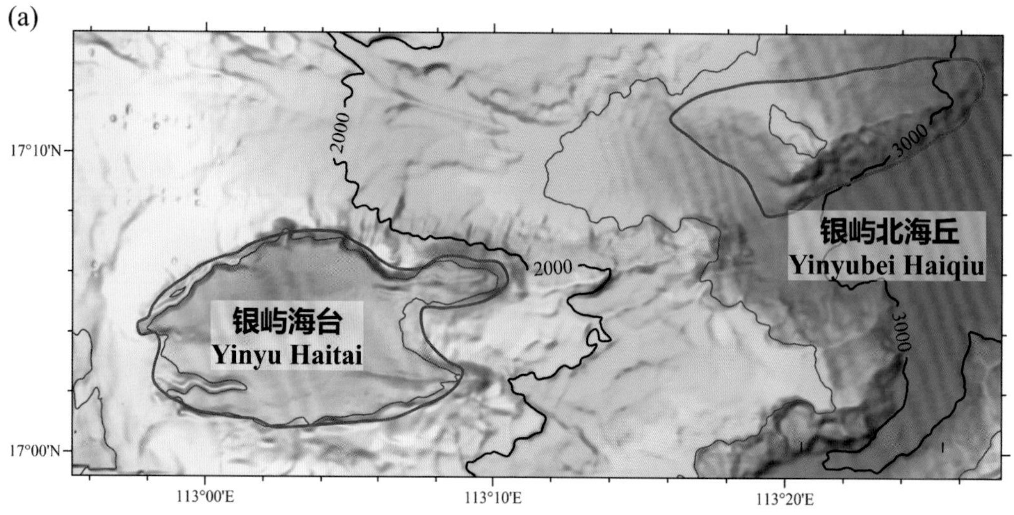

(b)

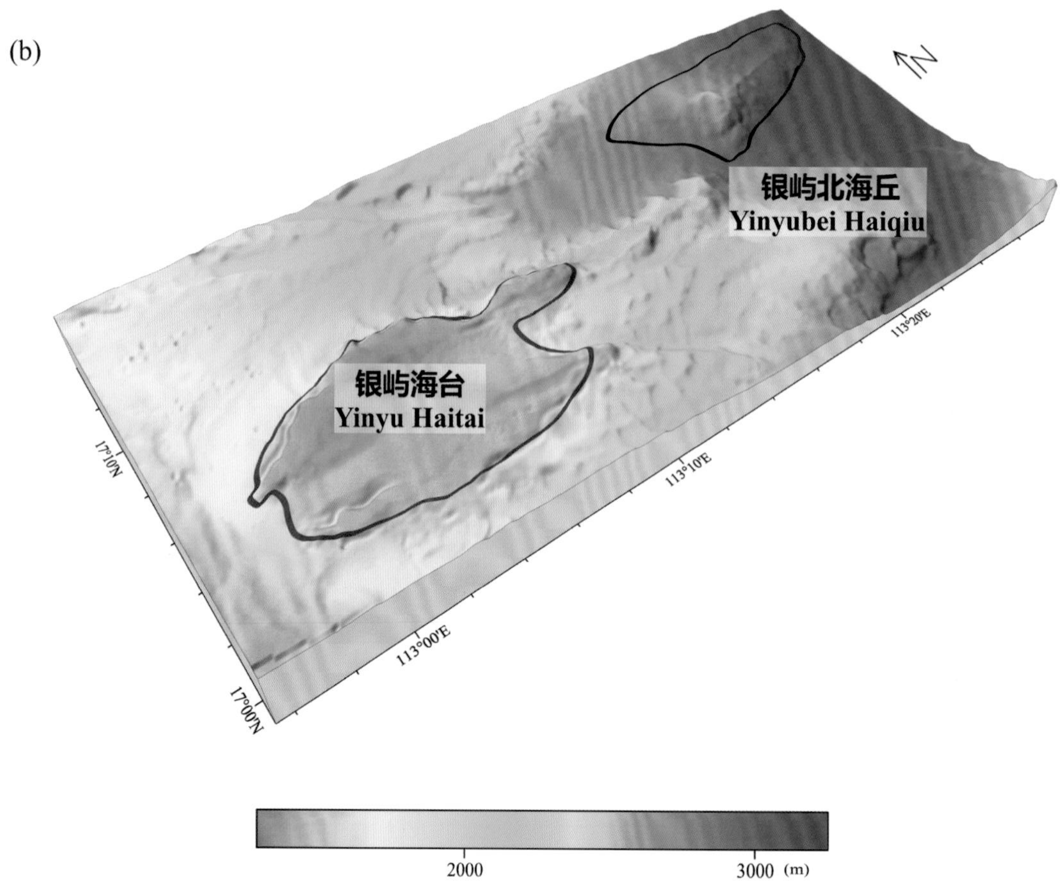

图 2-5　银屿海台、银屿北海丘

(a) 海底地形图（等深线间隔 500 米）；(b) 三维海底地形图

Fig.2-5　Yinyu Haitai, Yinyubei Haiqiu

(a) Seafloor topographic map (with contour interval of 500 m); (b) 3-D seafloor topographic map

2.15 永乐海底峡谷

标准名称 Standard Name	永乐海底峡谷 Yongle Haidixiagu	类别 Generic Term	海底峡谷 Canyon
中心点坐标 Center Coordinates	16°53.0′N, 113°00.0′E	规模（千米 × 千米） Dimension（km × km）	54 × 4
最小水深（米） Min Depth (m)	1220	最大水深（米） Max Depth (m)	3300
地理实体描述 Feature Description	永乐海底峡谷发育在西沙群岛东部，其头部位于西沙海台，从西向东延伸发育，最终汇入南海海盆（图 2–6）。 Yongle Haidixiagu is located in the east of Xisha Qundao, with its head on Xisha Haitai. The Canyon stretches from west to east until it ends at Nanhai Haipen (Fig.2–6).		
命名由来 Origin of Name	该海底峡谷位于永乐海隆附近，因此得名。 The Canyon is located near Yongle Hailong, so the word “Yongle” was used to name the Canyon.		

2.16 西渡海山

标准名称 Standard Name	西渡海山 Xidu Haishan	类别 Generic Term	海山 Seamount
中心点坐标 Center Coordinates	16°47.7′N, 113°15.1′E	规模（千米 × 千米） Dimension（km × km）	19 × 9
最小水深（米） Min Depth (m)	650	最大水深（米） Max Depth (m)	2030
地理实体描述 Feature Description	西渡海山发育在西沙海隆北部，平面形态呈北东—南西向的椭圆形。海山的西北侧坡较陡，东南侧坡较缓（图 2–6）。 Xidu Haishan is located in the north of Xisha Hailong, with an oval-shaped planform in the direction of NE–SW. Its northwest slope is relatively steeper while its southwest slope is relatively gentler (Fig.2–6).		
命名由来 Origin of Name	该海山位于我国西沙群岛附近，故以西沙群岛中的岛屿、暗礁、暗滩、环礁、水道、沙洲名进行地名的团组化命名。“西渡滩”属于西沙群岛的一个暗滩，取“西渡”两字命名该海山。 The Seamount is located near the Xisha Qundao in China, so it is named after a group of islands, reefs, shoals, atolls, channels and sandbars in Xisha Qundao. “Xidu Tan” is one of the shoals belonging to Xisha Qundao, so the word “Xidu” was used to name the Seamount.		

2.17 西新海丘

标准名称 Standard Name	西新海丘 Xixin Haiqiu	类别 Generic Term	海丘 Hill
中心点坐标 Center Coordinates	16°51.7′N, 113°29.9′E	规模（千米 × 千米） Dimension（km × km）	18 × 13
最小水深（米） Min Depth (m)	1730	最大水深（米） Max Depth (m)	2260
地理实体描述 Feature Description	西新海丘发育在西沙海隆北部，平面形态呈北西—南东向的椭圆形，其顶部发育多座峰（图 2-6）。 Xixin Haiqiu is located in the north of Xisha Hailong, with an oval-shaped planform in the direction of NW–SE, and has several peaks developed on its top (Fig.2–6).		
命名由来 Origin of Name	该海丘位于我国西沙群岛附近，故以西沙群岛中的岛屿、暗礁、暗滩、环礁、水道、沙洲名进行地名的团组化命名。“西新沙州”为西沙群岛的一个沙洲，取“西新”两字命名该海丘。 The Hill is located near the Xisha Qundao in China, so it is named after a group of islands, reefs, shoals, atolls, channels and sandbars in Xisha Qundao. “Xixin Shazhou” is one of the sandbars belonging to Xisha Qundao, so the word “Xixin” was used to name the Hill.		

(a)

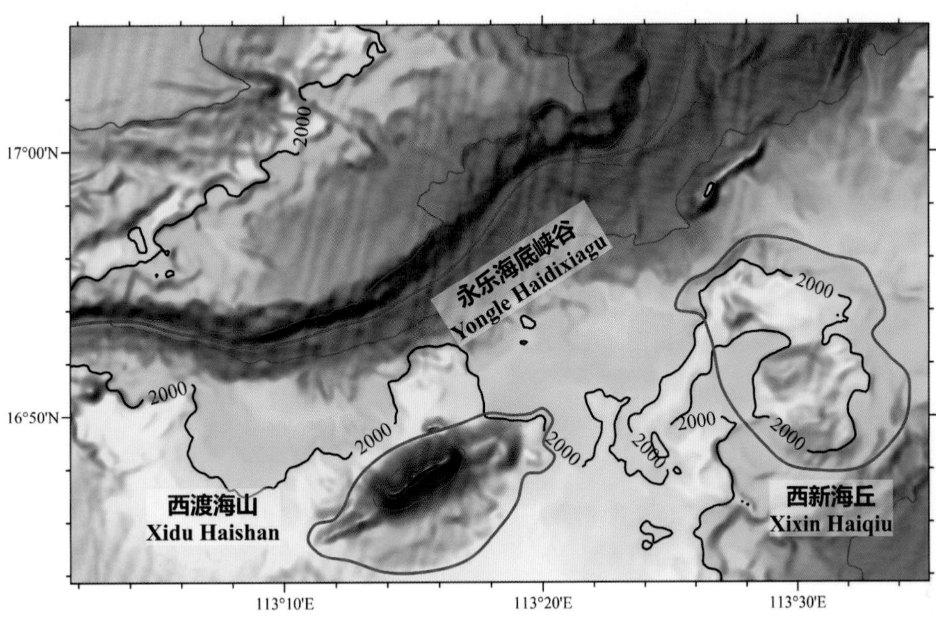

(b)

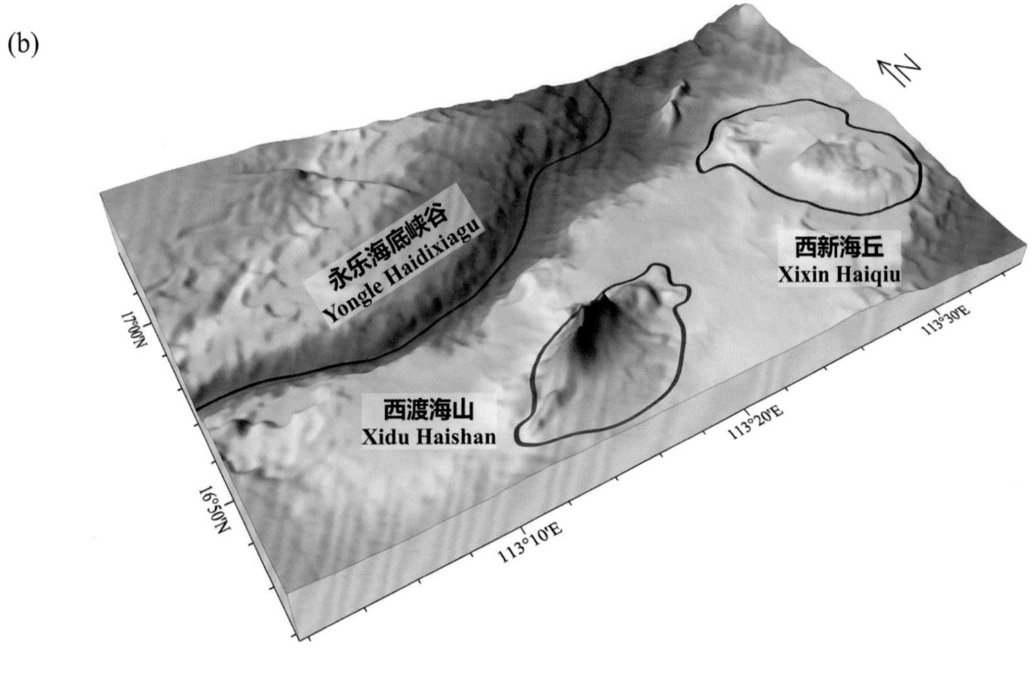

图 2-6 永乐海底峡谷、西渡海山、西新海丘

(a) 海底地形图（等深线间隔 500 米）；(b) 三维海底地形图

Fig.2-6 Yongle Haidixiagu, Xidu Haishan, Xixin Haiqiu

(a) Seafloor topographic map (with contour interval of 500 m); (b) 3-D seafloor topographic map

2.18 华光海山

标准名称 Standard Name	华光海山 Huaguang Haishan	类别 Generic Term	海山 Seamount
中心点坐标 Center Coordinates	17°03.4′N, 113°35.4′E	规模（千米 × 千米） Dimension（km × km）	32 × 20
最小水深（米） Min Depth (m)	2060	最大水深（米） Max Depth (m)	3200
地理实体描述 Feature Description	华光海山发育在西沙海隆北部，平面形态像一只倒置的靴子，其顶部发育两座峰（图 2-7）。 Huaguang Haishan is located in the north of Xisha Hailong, with an inverted-boot-shaped planform, and has two peaks developed on its top (Fig.2-7).		
命名由来 Origin of Name	该海山位于我国西沙群岛附近，故以西沙群岛中的岛屿、暗礁、暗滩、环礁、水道、沙洲名进行地名的团组化命名。“华光礁”属于西沙群岛的一个暗礁，取“华光”两字命名该海山。 The Seamount is located near the Xisha Qundao in China, so it is named after a group of islands, reefs, shoals, atolls, channels and sandbars in Xisha Qundao. “Huaguang Jiao” is one of the reefs belonging to Xisha Qundao, so the word “Huaguang” was used to name the Seamount.		

2.19 北边廊海丘

标准名称 Standard Name	北边廊海丘 Beibianlang Haiqiu	类别 Generic Term	海丘 Hill
中心点坐标 Center Coordinates	17°13.1'N, 113°39.2'E	规模（千米 × 千米） Dimension（km × km）	13 × 5
最小水深（米） Min Depth (m)	2600	最大水深（米） Max Depth (m)	3420
地理实体描述 Feature Description	北边廊海丘发育在西沙海隆北部，平面形态呈东—西向的纺锤形（图 2-7）。 Beibianlang Haiqiu is located in the north of Xisha Hailong, with a spindle-shaped planform in the direction of E-W (Fig.2-7).		
命名由来 Origin of Name	该海丘位于我国西沙群岛附近，故以西沙群岛中的岛屿、暗礁、暗滩、环礁、水道、沙洲名进行地名的团组化命名。“北边廊”属于西沙群岛的一个暗礁，取“北边廊”三字命名该海丘。 The Hill is located near the Xisha Qundao in China, so it is named after a group of islands, reefs, shoals, atolls, channels and sandbars in Xisha Qundao. “Beibianlang” is one of the reefs belonging to Xisha Qundao, so the word “Beibianlang” was used to name the Hill.		

2.20 盘石海山

标准名称 Standard Name	盘石海山 Panshi Haishan	类别 Generic Term	海山 Seamount
中心点坐标 Center Coordinates	17°04.9′N, 113°54.3′E	规模（千米 × 千米） Dimension（km × km）	16 × 7
最小水深（米） Min Depth (m)	2610	最大水深（米） Max Depth (m)	3440
地理实体描述 Feature Description	盘石海山发育在西沙海隆北部，平面形态呈北东—南西向的不规则多边形（图 2–7）。 Panshi Haishan is located in the north of Xisha Hailong, with a planform in the shape of an irregular polygon in the direction of NE–SW (Fig.2–7).		
命名由来 Origin of Name	该海山位于我国西沙群岛附近，故以西沙群岛中的岛屿、暗礁、暗滩、环礁、水道、沙洲名进行地名的团组化命名。“盘石屿”属于西沙群岛的一个环礁，取“盘石”两字命名该海山。 The Seamount is located near the Xisha Qundao in China, so it is named after a group of islands, reefs, shoals, atolls, channels and sandbars in Xisha Qundao. “Panshi Yu” is one of the atolls belonging to Xisha Qundao, so the word “Panshi” was used to name the Seamount.		

(a)

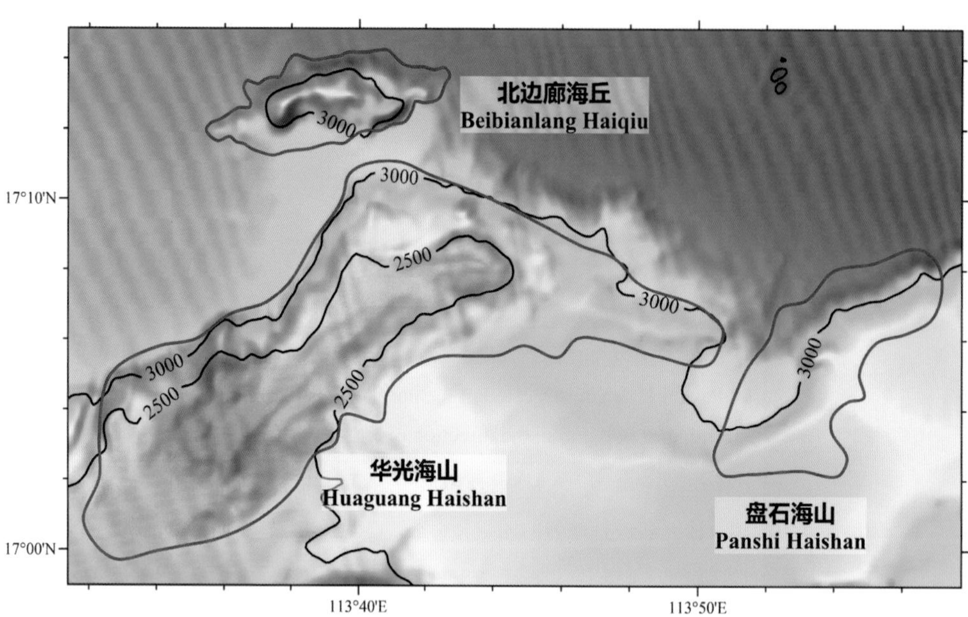

(b)

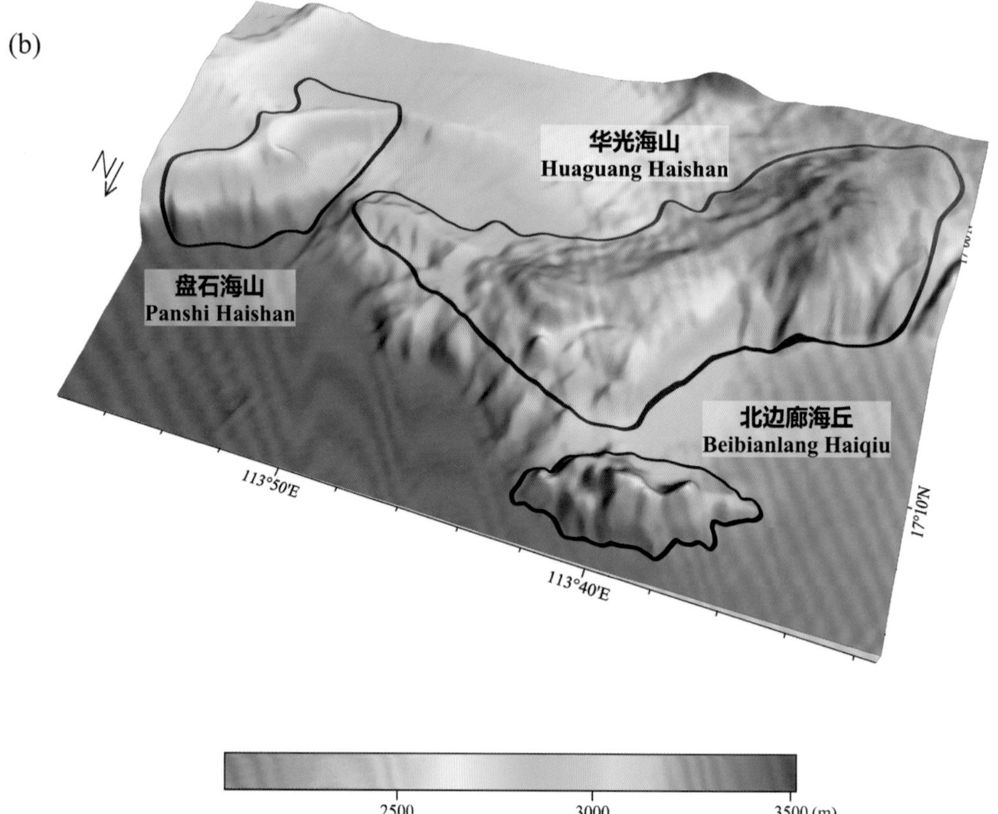

图 2-7　华光海山、北边廊海丘、盘石海山

(a) 海底地形图（等深线间隔 500 米）；(b) 三维海底地形图

Fig.2-7　Huaguang Haishan, Beibianlang Haiqiu, Panshi Haishan

(a) Seafloor topographic map (with contour interval of 500 m)；(b) 3-D seafloor topographic map

2.21 滨湄海丘

标准名称 Standard Name	滨湄海丘 Binmei Haiqiu	类别 Generic Term	海丘 Hill
中心点坐标 Center Coordinates	16°58.5'N, 113°40.0'E	规模（千米 × 千米） Dimension（km × km）	10 × 3
最小水深（米） Min Depth (m)	1950	最大水深（米） Max Depth (m)	2670
地理实体描述 Feature Description	滨湄海丘发育在西沙海隆东北部，平面形态呈东—西向的纺锤形（图 2–8）。 Binmei Haiqiu is located in the northeast of Xisha Hailong, with a spindle-shaped planform in the direction of E–W (Fig.2–8).		
命名由来 Origin of Name	该海丘位于我国西沙群岛附近，故以西沙群岛中的岛屿、暗礁、暗滩、环礁、水道、沙洲名进行地名的团组化命名。“滨湄滩”属于西沙群岛的一个暗滩，取“滨湄”两字命名该海丘。 The Hill is located near the Xisha Qundao in China, so it is named after a group of islands, reefs, shoals, atolls, channels and sandbars in Xisha Qundao. “Binmei Tan” is one of the shoals belonging to Xisha Qundao, so the word “Binmei” was used to name the Hill.		

2.22 嵩焘海丘

标准名称 Standard Name	嵩焘海丘 Songtao Haiqiu	类别 Generic Term	海丘 Hill
中心点坐标 Center Coordinates	16°52.2'N, 113°52.5'E	规模（千米 × 千米） Dimension（km × km）	20 × 10
最小水深（米） Min Depth (m)	2610	最大水深（米） Max Depth (m)	3150
地理实体描述 Feature Description	嵩焘海丘发育在西沙海隆东部，平面形态呈东—西向的不规则多边形（图 2–8）。 Songtao Haiqiu is located in the east of Xisha Hailong, with a planform in the shape of an irregular polygon in the direction of E–W (Fig.2–8).		
命名由来 Origin of Name	该海丘位于我国西沙群岛附近，故以西沙群岛中的岛屿、暗礁、暗滩、环礁、水道、沙洲名进行地名的团组化命名。“嵩焘滩”属于西沙群岛的一个暗礁，取“嵩焘”两字命名该海丘。 The Hill is located near the Xisha Qundao in China, so it is named after a group of islands, reefs, shoals, atolls, channels and sandbars in Xisha Qundao. “Songtao Tan” is one of the reefs belonging to Xisha Qundao, so the word “Songtao” was used to name the Hill.		

2.23 东新海丘

标准名称 Standard Name	东新海丘 Dongxin Haiqiu	类别 Generic Term	海丘 Hill
中心点坐标 Center Coordinates	16°47.3′N, 113°52.9′E	规模（千米 × 千米） Dimension（km × km）	7 × 7
最小水深（米） Min Depth (m)	2510	最大水深（米） Max Depth (m)	3260
地理实体描述 Feature Description	东新海丘发育在西沙海隆东部，平面形态近似三角形（图 2−8）。 Dongxin Haiqiu is located in the east of Xisha Hailong, with a subtriangular planform (Fig.2−8).		
命名由来 Origin of Name	该海丘位于我国西沙群岛附近，故以西沙群岛中的岛屿、暗礁、暗滩、环礁、水道、沙洲名进行地名的团组化命名。“东新沙州”属于西沙群岛的一个沙洲，取“东新”两字命名该海丘。 The Hill is located near the Xisha Qundao in China, so it is named after a group of islands, reefs, shoals, atolls, channels and sandbars in Xisha Qundao. “Dongxin Shazhou” is one of the sandbars belonging to Xisha Qundao, so the word “Dongxin” was used to name the Hill.		

(a)

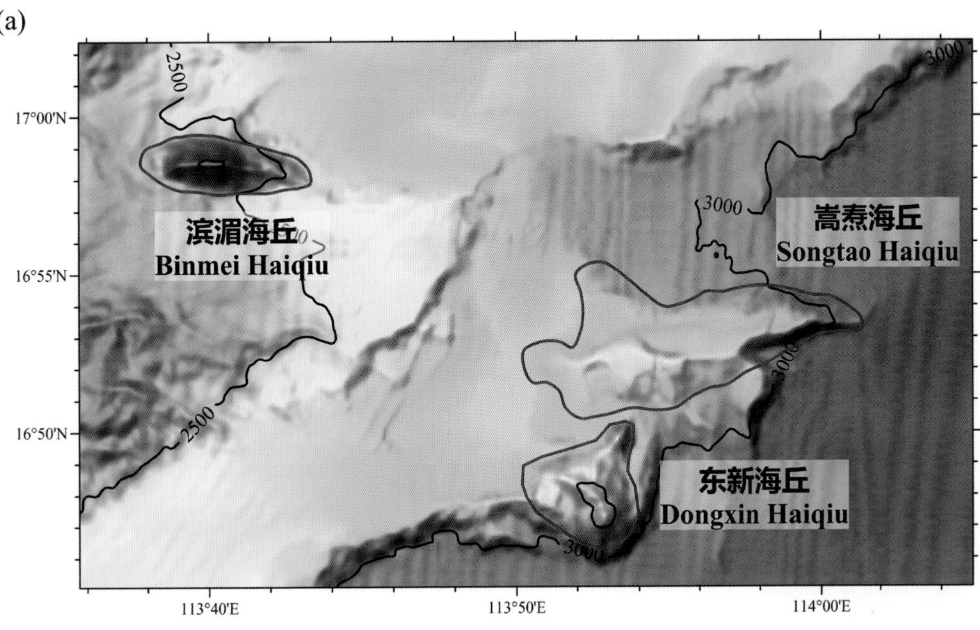

(b)

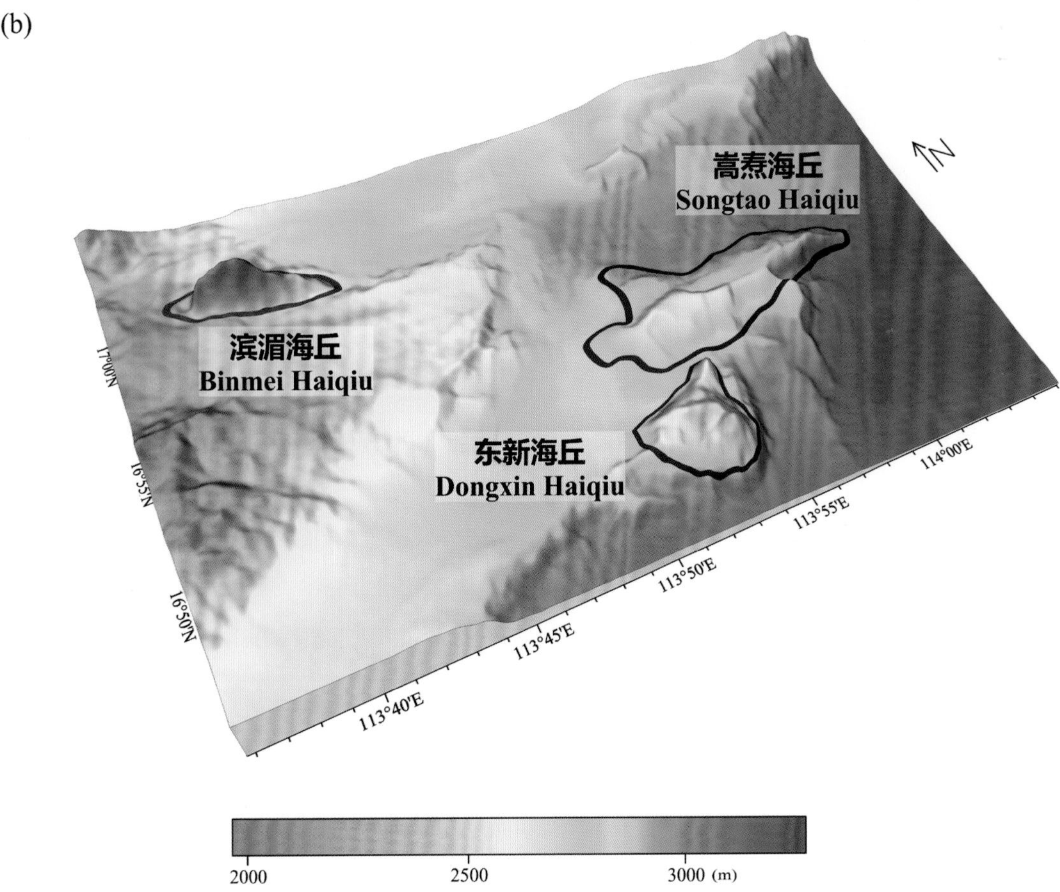

图 2-8 滨湄海丘、嵩焘海丘、东新海丘

(a) 海底地形图（等深线间隔 500 米）；(b) 三维海底地形图

Fig.2-8 Binmei Haiqiu, Songtao Haiqiu, Dongxin Haiqiu

(a) Seafloor topographic map (with contour interval of 500 m)；(b) 3-D seafloor topographic map

2.24 鸭公海丘

标准名称 Standard Name	鸭公海丘 Yagong Haiqiu	类别 Generic Term	海丘 Hill
中心点坐标 Center Coordinates	16°37.5'N, 113°41.9'E	规模（千米 × 千米） Dimension（km × km）	13 × 8
最小水深（米） Min Depth (m)	2460	最大水深（米） Max Depth (m)	3050
地理实体描述 Feature Description	鸭公海丘发育在西沙海隆东南部，平面形态呈北东—南西向的三角形。海丘的西南侧水深较浅，东北侧水深较深（图 2−9）。 Yagong Haiqiu is located in the southeast of Xisha Hailong, with a triangle-shaped planform in the direction of NE−SW. The depth of the Hill is shallower on the southwest side and is deeper on the northeast side (Fig.2−9).		
命名由来 Origin of Name	该海丘位于我国西沙群岛附近，故以西沙群岛中的岛屿、暗礁、暗滩、环礁、水道、沙洲名进行地名的团组化命名。"鸭公岛"属于西沙群岛的一个岛屿，取"鸭公"两字命名该海丘。 The Hill is located near the Xisha Qundao in China, so it is named after a group of islands, reefs, shoals, atolls, channels and sandbars in Xisha Qundao. "Yagong Dao" is one of the islands belonging to Xisha Qundao, so the word "Yagong" was used to name the Hill.		

2.25 咸舍海丘

标准名称 Standard Name	咸舍海丘 Xianshe Haiqiu	类别 Generic Term	海丘 Hill
中心点坐标 Center Coordinates	16°32.5'N, 113°45.0'E	规模（千米 × 千米） Dimension（km × km）	20 × 15
最小水深（米） Min Depth (m)	2160	最大水深（米） Max Depth (m)	3020
地理实体描述 Feature Description	咸舍海丘发育在西沙海隆东部，平面形态呈不规则多边形。海丘的西南侧水深较浅，东北侧水深较深（图 2−9）。 Xianshe Haiqiu is located in the east of Xisha Hailong, with a planform in the shape of an irregular polygon. The depth of the Hill is shallower on the southwest side and is deeper on the northeast side (Fig.2−9).		
命名由来 Origin of Name	该海丘位于我国西沙群岛附近，故以西沙群岛中的岛屿、暗礁、暗滩、环礁、水道、沙洲名进行地名的团组化命名。"咸舍屿"属于西沙群岛的一个岛屿，取"咸舍"两字命名该海丘。 The Hill is located near the Xisha Qundao in China, so it is named after a group of islands, reefs, shoals, atolls, channels and sandbars in Xisha Qundao. "Xianshe Yu" is one of the islands belonging to Xisha Qundao, so the word "Xianshe" was used to name the Hill.		

2.26 全富海丘

标准名称 Standard Name	全富海丘 Quanfu Haiqiu	类别 Generic Term	海丘 Hill
中心点坐标 Center Coordinates	16°27.2'N, 113°51.7'E	规模（千米 × 千米） Dimension（km × km）	10 × 6
最小水深（米） Min Depth (m)	2410	最大水深（米） Max Depth (m)	2910
地理实体描述 Feature Description	全富海丘发育在中沙海槽内，平面形态近似东—西向的椭圆形。海丘的北坡较陡，南坡较缓（图 2–9）。 Quanfu Haiqiu is located in Zhongsha Haicao, with an oval-shaped planform in the direction of E–W. Its north slope is relatively steeper and the south slope is relatively gentler (Fig.2–9).		
命名由来 Origin of Name	该海丘位于我国西沙群岛附近，故以西沙群岛中的岛屿、暗礁、暗滩、环礁、水道、沙洲名进行地名的团组化命名。“全富岛”属于西沙群岛的一个岛屿，取“全富”两字命名该海丘。 The Hill is located near the Xisha Qundao in China, so it is named after a group of islands, reefs, shoals, atolls, channels and sandbars in Xisha Qundao. “Quanfu Dao” is one of the islands belonging to Xisha Qundao, so the word “Quanfu” was used to name the Hill.		

2.27 七连海丘

标准名称 Standard Name	七连海丘 Qilian Haiqiu	类别 Generic Term	海丘 Hill
中心点坐标 Center Coordinates	16°31.4'N, 113°56.9'E	规模（千米 × 千米） Dimension（km × km）	9 × 6
最小水深（米） Min Depth (m)	2460	最大水深（米） Max Depth (m)	3140
地理实体描述 Feature Description	七连海丘发育在中沙海槽内，平面形态近似南—北向的椭圆形。海丘的北坡较陡，南坡较缓（图 2–9）。 Qilian Haiqiu is located in Zhongsha Haicao, with an oval-shaped planform in the direction of S–N. Its north slope is relatively steeper and the south slope is relatively gentler (Fig.2–9).		
命名由来 Origin of Name	该海丘位于我国西沙群岛附近，故以西沙群岛中的岛屿、暗礁、暗滩、环礁、水道、沙洲名进行地名的团组化命名。“七连屿”属于西沙群岛的一个岛屿，取“七连”两字命名该海丘。 The Hill is located near the Xisha Qundao in China, so it is named after a group of islands, reefs, shoals, atolls, channels and sandbars in Xisha Qundao. “Qilian Yu” is one of the islands belonging to Xisha Qundao, so the word “Qilian” was used to name the Hill.		

(a)

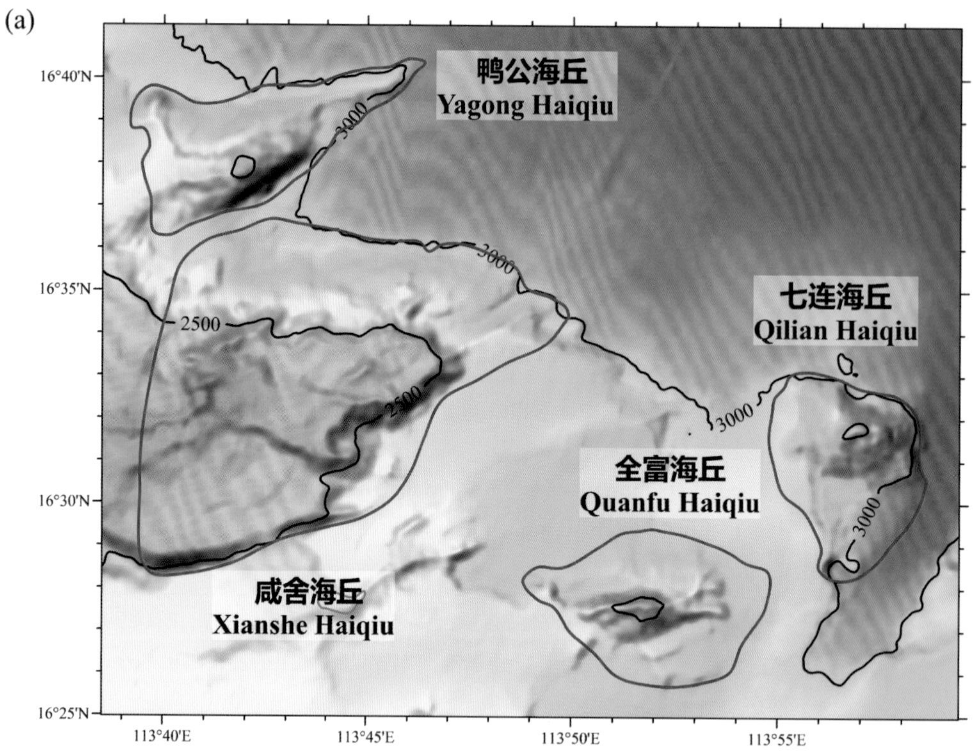

(b)

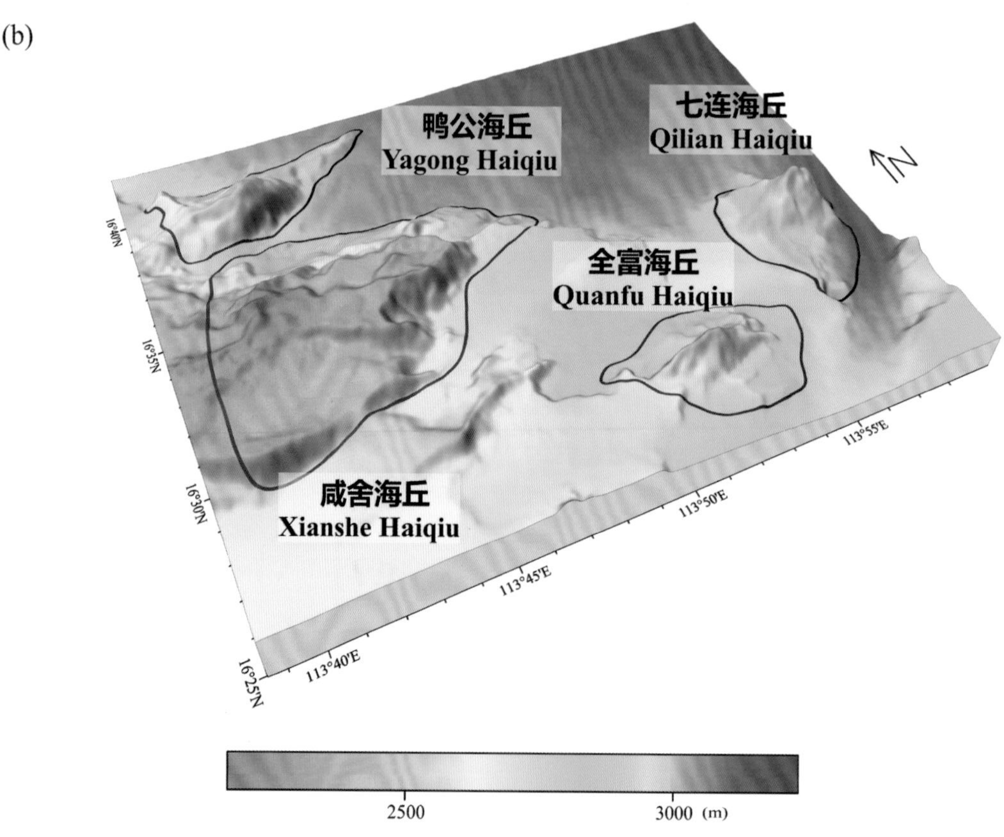

图 2-9　鸭公海丘、咸舍海丘、全富海丘、七连海丘

(a) 海底地形图（等深线间隔 500 米）；(b) 三维海底地形图

Fig.2-9　Yagong Haiqiu, Xianshe Haiqiu, Quanfu Haiqiu, Qilian Haiqiu

(a) Seafloor topographic map (with contour interval of 500 m); (b) 3-D seafloor topographic map

2.28 湛涵海台

标准名称 Standard Name	湛涵海台 Zhanhan Haitai	类别 Generic Term	海台 Plateau
中心点坐标 Center Coordinates	16°30.4′N, 112°57.0′E	规模（千米 × 千米） Dimension（km × km）	20 × 20
最小水深（米） Min Depth (m)	300	最大水深（米） Max Depth (m)	1240
地理实体描述 Feature Description	湛涵海台发育在西沙海隆西部，平面形态呈圆形（图 2–10）。 Zhanhan Haitai is located in the west of Xisha Hailong, with a circular planform (Fig.2–10).		
命名由来 Origin of Name	该海丘位于我国西沙群岛附近，故以西沙群岛中的岛屿、暗礁、暗滩、环礁、水道、沙洲名进行地名的团组化命名。“湛涵滩”属于西沙群岛的一个暗礁，取“湛涵”两字命名该海丘。 The Plateau is located near the Xisha Qundao in China, so it is named after a group of islands, reefs, shoals, atolls, channels and sandbars in Xisha Qundao. “Zhanhan Tan” is one of the reefs belonging to Xisha Qundao, so the word “Zhanhan” was used to name the Plateau.		

2.29 浪花海谷

标准名称 Standard Name	浪花海谷 Langhua Haigu	类别 Generic Term	海谷 Valley
中心点坐标 Center Coordinates	16°12.0′N, 112°52.0′E	规模（千米 × 千米） Dimension（km × km）	53 × 4
最小水深（米） Min Depth (m)	1200	最大水深（米） Max Depth (m)	2520
地理实体描述 Feature Description	浪花海谷发源于西沙群岛的东岛环礁，整体呈北西—南东向延伸发育，最终汇入到中沙海槽（图 2–10）。 Langhua Haigu is originated from Dongdao Huanjiao in Xisha Qundao. The Valley stretches in the direction of NW–SE, and finally ends at Zhongsha Haicao (Fig.2–10).		
命名由来 Origin of Name	该海谷位于我国西沙群岛附近，故以西沙群岛中的岛屿、暗礁、暗滩、环礁、水道、沙洲名进行地名的团组化命名。“浪花礁”属于西沙群岛的一个环礁，取“浪花”两字命名该海谷。 The Valley is located near the Xisha Qundao in China, so it is named after a group of islands, reefs, shoals, atolls, channels and sandbars in Xisha Qundao. “Langhua Jiao” is one of the atolls belonging to Xisha Qundao, so the word “Langhua” was used to name the Valley.		

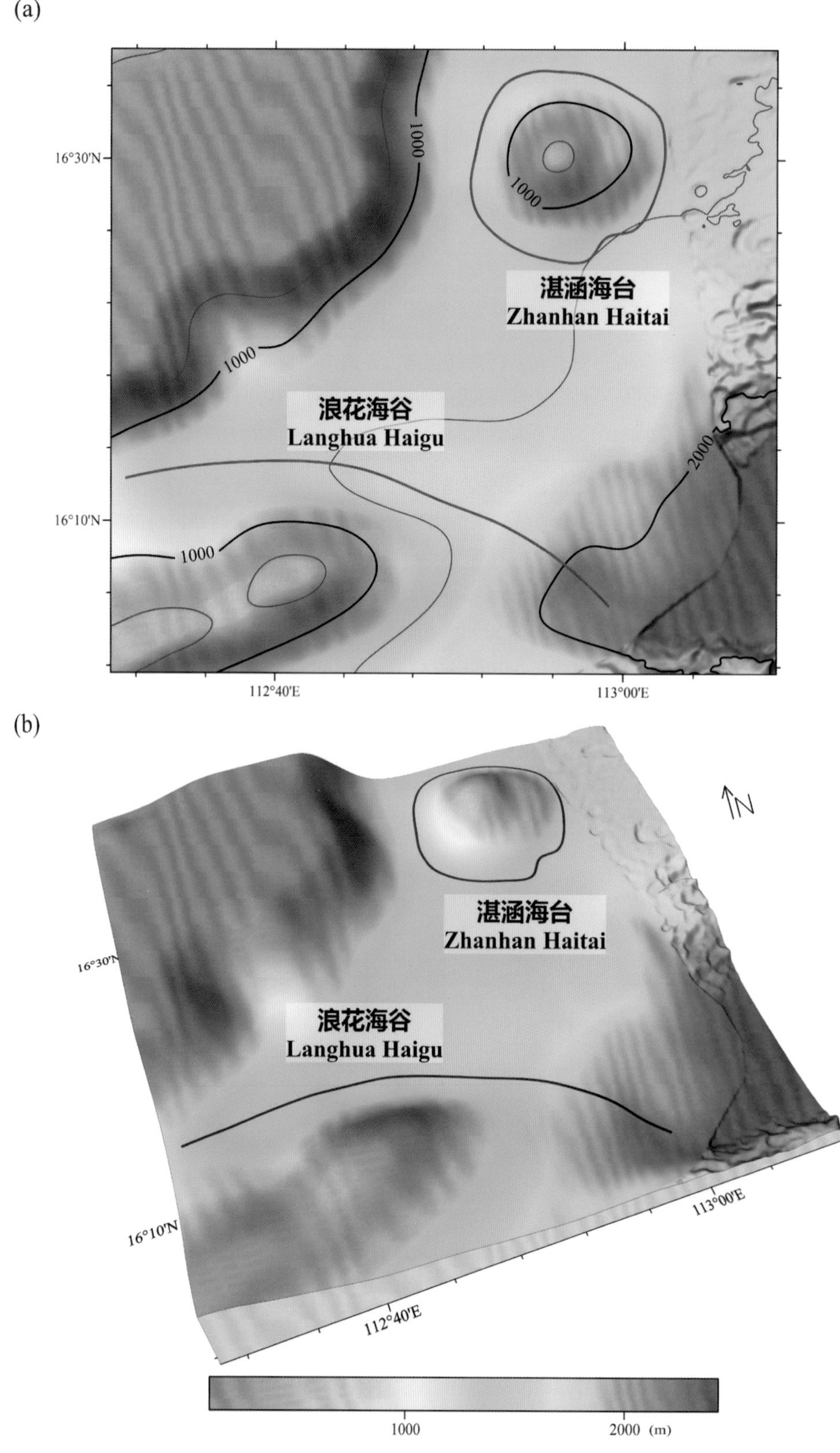

图 2-10 湛涵海台、浪花海谷

(a) 海底地形图（等深线间隔 500 米）；(b) 三维海底地形图

Fig.2-10 Zhanhan Haitai, Langhua Haigu

(a) Seafloor topographic map (with contour interval of 500 m); (b) 3-D seafloor topographic map

2.30 浪花海山

标准名称 Standard Name	浪花海山 Langhua Haishan	类别 Generic Term	海山 Seamount
中心点坐标 Center Coordinates	16°00.0′N, 112°59.9′E	规模（千米 × 千米） Dimension（km × km）	68 × 15
最小水深（米） Min Depth (m)	880	最大水深（米） Max Depth (m)	2660
地理实体描述 Feature Description	浪花海山发育在西沙群岛东南部，平面形态整体呈近东—西向的长条形。海山的北坡较陡，南坡较缓，其顶部发育 4 座峰（图 2-11）。 Langhua Haishan is located in the southeast of Xisha Qundao, with a planform in the shape of a long strip nearly in the direction of E–W. Its north slope is relative steeper while the south slope is relative gentler, with four peaks developed on its top (Fig.2–11).		
命名由来 Origin of Name	该海山位于浪花海谷附近，因此得名。 The Seamount is located near Langhua Haigu, so the word “Langhua” was used to name the Seamount.		

2.31 宣德海丘

标准名称 Standard Name	宣德海丘 Xuande Haiqiu	类别 Generic Term	海丘 Hill
中心点坐标 Center Coordinates	15°55.4'N, 113°05.5'E	规模（千米 × 千米） Dimension（km × km）	10 × 8
最小水深（米） Min Depth (m)	1760	最大水深（米） Max Depth (m)	2640
地理实体描述 Feature Description	宣德海丘位于西沙群岛东南部、浪花海山南侧，平面形态近似三角形（图 2-11）。 Xuande Haiqiu is located in the southeast of Xisha Qundao and is on the south of Langhua Haishan, with a subtriangular planform (Fig.2–11).		
命名由来 Origin of Name	该海丘位于我国西沙群岛附近，故以西沙群岛中的岛屿、暗礁、暗滩、环礁、水道、沙洲名进行地名的团组化命名。“宣德群岛”属于西沙群岛的一组岛屿，取“宣德”两字命名该海丘。 The Hill is located near the Xisha Qundao in China, so it is named after a group of islands, reefs, shoals, atolls, channels and sandbars in Xisha Qundao. “Xuande Qundao” is one of the islands belonging to Xisha Qundao, so the word “Xuande” was used to name the Hill.		

2.32 玉琢海底峡谷

标准名称 Standard Name	玉琢海底峡谷 Yuzhuo Haidixiagu	类别 Generic Term	海底峡谷 Canyon
中心点坐标 Center Coordinates	15°53.9'N, 112°44.3'E	规模（千米 × 千米） Dimension（km × km）	40 × 2
最小水深（米） Min Depth (m)	1600	最大水深（米） Max Depth (m)	2540
地理实体描述 Feature Description	玉琢海底峡谷发源于西沙群岛的浪花礁，其头部由两条分支海底峡谷分别沿着南方和东北方向发育，在15°53′N, 112°41′E处汇合形成一条主干峡谷并继续向东侧延伸发育，峡谷总长约40千米，峡谷头部和尾部的水深差达1000米（图2−11）。 Yuzhuo Haidixiagu originates from Langhua Jiao in Xisha Qundao. Two canyons develop from the head towards south and northeast respectively and combine into one canyon at 15°53′N, 112°41′E, which continues to stretches towards east. The total length of the Canyon of about 40 km, and the depth difference of the head and the tail of the Canyon is up to 1000 m (Fig.2−11).		
命名由来 Origin of Name	该海底峡谷位于我国西沙群岛附近，故以西沙群岛中的岛屿、暗礁、暗滩、环礁、水道、沙洲名进行地名的团组化命名。“玉琢礁”属于西沙群岛的一个环礁，取“玉琢”两字命名该海山。 The Canyon is located near the Xisha Qundao in China, so it is named after a group of islands, reefs, shoals, atolls, channels and sandbars in Xisha Qundao. “Yuzhuo Jiao” is one of the atolls belonging to Xisha Qundao, so the word “Yuzhuo” was used to name the Canyon.		

2.33 银砾海山

标准名称 Standard Name	银砾海山 Yinli Haishan	类别 Generic Term	海山 Seamount
中心点坐标 Center Coordinates	15°47.5′N, 112°42.5′E	规模（千米 × 千米） Dimension（km × km）	25 × 15
最小水深（米） Min Depth (m)	290	最大水深（米） Max Depth (m)	2660
地理实体描述 Feature Description	银砾海山位于玉琢海底峡谷南侧，顶部平坦。海山的北坡较缓，南坡较陡（图2−11）。 Yinli Haishan is located on the south of Yuzuo Haidixiagu, with a flat top. Its north slope is relative gentler and the south slope is relative steeper (Fig.2−11).		
命名由来 Origin of Name	该海山位于我国西沙群岛附近，故以西沙群岛中的岛屿、暗礁、暗滩、环礁、水道、沙洲名进行地名的团组化命名。“银砾滩”属于西沙群岛的一个暗滩，取“银砾”两字命名该海山。 The Seamount is located near the Xisha Qundao in China, so it is named after a group of islands, reefs, shoals, atolls, channels and sandbars in Xisha Qundao. “Yinli Tan” is one of the shoals belonging to Xisha Qundao, so the word “Yinli” was used to name the Seamount.		

2.34 筐仔海山

标准名称 Standard Name	筐仔海山 Kuangzai Haishan	类别 Generic Term	海山 Seamount
中心点坐标 Center Coordinates	15°49.2'N, 112°49.5'E	规模（千米 × 千米） Dimension（km × km）	12 × 12
最小水深（米） Min Depth (m)	820	最大水深（米） Max Depth (m)	2500
地理实体描述 Feature Description	筐仔海山位于玉琢海底峡谷南侧，与银砾海山相邻。海山的北坡较缓，南坡较陡（图 2–11）。 Kuangzai Haishan is located on the south of Yuzuo Haidixiagu and is adjacent to Yinli Haishan. Its north slope is relative gentler and the south slope is relative steeper (Fig.2–11).		
命名由来 Origin of Name	该海山位于我国西沙群岛附近，故以西沙群岛中的岛屿、暗礁、暗滩、环礁、水道、沙洲名进行地名的团组化命名。“筐仔沙洲”属于西沙群岛的一个沙洲，取“筐仔”两字命名该海山。 The Seamount is located near the Xisha Qundao in China, so it is named after a group of islands, reefs, shoals, atolls, channels and sandbars in Xisha Qundao. “Kuangzai Shazhou” is one of the sandbars belonging to Xisha Qundao, so the word “Kuangzai” was used to name the Seamount.		

2.35 金银海脊

标准名称 Standard Name	金银海脊 Jinyin Haiji	类别 Generic Term	海脊 Ridge
中心点坐标 Center Coordinates	15°44.7'N, 113°01.0'E	规模（千米 × 千米） Dimension（km × km）	30 × 3
最小水深（米） Min Depth (m)	1870	最大水深（米） Max Depth (m)	2770
地理实体描述 Feature Description	金银海脊位于西沙群岛东南部，其脊线呈北西—南东走向（图 2–11）。 Jinyin Haiji is located in the southeast of Xisha Qundao and stretches in the direction of NW—SE (Fig.2–11).		
命名由来 Origin of Name	该海脊位于我国西沙群岛附近，故以西沙群岛中的岛屿、暗礁、暗滩、环礁、水道、沙洲名进行地名的团组化命名。“金银岛”属于西沙群岛的一个岛屿，取“金银”两字命名该海脊。 The Ridge is located near the Xisha Qundao in China, so it is named after a group of islands, reefs, shoals, atolls, channels and sandbars in Xisha Qundao. “Jinyin Dao” is one of the islands belonging to Xisha Qundao, so the word “Jinyin” was used to name the Ridge.		

(a)

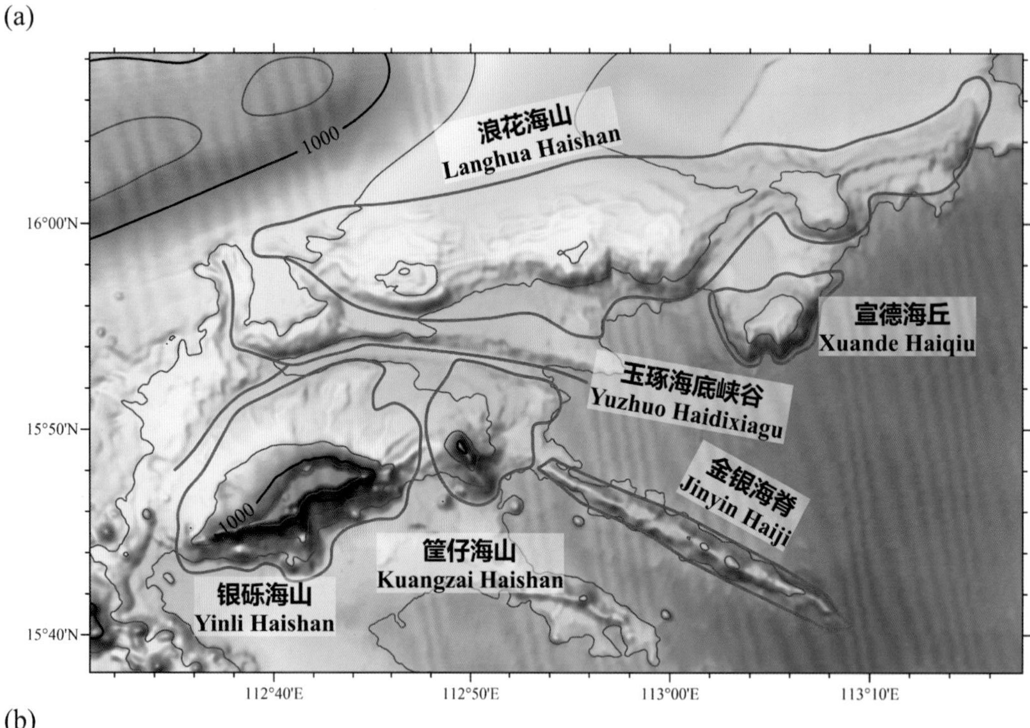

(b)

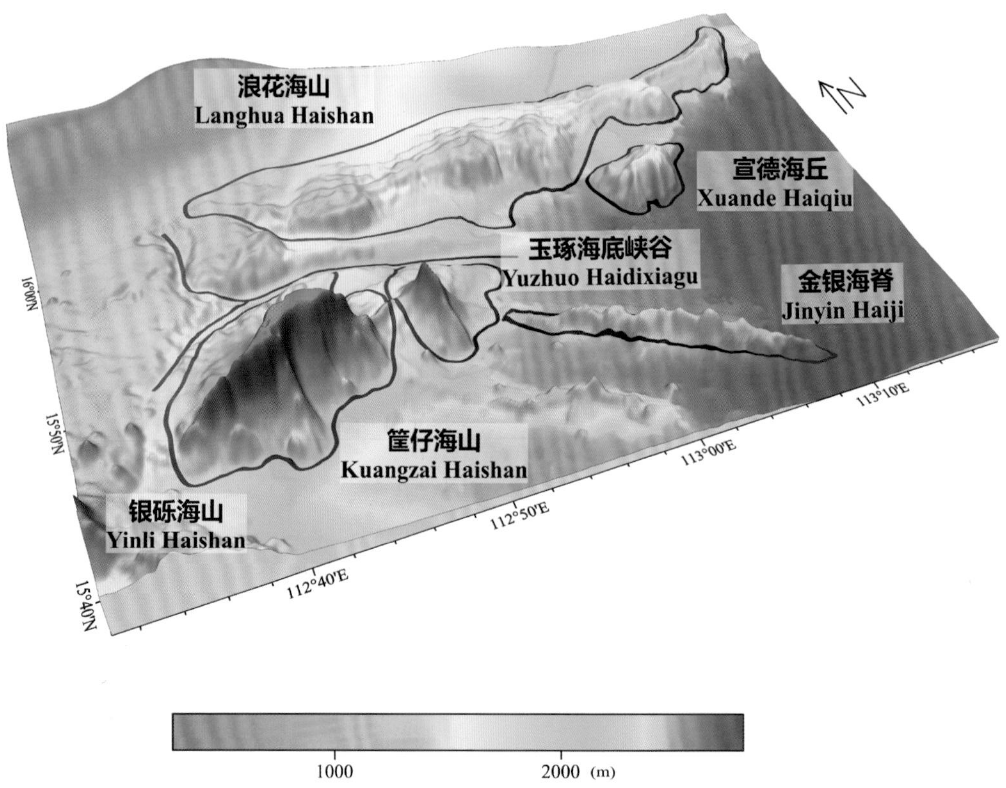

图 2-11　浪花海山、宣德海丘、玉琢海底峡谷、银砾海山、筐仔海山、金银海脊

(a) 海底地形图（等深线间隔 500 米）；(b) 三维海底地形图

Fig.2-11　Langhua Haishan, Xuande Haiqiu, Yuzhuo Haidixiagu, Yinli Haishan, Kuangzai Haishan, Jinyin Haiji

(a) Seafloor topographic map (with contour interval of 500 m); (b) 3-D seafloor topographic map

2.36 广金海脊

标准名称 Standard Name	广金海脊 Guangjin Haiji	类别 Generic Term	海脊 Ridge
中心点坐标 Center Coordinates	15°40.5′N, 112°29.2′E	规模（千米 × 千米） Dimension（km × km）	35 × 8
最小水深（米） Min Depth (m)	440	最大水深（米） Max Depth (m)	2560
地理实体描述 Feature Description	广金海脊位于西沙群岛东南部，其脊线呈北西—南东走向（图 2–12）。 Guangjin Haiji is located in the southeast of Xisha Qundao and stretches in the direction of NW–SE (Fig.2–12).		
命名由来 Origin of Name	该海脊位于我国西沙群岛附近，故以西沙群岛中的岛屿、暗礁、暗滩、环礁、水道、沙洲名进行地名的团组化命名。“广金岛”属于西沙群岛的一个岛屿，取“广金”两字命名该海脊。 The Ridge is located near the Xisha Qundao in China, so it is named after a group of islands, reefs, shoals, atolls, channels and sandbars in Xisha Qundao. “Guangjin Dao” is one of the islands belonging to Xisha Qundao, so the word “Guangjin” was used to name the Ridge.		

2.37 湛涵海脊

标准名称 Standard Name	湛涵海脊 Zhanhan Haiji	类别 Generic Term	海脊 Ridge
中心点坐标 Center Coordinates	15°29.5′N, 112°28.2′E	规模（千米 × 千米） Dimension（km × km）	32 × 5
最小水深（米） Min Depth (m)	1410	最大水深（米） Max Depth (m)	2190
地理实体描述 Feature Description	湛涵海脊位于西沙群岛南部，其脊线呈北东东—南西西走向（图 2–12）。 Zhanhan Haiji is located in the south of Xisha Qundao and stretches in the direction of NEE–SWW (Fig.2–12).		
命名由来 Origin of Name	该海脊位于我国西沙群岛附近，故以西沙群岛中的岛屿、暗礁、暗滩、环礁、水道、沙洲名进行地名的团组化命名。“湛涵滩”属于西沙群岛的一个暗礁，取“湛涵”两字命名该海脊。 The Ridge is located near the Xisha Qundao in China, so it is named after a group of islands, reefs, shoals, atolls, channels and sandbars in Xisha Qundao. “Zhanhan Tan” is one of the reefs belonging to Xisha Qundao, so the word “Zhanhan” was used to name the Ridge.		

2.38 湛涵北海丘

标准名称 Standard Name	湛涵北海丘 Zhanhanbei Haiqiu	类别 Generic Term	海丘 Hill
中心点坐标 Center Coordinates	15°32.0′N, 112°12.4′E	规模（千米 × 千米） Dimension（km × km）	13 × 9
最小水深（米） Min Depth (m)	1600	最大水深（米） Max Depth (m)	2000
地理实体描述 Feature Description	湛涵北海丘位于西沙群岛南部，平面形态不规则。海丘的西北水深较浅，东南水深较深（图 2–12）。 Zhanhanbei Haiqiu is located in the south of Xisha Qundao with an irregular planform. The depth of the Hill is shallower on the northwest side and is deeper on the southeast side (Fig.2–12).		
命名由来 Origin of Name	该海丘位于湛涵海脊以北海域，因此得名。 The Hill is located in the sea area to the north of Zhanhan Haiji, and “Bei” means north in Chinese, so the word “Zhanhanbei” was used to name the Hill.		

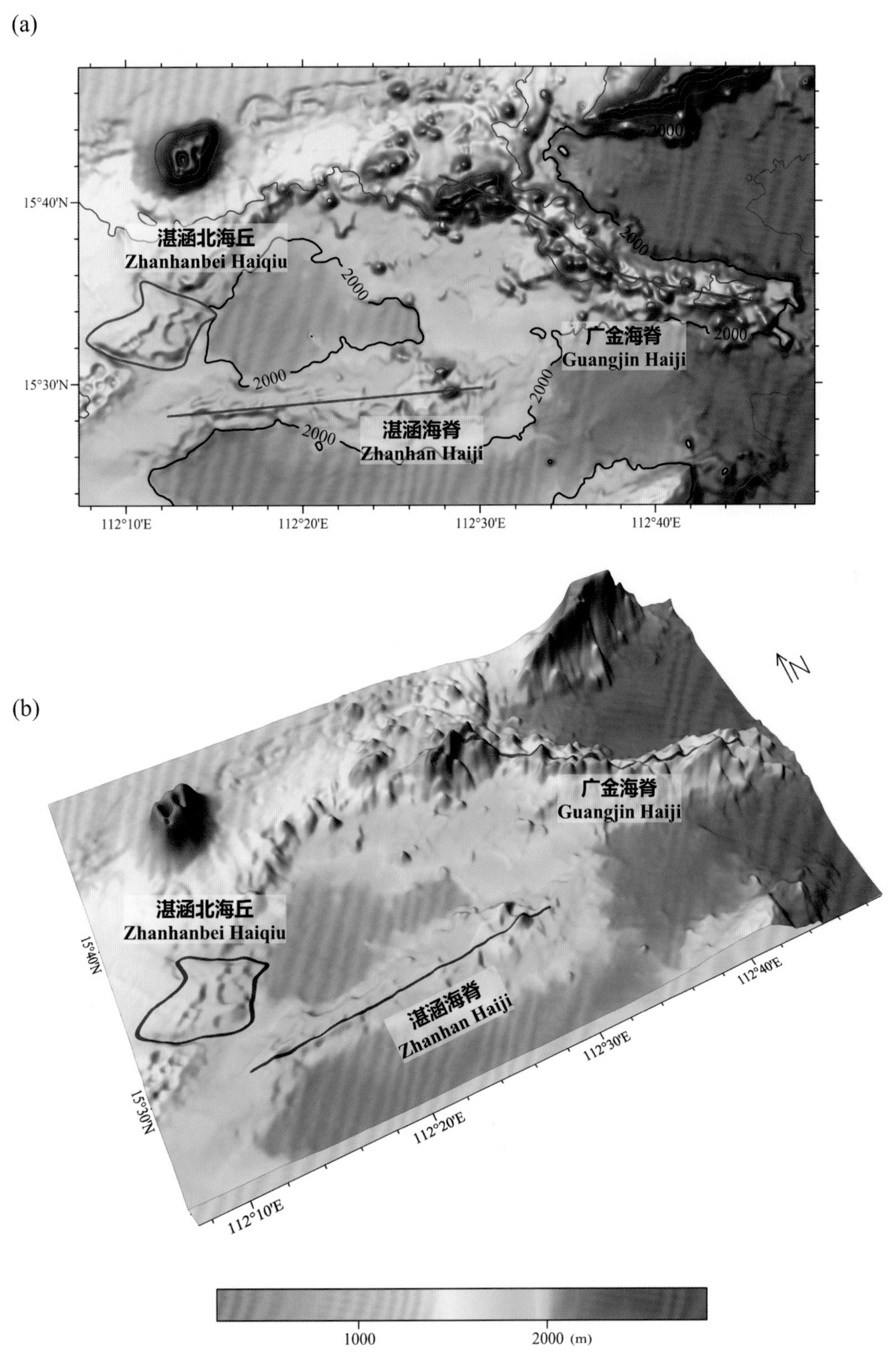

图 2-12　广金海脊、湛涵海脊、湛涵北海丘

(a) 海底地形图（等深线间隔 500 米）；(b) 三维海底地形图

Fig.2-12　Guangjin Haiji, Zhanhan Haiji, Zhanhanbei Haiqiu

(a) Seafloor topographic map (with contour interval of 500 m); (b) 3-D seafloor topographic map

2.39 羚羊海山

标准名称 Standard Name	羚羊海山 Lingyang Haishan	类别 Generic Term	海山 Seamount
中心点坐标 Center Coordinates	15°26.5′N, 112°04.2′E	规模（千米 × 千米） Dimension（km × km）	22 × 10
最小水深（米） Min Depth (m)	396	最大水深（米） Max Depth (m)	1732
地理实体描述 Feature Description	羚羊海山位于西沙群岛南部，平面形态呈北东—南西向的纺锤形。海山的西北坡较缓，东南坡较陡（图 2−13）。 Lingyang Haishan is located in the south of Xisha Qundao, with a spindle-shaped planform in the direction of NE−SW. Its northwest slope is relative gentler while its southeast slope is relatively steeper (Fig.2−13).		
命名由来 Origin of Name	该海山位于我国西沙群岛附近，故以西沙群岛中的岛屿、暗礁、暗滩、环礁、水道、沙洲名进行地名的团组化命名。“羚羊礁”属于西沙群岛的一个环礁，取“羚羊”两字命名该海山。 The Seamount is located near the Xisha Qundao in China, so it is named after a group of islands, reefs, shoals, atolls, channels and sandbars in Xisha Qundao. “Lingyang Jiao” is one of the atolls belonging to Xisha Qundao, so the word “Lingyang” was used to name the Seamount.		

2.40 羚羊南海山

标准名称 Standard Name	羚羊南海山 Lingyangnan Haishan	类别 Generic Term	海山 Seamount
中心点坐标 Center Coordinates	15°09.6'N, 111°54.2'E	规模（千米 × 千米） Dimension（km × km）	17 × 9
最小水深（米） Min Depth (m)	240	最大水深（米） Max Depth (m)	1350
地理实体描述 Feature Description	羚羊南海山位于西沙群岛南部，平面形态呈北西—南东向的纺锤形（图 2−13）。 Lingyangnan Haishan is located in the south of Xisha Qundao, with a spindle-shaped planform in the direction of NW−SE (Fig.2−13).		
命名由来 Origin of Name	该海山位于羚羊海山以南，因此得名。 Named from its geographical location. The Seamount is located to the south of Lingyang Haishan, and “Nan” means south in Chinese, so the word “Lingyangnan” was used to name the Seamount.		

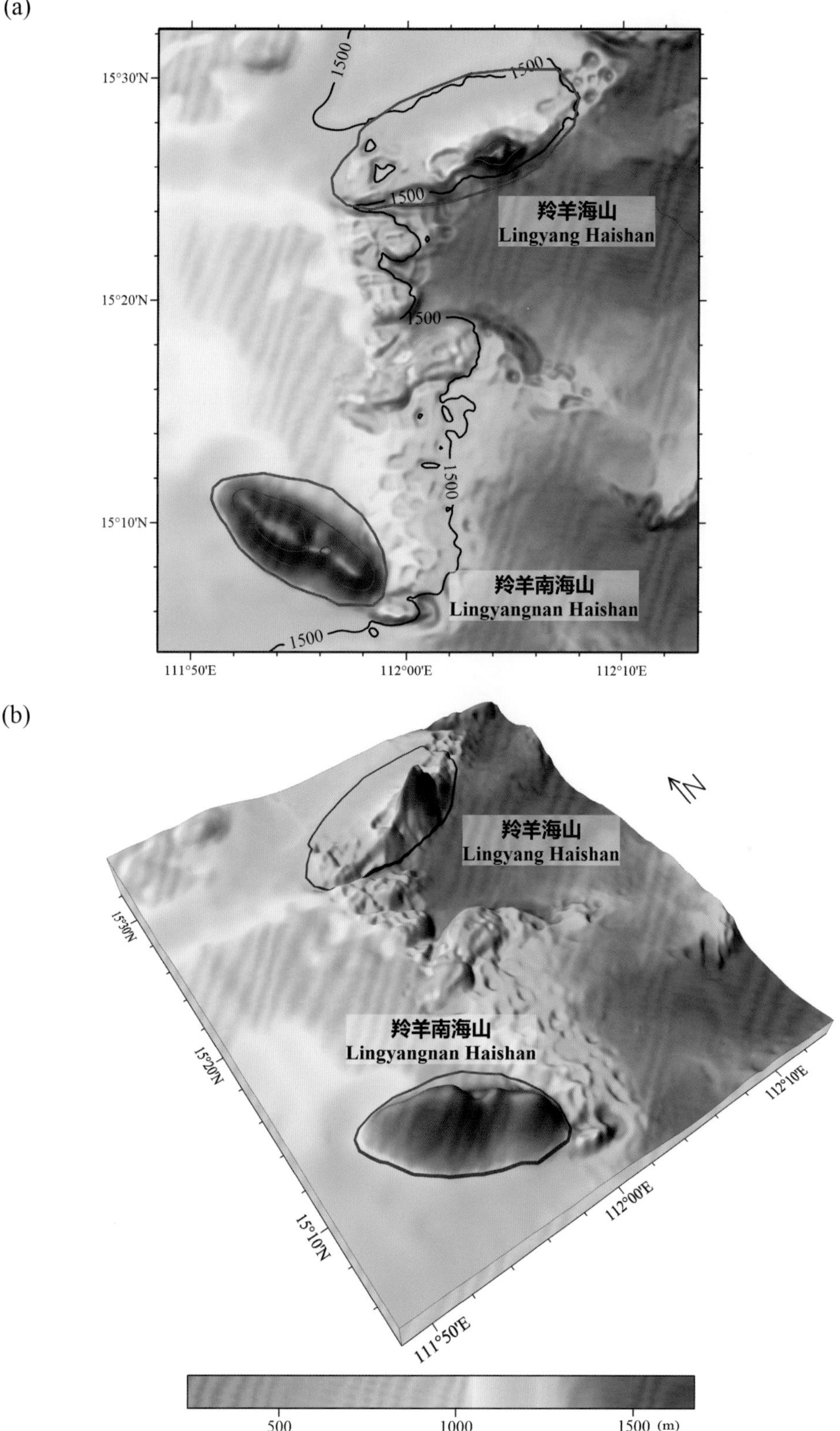

图 2-13　羚羊海山、羚羊南海山

(a) 海底地形图（等深线间隔 500 米）；(b) 三维海底地形图

Fig.2-13　Lingyang Haishan, Lingyangnan Haishan

(a) Seafloor topographic map (with contour interval of 500 m)；(b) 3-D seafloor topographic map

2.41 杨柳西海丘

标准名称 Standard Name	杨柳西海丘 Yangliuxi Haiqiu	类别 Generic Term	海丘 Hill
中心点坐标 Center Coordinates	14°48.5'N, 111°38.9'E	规模（千米 × 千米） Dimension（km × km）	15 × 6
最小水深（米） Min Depth (m)	360	最大水深（米） Max Depth (m)	2020
地理实体描述 Feature Description	杨柳西海丘发育在中建南斜坡东部，平面形态呈北东—南西向的椭圆形（图 2−14）。 Yangliuxi Haiqiu is located in the east of Zhongjiannan Xiepo, with an oval-shaped planform in the direction of NE−SW (Fig.2−14).		
命名由来 Origin of Name	该海丘位于杨柳海山以西，因此得名。 The Hill is located to the west of Yangliu Haishan, and ‘Xi’ means west in Chinese, so the word “Yangliuxi” was used to name the Hill.		

(a)

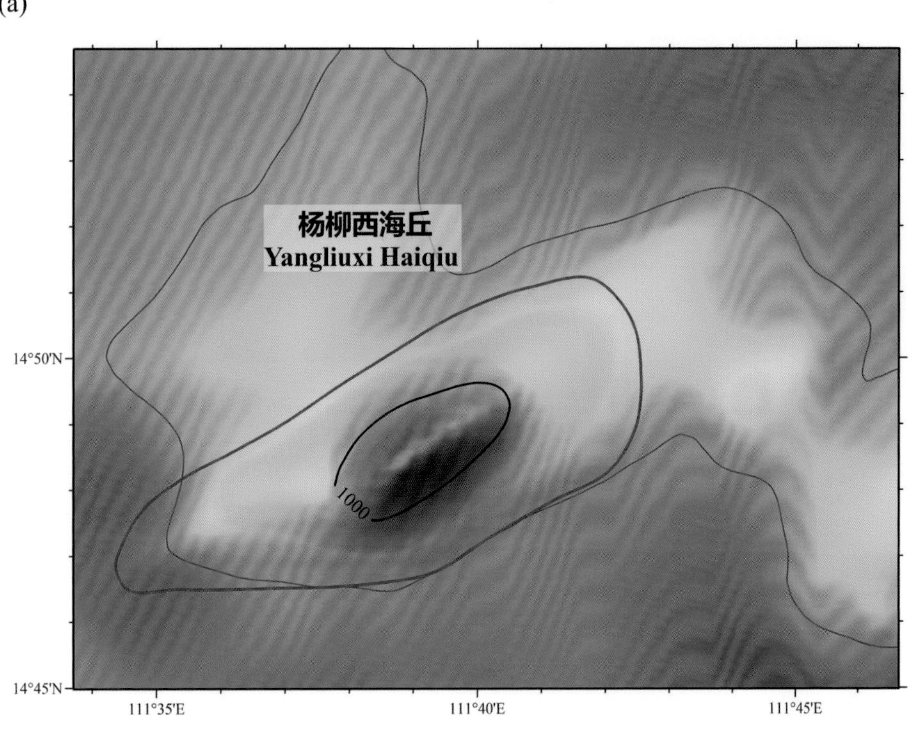

(b)

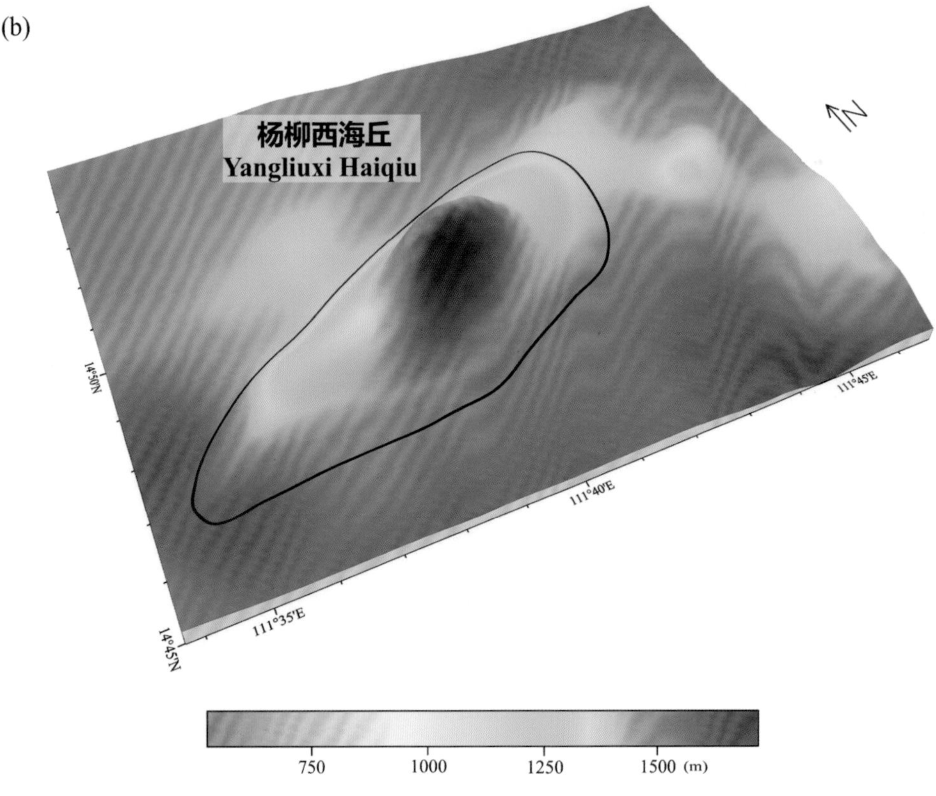

图 2-14 杨柳西海丘

(a) 海底地形图（等深线间隔 500 米）；(b) 三维海底地形图

Fig.2-14 Yangliuxi Haiqiu

(a) Seafloor topographic map (with contour interval of 500 m); (b) 3-D seafloor topographic map

2.42 晋卿海山

标准名称 Standard Name	晋卿海山 Jinqing Haishan	类别 Generic Term	海山 Seamount
中心点坐标 Center Coordinates	15°23.6′N, 112°39.8′E	规模（千米 × 千米） Dimension（km × km）	25 × 13
最小水深（米） Min Depth (m)	1610	最大水深（米） Max Depth (m)	3030
地理实体描述 Feature Description	晋卿海山位于西沙群岛南部，兰宫洼地北侧，平面形态呈北东东—南西西向的长条形。海山的西北坡较缓，东南坡较陡（图 2–15）。 Jinqing Haishan is located to the south of Xisha Qundao and is on the north of Langong Wadi, with a planform in the shape of a long strip in the direction of NEE–SWW. Its northwest slope is relatively gentler, while its southeast slope is relatively steeper (Fig.2–15).		
命名由来 Origin of Name	该海山位于我国西沙群岛附近，故以西沙群岛中的岛屿、暗礁、暗滩、环礁、水道、沙洲名进行地名的团组化命名。"晋卿岛"属于西沙群岛的一个岛屿，取"晋卿"两字命名该海山。 The Seamount is located near the Xisha Qundao in China, so it is named after a group of islands, reefs, shoals, atolls, channels and sandbars in Xisha Qundao. "Jinqing Dao" is one of the islands belonging to Xisha Qundao, so the word "Jinqing" was used to name the Seamount.		

2.43 晋卿西海山

标准名称 Standard Name	晋卿西海山 Jinqingxi Haishan	类别 Generic Term	海山 Seamount
中心点坐标 Center Coordinates	15°17.8′N, 112°29.2′E	规模（千米 × 千米） Dimension（km × km）	18 × 10
最小水深（米） Min Depth (m)	560	最大水深（米） Max Depth (m)	2890
地理实体描述 Feature Description	晋卿西海山位于西沙群岛南部，兰宫洼地北侧，平面形态呈北东—南西向的不规则多边形。海山的西北坡较缓，东南坡较陡（图 2–15）。 Jinqingxi Haishan is located to the south of Xisha Qundao and is on the north of Langong Wadi, with a planform in the shape of an irregular polygon in the direction of NE–SW. Its northwest slope is relatively gentler, while its southeast slope is relatively steeper (Fig.2–15).		
命名由来 Origin of Name	该海山位于晋卿海山以西，因此得名。 The Seamount is located on the west of Jinqing Haishan, and "Xi" means west in Chinese, so the word "Jinqingxi" was used to name the Seamount.		

(a)

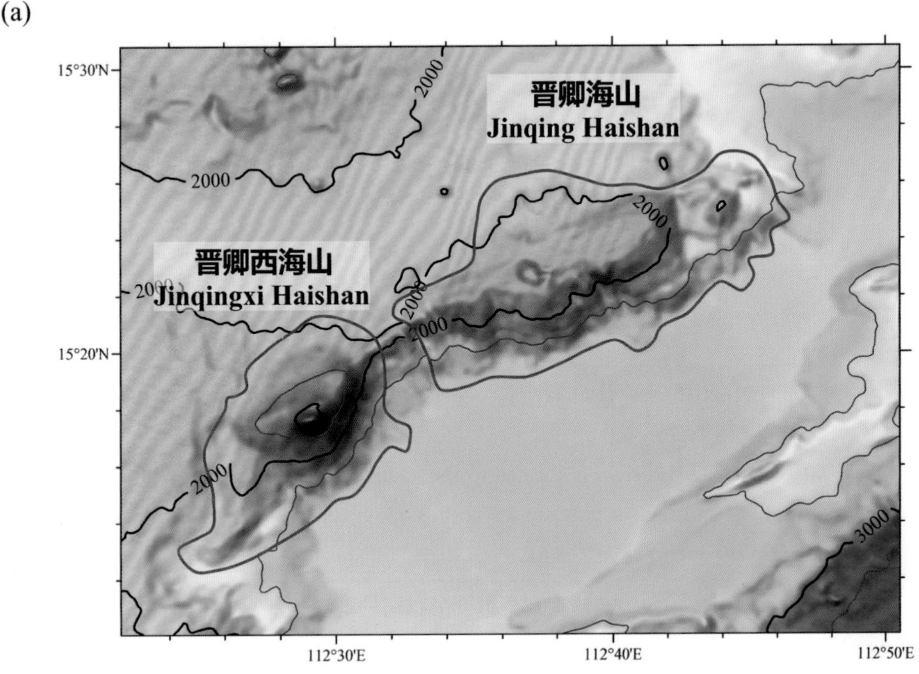

(b)

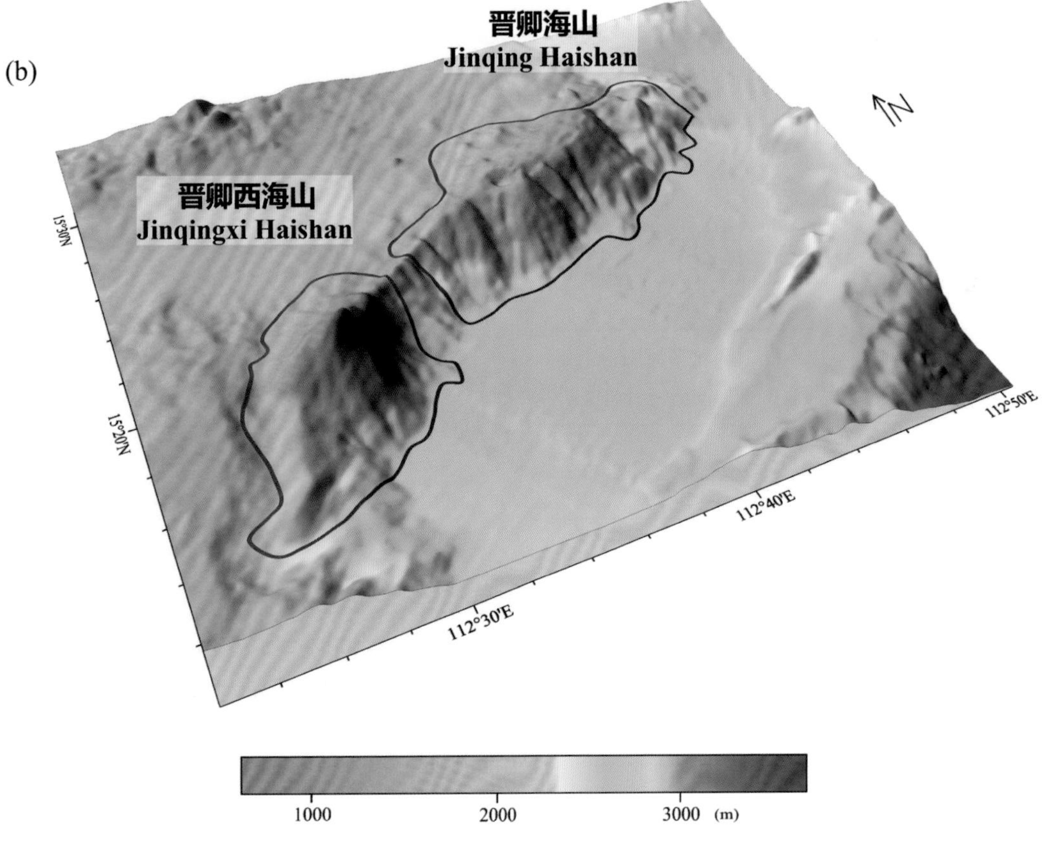

图 2-15　晋卿海山、晋卿西海山

(a) 海底地形图（等深线间隔 500 米）；(b) 三维海底地形图

Fig.2-15　Jinqing Haishan, Jinqingxi Haishan

(a) Seafloor topographic map (with contour interval of 500 m); (b) 3-D seafloor topographic map

2.44 琛航海山

标准名称 Standard Name	琛航海山 Chenhang Haishan	类别 Generic Term	海山 Seamount
中心点坐标 Center Coordinates	15°00.0′N, 112°16.7′E	规模（千米 × 千米） Dimension（km × km）	31 × 16
最小水深（米） Min Depth (m)	384	最大水深（米） Max Depth (m)	2444
地理实体描述 Feature Description	琛航海山位于西沙群岛南部，平面形态呈不规则多边形。海山的北坡较缓，南坡较陡（图 2–16）。 Chenhang Haishan is located to the south of Xisha Qundao, with a planform in the shape of an irregular polygon. Its north slope is relatively gentler, while its south slope is relatively steeper (Fig.2–16).		
命名由来 Origin of Name	该海山位于我国西沙群岛附近，故以西沙群岛中的岛屿、暗礁、暗滩、环礁、水道、沙洲名进行地名的团组化命名。“琛航岛”属于西沙群岛的一个岛屿，取“琛航”两字命名该海山。 The Seamount is located near the Xisha Qundao in China, so it is named after a group of islands, reefs, shoals, atolls, channels and sandbars in Xisha Qundao. “Chenhang Dao” is one of the islands belonging to Xisha Qundao, so the word “Chenhang” was used to name the Seamount.		

2.45 琛航东海山

标准名称 Standard Name	琛航东海山 Chenhangdong Haishan	类别 Generic Term	海山 Seamount
中心点坐标 Center Coordinates	15°04.9'N, 112°22.4'E	规模（千米 × 千米） Dimension（km × km）	18 × 9
最小水深（米） Min Depth (m)	1200	最大水深（米） Max Depth (m)	2700
地理实体描述 Feature Description	琛航东海山位于西沙群岛南部，平面形态呈北东—南西向的长条形。海山的西北坡较缓，东南坡较陡（图 2–16）。 Chenhangdong Haishan is located to the south of Xisha Qundao, with a planform of long strip in the direction of NE–SW. Its northwest slope is a gentler, while its southeast slope is steeper (Fig.2–16).		
命名由来 Origin of Name	该海山位于琛航海山以东海域，因此得名。 Named from its geographical location. The Seamount is located in the sea area to the east of Chenhang Haishan, and “Dong” means east in Chinese, so the word “Chenhangdong” was used to name the Seamount.		

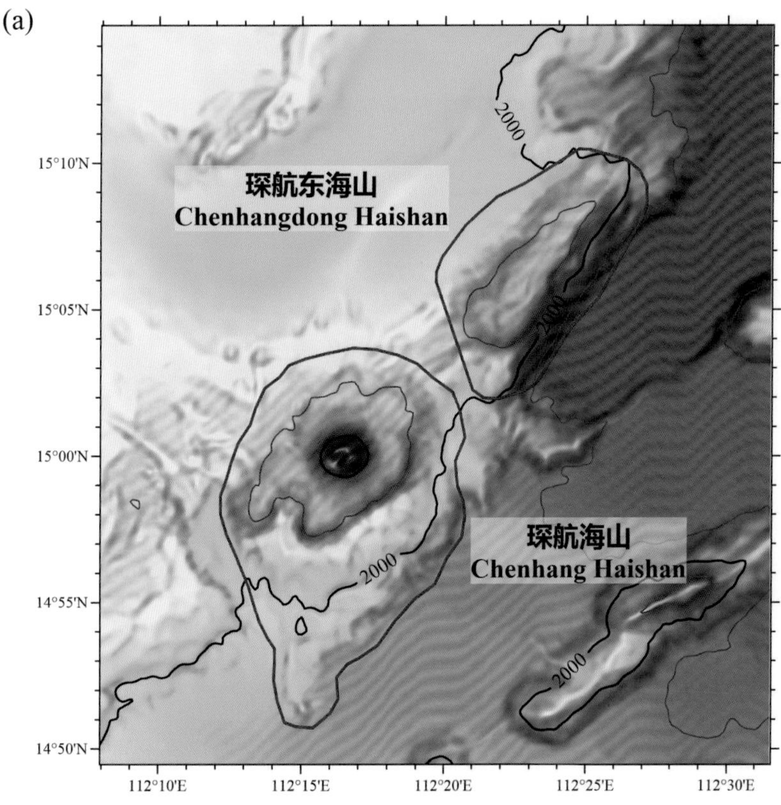

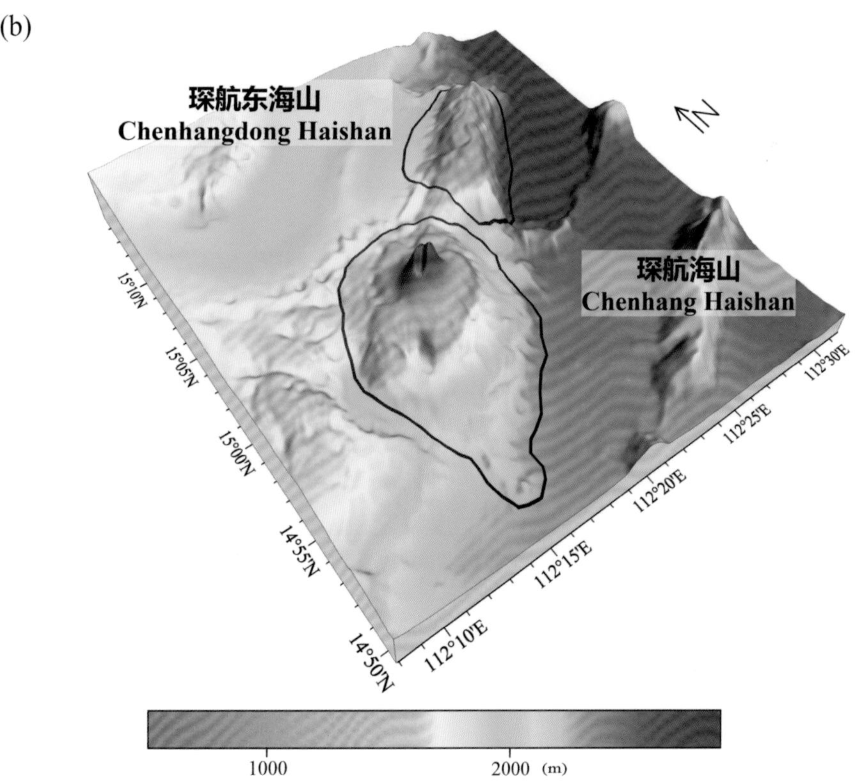

图 2-16 琛航海山、琛航东海山

(a) 海底地形图（等深线间隔 500 米）；(b) 三维海底地形图

Fig.2-16 Chenhang Haishan, Chenhangdong Haishan

(a) Seafloor topographic map (with contour interval of 500 m); (b) 3-D seafloor topographic map

2.46 甘泉海台

标准名称 Standard Name	甘泉海台 Ganquan Haitai	类别 Generic Term	海台 Plateau
中心点坐标 Center Coordinates	16°42.0′N, 110°56.0′E	规模（千米 × 千米） Dimension（km × km）	53 × 21
最小水深（米） Min Depth (m)	570	最大水深（米） Max Depth (m)	1490
地理实体描述 Feature Description	甘泉海台位于西沙群岛西部，平面形态呈北东—南西向的长条形。海台的顶部较平坦，平均水深 600 米，向四周水深迅速变大（图 2–17）。 Ganquan Haitai is located in the west of Xisha Qundao, with a planform in the shape of a long strip in the direction of NE–SW. The Plateau has a flat top with an average depth of 600 m, and depth increases rapidly from the center to all directions (Fig.2–17).		
命名由来 Origin of Name	该海台位于我国西沙群岛附近，故以西沙群岛中的岛屿、暗礁、暗滩、环礁、水道、沙洲名进行地名的团组化命名。“甘泉岛”属于西沙群岛的一个岛屿，取“甘泉”两字命名该海台。 The Plateau is located near the Xisha Qundao in China, so it is named after a group of islands, reefs, shoals, atolls, channels and sandbars in Xisha Qundao. “Ganquan Dao” is one of the islands belonging to Xisha Qundao, so the word “Ganquan” was used to name the Plateau.		

2.47 甘泉海谷

标准名称 Standard Name	甘泉海谷 Ganquan Haigu	类别 Generic Term	海谷 Valley
中心点坐标 Center Coordinates	16°18.0′N, 111°00.0′E	规模（千米 × 千米） Dimension（km × km）	120 × 20
最小水深（米） Min Depth (m)	1130	最大水深（米） Max Depth (m)	1480
地理实体描述 Feature Description	甘泉海谷位于西沙群岛西南部，平面形态呈北西—南东向的折线，其北端与西沙海槽相连（图 2–17）。 Ganquan Haigu is located in the southwest of Xisha Qundao, with a planform in the shape of a polyline in the direction of NW–SE, and connects to Xisha Haicao on the north (Fig.2–17).		
命名由来 Origin of Name	该海谷位于甘泉海台附近，因此得名。 The Valley is located near Ganquan Haitai, so the word “Ganquan” was used to name the Valley.		

2.48 中建北海台

标准名称 Standard Name	中建北海台 Zhongjianbei Haitai	类别 Generic Term	海台 Plateau
中心点坐标 Center Coordinates	16°01.9′N, 110°57.0′E	规模（千米 × 千米）Dimension（km × km）	23 × 20
最小水深（米）Min Depth (m)	380	最大水深（米）Max Depth (m)	1380
地理实体描述 Feature Description	中建北海台位于西沙群岛西南部，其顶部较平坦，平均水深 400 米，向四周水深迅速变大（图 2-17）。 Zhongjianbei Haitai is located in the southwest of Xisha Qundao. The Plateau has a flat top, with an average depth of 400 m, and the depth increases rapidly from the center to all directions (Fig.2-17).		
命名由来 Origin of Name	该海台位于中建岛以北，因此得名。 The Plateau is located on the north of Zhongjian Dao, and “Bei” means north in Chinese, so the word “Zhongjianbei” was used to name the Plateau.		

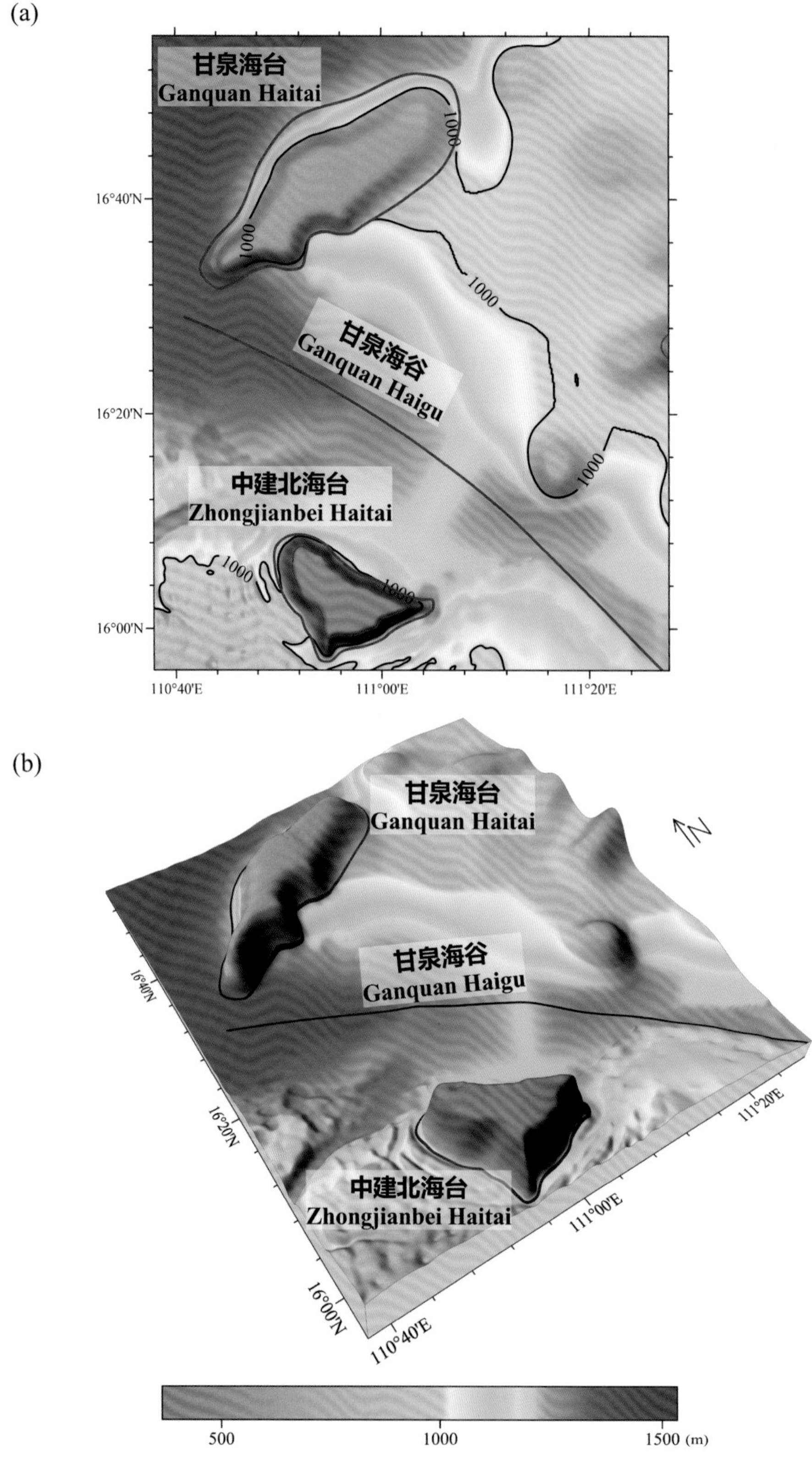

图 2-17 甘泉海台、甘泉海谷、中建北海台

(a) 海底地形图（等深线间隔 500 米）；(b) 三维海底地形图

Fig.2-17 Ganquan Haitai, Ganquan Haigu, Zhongjianbei Haitai

(a) Seafloor topographic map (with contour interval of 500 m); (b) 3-D seafloor topographic map

2.49 中建阶地

标准名称 Standard Name	中建阶地 Zhongjian Jiedi	类别 Generic Term	海底阶地 Terrace
中心点坐标 Center Coordinates	15°30.0′N, 110°30.0′E	规模（千米 × 千米） Dimension（km × km）	190 × 55
最小水深（米） Min Depth (m)	250	最大水深（米） Max Depth (m)	700
地理实体描述 Feature Description	中建阶地北接西沙海槽和西沙海台，东邻中建南斜坡，南至中建南海底峡谷群，南北向长约 170 千米，东西向宽约 80 千米。阶地的平面形态整体呈南—北走向，其东北侧有一块凸出区域（图 2–18）。 Zhongjian Jiedi borders Xisha Haicao and Xisha Haitai on the north, is adjacent to Zhongjiannan Xiepo on the east and reaches Zhongjiannan Haidixiaguqun on the south, with a length of about 170 km from south to north, a width of about 80 km from east to west. The whole Terrace is NS-oriented with a protruding area on the northeast side (Fig.2–18).		
命名由来 Origin of Name	该海底阶地邻近中建岛，因此得名。 The Terrace is adjacent to Zhongjian Dao, so the word "Zhongjian" was used to name the Terrace.		

2.50 中建南斜坡

标准名称 Standard Name	中建南斜坡 Zhongjiannan Xiepo	类别 Generic Term	海底斜坡 Slope
中心点坐标 Center Coordinates	14°50.0′N, 111°30.0′E	规模（千米 × 千米） Dimension（km × km）	250 × 170
最小水深（米） Min Depth (m)	50	最大水深（米） Max Depth (m)	2600
地理实体描述 Feature Description	中建南斜坡位于南海西部陆坡中部、中建阶地以东，水深由西北向东南方向逐渐增大。斜坡上发育多个海底峡谷群、海丘等（图 2–18）。 Zhongjiannan Xiepo is located in the middle of Nanhaixibu Lupo and is to the east of Zhongjian Jiedi, with depth increasing gradually from northwest to southeast. Several canyons and hills have developed on it (Fig.2–18).		
命名由来 Origin of Name	该海底斜坡位于中建岛以南，因此得名。 The Slope is located to the south of Zhongjian Dao, and "Nan" means south in Chinese, so the word "Zhongjiannan" was used to name the Slope.		

2.51 中建西海底峡谷群

标准名称 Standard Name	中建西海底峡谷群 Zhongjianxi Haidixiaguqun	类别 Generic Term	海底峡谷群 Canyons
中心点坐标 Center Coordinates	15°40.0′N, 110°40.0′E	规模（千米 × 千米） Dimension（km × km）	132 × 49
最小水深（米） Min Depth (m)	596	最大水深（米） Max Depth (m)	1296
地理实体描述 Feature Description	中建西海底峡谷群发育在中建南斜坡上、中建阶地东部，由一条南—北向的大型峡谷和数十条北西—南东向的小型峡谷组成（图 2–18）。 Zhongjianxi Haidixiaguqun are located in Zhongjiannan Xiepo and is in the east of Zhongjian Jiedi. It is composed of a larger canyon in the direction of S–N and dozens of smaller canyons in the direction of NW–SE (Fig.2–18).		
命名由来 Origin of Name	该海底峡谷群位于中建岛以西，因此得名。 These Canyons are located to the west of Zhongjian Dao, and “Xi” means west in Chinese, so the word “Zhongjianxi” was used to name these Canyons.		

2.52 中建南海底峡谷群

标准名称 Standard Name	中建南海底峡谷群 Zhongjiannan Haidixiaguqun	类别 Generic Term	海底峡谷群 Canyons
中心点坐标 Center Coordinates	14°12.0′N, 110°12.0′E	规模（千米 × 千米） Dimension（km × km）	90 × 25
最小水深（米） Min Depth (m)	500	最大水深（米） Max Depth (m)	1650
地理实体描述 Feature Description	中建南海底峡谷群发育在中建南斜坡上、中建阶地南部，由数十条北西—南东向相间排列的海底峡谷组成，这些峡谷的头部水深约 500 米，尾部水深约 1600 米（图 2–18）。 Zhongjiannan Haidixiaguqun are located in the Zhongjiannan Xiepo and is in the south of Zhongjian Jiedi. It is composed of dozens of alternately arranged canyons in the direction of NW–SE. The depth of the Canyons is about 500 m and about 1600 m at the head and at the tail respectively (Fig.2–18).		
命名由来 Origin of Name	该海底峡谷群位于中建阶地以南，因此得名。 These Canyons are located to the south of Zhongjian Jiedi, and “Nan” means south in Chinese, so the word “Zhongjiannan” was used to name these Canyons.		

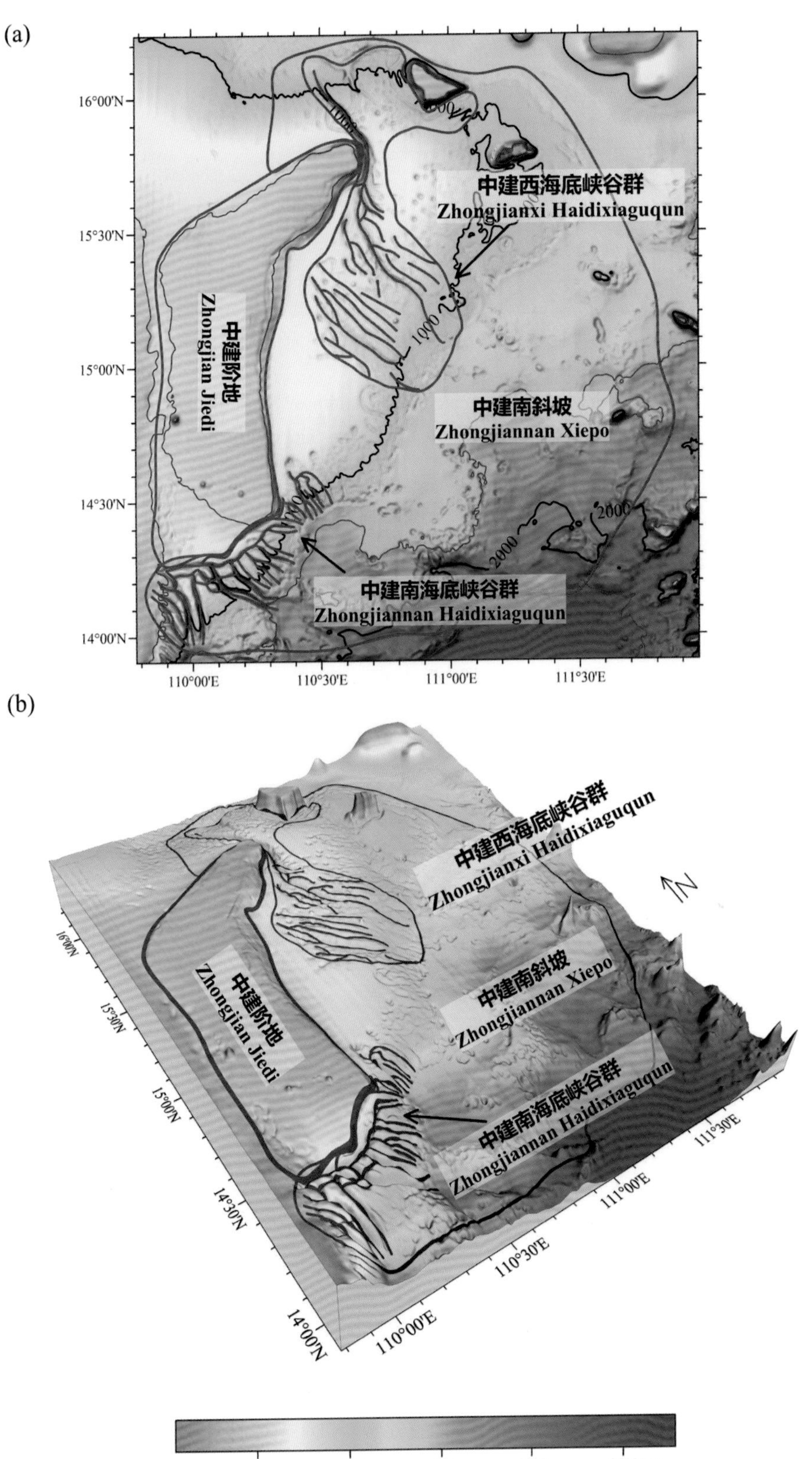

图 2-18　中建阶地、中建南斜坡、中建西海底峡谷群、中建南海底峡谷群

(a) 海底地形图（等深线间隔 500 米）；(b) 三维海底地形图

Fig.2-18　Zhongjian Jiedi, Zhongjiannan Xiepo, Zhongjianxi Haidixiaguqun, Zhongjiannan Haidixiaguqun

(a) Seafloor topographic map (with contour interval of 500 m)；(b) 3-D seafloor topographic map

2.53 金吒海丘

标准名称 Standard Name	金吒海丘 Jinzha Haiqiu	类别 Generic Term	海丘 Hill
中心点坐标 Center Coordinates	14°48.7'N, 109°56.1'E	规模（千米 × 千米） Dimension (km × km)	6 × 5
最小水深（米） Min Depth (m)	150	最大水深（米） Max Depth (m)	560
地理实体描述 Feature Description	金吒海丘发育在中建阶地南部，平面形态呈圆形。海丘的东坡平缓，西坡陡峭（图 2-19）。 Jinzha Haiqiu is located in the south of Zhongjian Jiedi and has a circular planform, with a gentle east slope and a steep west slope (Fig.2-19).		
命名由来 Origin of Name	取自中国古代神话传说人物“金吒”，因此得名。 The Hill is named after “Jinzha”, a legendary figure in the ancient Chinese myths and legends.		

2.54 木吒海丘

标准名称 Standard Name	木吒海丘 Muzha Haiqiu	类别 Generic Term	海丘 Hill
中心点坐标 Center Coordinates	14°34.2'N, 110°02.5'E	规模（千米 × 千米） Dimension (km × km)	5 × 4
最小水深（米） Min Depth (m)	250	最大水深（米） Max Depth (m)	550
地理实体描述 Feature Description	木吒海丘发育在中建阶地南部，平面形态呈椭圆形（图 2-19）。 Muzha Haiqiu is located in the south of Zhongjian Jiedi and has an oval-shaped planform (Fig.2-19).		
命名由来 Origin of Name	取自中国古代神话传说人物“木吒”，因此得名。 The Hill is named after “Muzha”, a legendary figure in the ancient Chinese myths and legends.		

2.55 哪吒海丘

标准名称 Standard Name	哪吒海丘 Nezha Haiqiu	类别 Generic Term	海丘 Hill
中心点坐标 Center Coordinates	14°32.1'N, 110°11.3'E	规模（千米 × 千米） Dimension (km × km)	4 × 3
最小水深（米） Min Depth (m)	330	最大水深（米） Max Depth (m)	610
地理实体描述 Feature Description	哪吒海丘发育在中建阶地南部，平面形态呈圆形（图 2-19）。 Nezha Haiqiu is located in the south of Zhongjian Jiedi and has a circular planform (Fig.2-19).		
命名由来 Origin of Name	取自中国古代神话传说人物“哪吒”，因此得名。 The Hill is named after “Nezha”, a legendary figure in the ancient Chinese myths and legends.		

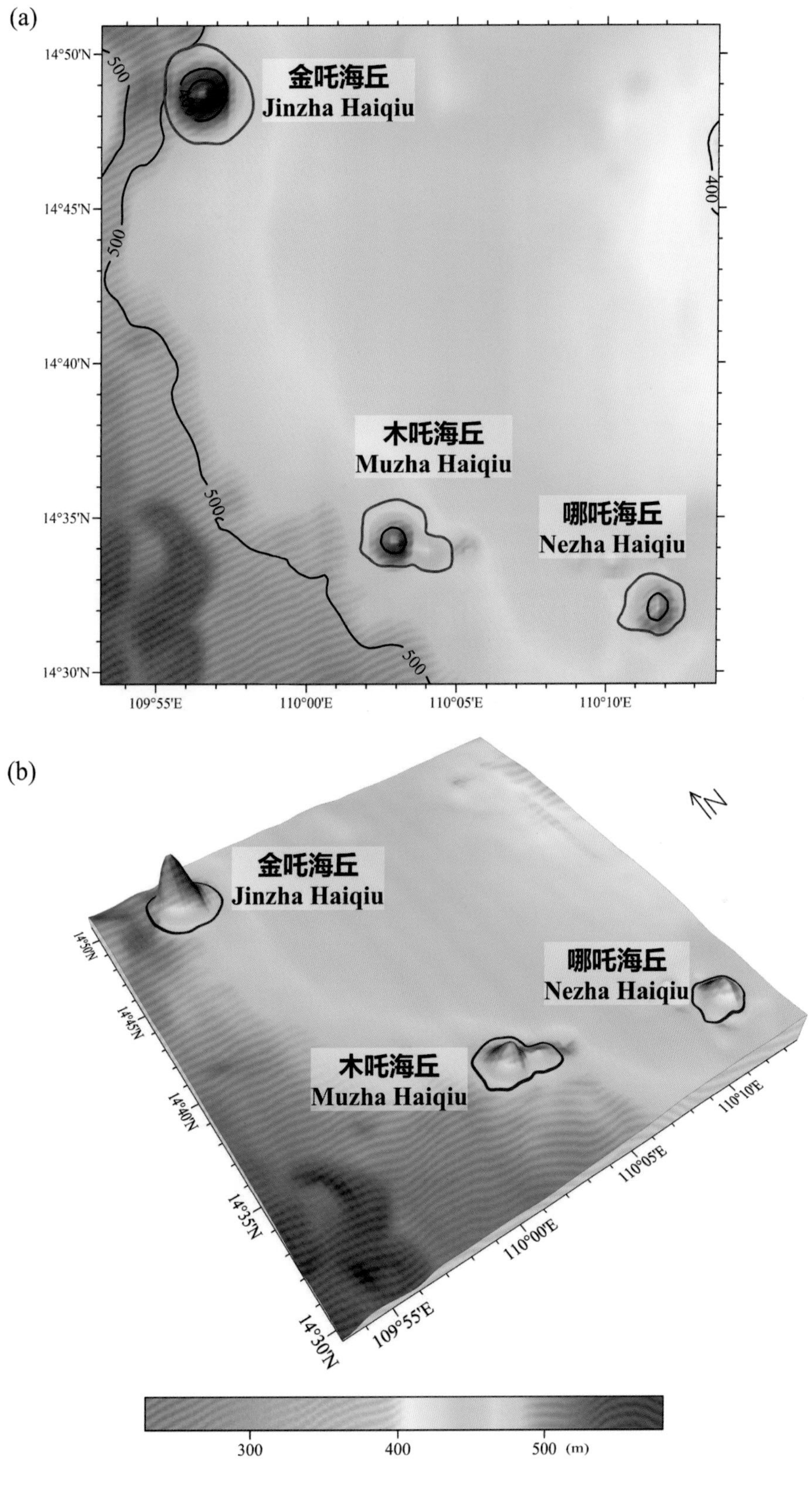

图 2-19　金吒海丘、木吒海丘、哪吒海丘

(a) 海底地形图（等深线间隔 100 米）；(b) 三维海底地形图

Fig.2-19　Jinzha Haiqiu, Muzha Haiqiu, Nezha Haiqiu

(a) Seafloor topographic map (with contour interval of 100 m); (b) 3-D seafloor topographic map

南海北部海域海底地理实体图集

Atlas of Undersea Features of Northern South China Sea

3

中沙北海隆周边海区海底地理实体

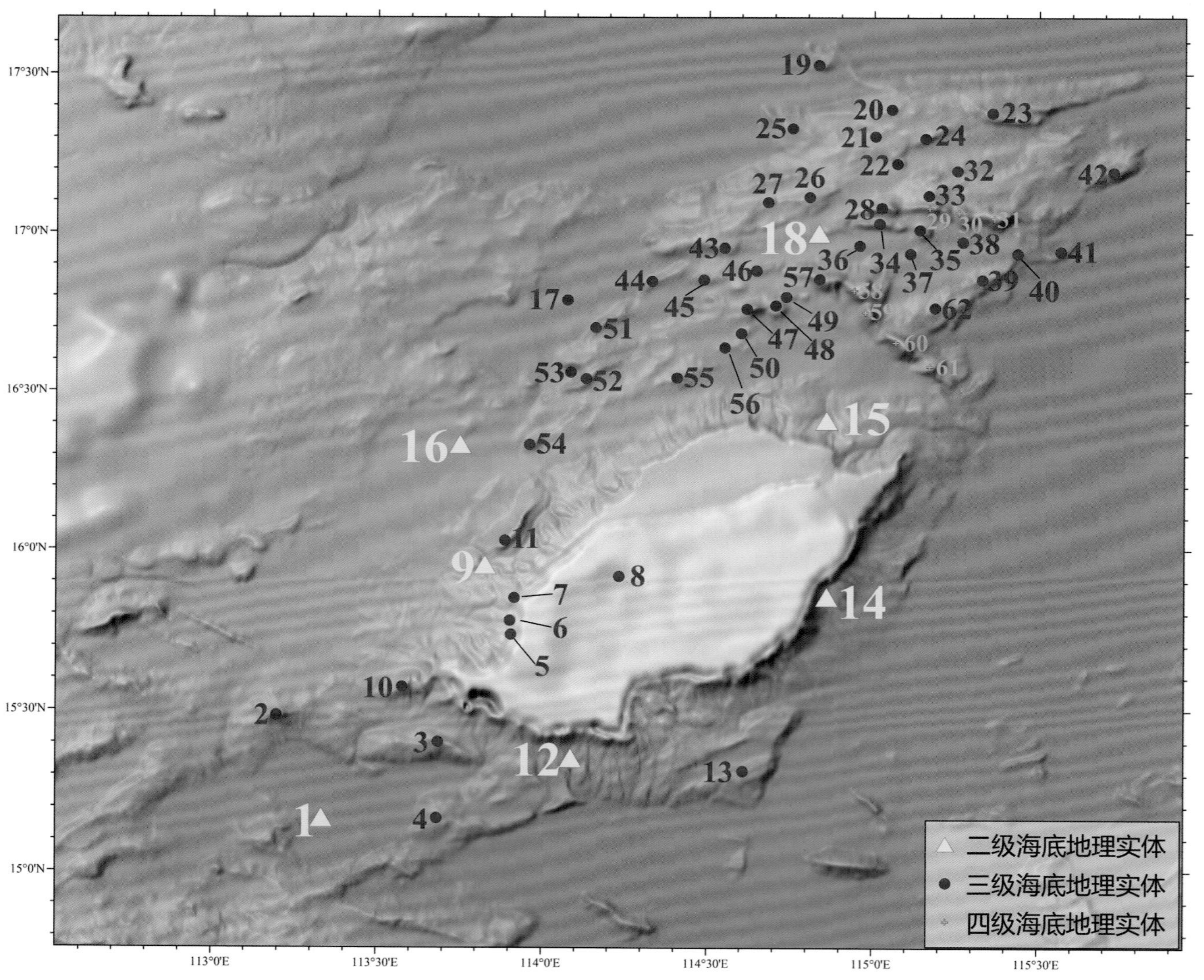

图 3-1　中沙北海隆周边海区海底地理实体中心点位置示意图，序号含义见表 3-1

Fig. 3-1　Location of center coordinates of undersea features in the sea area off Zhongshabei Rise, with the meanings of the serial numbers shown in Tab. 3-1

表 3-1　中沙北海隆周边海区海底地理实体列表

Tab.3-1　List of undersea features in the sea area off Zhongshabei Rise

序号 No.	标准名称 Standard Name	汉语拼音 Chinese Phonetic Alphabet	类别 Generic Term	中心点坐标 Center Coordinates		实体等级 Order
				纬度 Latitude	经度 Longitude	
1	中沙南海盆 Zhongshanan Haipen	Zhōngshānán Hǎipén	海盆 Basin	15°10.0'N	113°20.0'E	2
2	果淀海山 Guodian Haishan	Guǒdiàn Hǎishān	海山 Seamount	15°29.1'N	113°11.8'E	3
3	波洑海山 Bofu Haishan	Bōfú Hǎishān	海山 Seamount	15°24.2'N	113°41.1'E	3
4	布德海山 Bude Haishan	Bùdé Hǎishān	海山 Seamount	15°10.0'N	113°40.9'E	3
5	南安暗沙 Nan'an Ansha	Nán'ān Ànshā	暗沙 Shoal	15°44.2'N	113°54.3'E	3
6	钟灵暗沙 Zhongling Ansha	Zhōnglíng Ànshā	暗沙 Shoal	15°46.8'N	113°54.1'E	3
7	炎黄暗沙 Yanhuang Ansha	Yánhuáng Ànshā	暗沙 Shoal	15°51.0'N	113°54.9'E	3
8	明珠暗沙 Mingzhu Ansha	Míngzhū Ànshā	暗沙 Shoal	15°55.0'N	114°13.9'E	3
9	中沙西海底峡谷群 Zhongshaxi Haidixiaguqun	Zhōngshāxī Hǎidǐxiágǔqún	海底峡谷群 Canyons	15°57.1'N	113°49.6'E	2
10	排洪海山 Paihong Haishan	Páihóng Hǎishān	海山 Seamount	15°34.4'N	113°34.6'E	3
11	排波海脊 Paibo Haiji	Páibō Hǎijǐ	海脊 Ridge	16°01.8'N	113°53.2'E	3
12	中沙南海底峡谷群 Zhongshanan Haidixiaguqun	Zhōngshānán Hǎidǐxiágǔqún	海底峡谷群 Canyons	15°21.2'N	114°05.2'E	2
13	美溪海山 Meixi Haishan	Měixī Hǎishān	海山 Seamount	15°18.8'N	114°36.5'E	3
14	中沙东海底峡谷群 Zhongshadong Haidixiaguqun	Zhōngshādōng Hǎidǐxiágǔqún	海底峡谷群 Canyons	15°51.3'N	114°51.6'E	2
15	中沙北海底峡谷群 Zhongshabei Haidixiaguqun	Zhōngshāběi Hǎidǐxiágǔqún	海底峡谷群 Canyons	16°24.6'N	114°51.5'E	2
16	中沙海槽 Zhongsha Haicao	Zhōngshā Hǎicáo	海槽 Trough	16°20.0'N	113°45.0'E	2
17	中沙北海盆 Zhongshabei Haipen	Zhōngshāběi Hǎipén	海盆 Basin	16°47.3'N	114°04.4'E	3

续表

序号 No.	标准名称 Standard Name	汉语拼音 Chinese Phonetic Alphabet	类别 Generic Term	中心点坐标 Center Coordinates		实体等级 Order
				纬度 Latitude	经度 Longitude	
18	中沙北海隆 Zhongshabei Hailong	Zhōngshāběi Hǎilóng	海隆 Rise	17°00.0'N	114°50.0'E	2
19	指掌海脊 Zhizhang Haiji	Zhǐzhǎng Hǎijǐ	海脊 Ridge	17°31.7'N	114°49.9'E	3
20	天鸡海丘 Tianji Haiqiu	Tiānjī Hǎiqiū	海丘 Hill	17°23.4'N	115°03.2'E	3
21	青云海丘 Qingyun Haiqiu	Qīngyún Hǎiqiū	海丘 Hill	17°18.3'N	115°00.2'E	3
22	谢公海丘 Xiegong Haiqiu	Xiègōng Hǎiqiū	海丘 Hill	17°13.1'N	115°04.2'E	3
23	美滨海脊 Meibin Haiji	Měibīn Hǎijǐ	海脊 Ridge	17°22.7'N	115°21.5'E	3
24	美滨海山 Meibin Haishan	Měibīn Hǎishān	海山 Seamount	17°17.9'N	115°09.3'E	3
25	中北海脊 Zhongbei Haiji	Zhōngběi Hǎijǐ	海脊 Ridge	17°19.8'N	114°45.2'E	3
26	鲁班海丘 Luban Haiqiu	Lǔbān Hǎiqiū	海丘 Hill	17°06.8'N	114°48.3'E	3
27	湖月圆丘 Huyue Yuanqiu	Húyuè Yuánqiū	圆丘 Knoll	17°05.8'N	114°40.8'E	3
28	隐矶海山 Yinji Haishan	Yǐnjī Hǎishān	海山 Seamount	17°04.8'N	115°01.4'E	3
29	烟霞海底峰 Yanxia Haidifeng	Yānxiá Hǎidǐfēng	海底峰 Peak	17°04.7'N	115°10.1'E	4
30	济猛海底峰 Jimeng Haidifeng	Jìměng Hǎidǐfēng	海底峰 Peak	17°03.9'N	115°15.4'E	4
31	天姥海底峰 Tianmu Haidifeng	Tiānmǔ Hǎidǐfēng	海底峰 Peak	17°02.8'N	115°21.9'E	4
32	海鸠海丘 Haijiu Haiqiu	Hǎijiū Hǎiqiū	海丘 Hill	17°11.8'N	115°15.1'E	3
33	本固海丘 Bengu Haiqiu	Běngù Hǎiqiū	海丘 Hill	17°07.1'N	115°09.9'E	3
34	隐矶南海山 Yinjinan Haishan	Yǐnjīnán Hǎishān	海山 Seamount	17°01.8'N	115°01.0'E	3
35	越人海山 Yueren Haishan	Yuèrén Hǎishān	海山 Seamount	17°00.6'N	115°08.3'E	3

续表

序号 No.	标准名称 Standard Name	汉语拼音 Chinese Phonetic Alphabet	类别 Generic Term	中心点坐标 Center Coordinates		实体等级 Order
				纬度 Latitude	经度 Longitude	
36	天台海丘 Tiantai Haiqiu	Tiāntái Hǎiqiū	海丘 Hill	16°57.6'N	114°57.4'E	3
37	吴越海丘 Wuyue Haiqiu	Wúyuè Hǎiqiū	海丘 Hill	16°56.1'N	115°06.6'E	3
38	天横圆丘 Tianheng Yuanqiu	Tiānhéng Yuánqiū	圆丘 Knoll	16°58.3'N	115°16.1'E	3
39	指掌海丘 Zhizhang Haiqiu	Zhǐzhǎng Hǎiqiū	海丘 Hill	16°51.2'N	115°19.7'E	3
40	南扉海脊 Nanfei Haiji	Nánfēi Hǎijǐ	海脊 Ridge	16°56.1'N	115°26.1'E	3
41	东流海丘 Dongliu Haiqiu	Dōngliú Hǎiqiū	海丘 Hill	16°56.5'N	115°33.9'E	3
42	比微海山 Biwei Haishan	Bǐwēi Hǎishān	海山 Seamount	17°11.5'N	115°43.5'E	3
43	西门海山 Ximen Haishan	Xīmén Hǎishān	海山 Seamount	16°57.2'N	114°32.9'E	3
44	乐西海脊 Lexi Haiji	Lèxī Hǎijǐ	海脊 Ridge	16°50.9'N	114°19.7'E	3
45	霹雳海丘 Pili Haiqiu	Pīlì Hǎiqiū	海丘 Hill	16°51.2'N	114°29.1'E	3
46	青崖海丘 Qingya Haiqiu	Qīngyá Hǎiqiū	海丘 Hill	16°52.9'N	114°38.7'E	3
47	漫步海山 Manbu Haishan	Mànbù Hǎishān	海山 Seamount	16°45.7'N	114°37.0'E	3
48	漫步东海丘 Manbudong Haiqiu	Mànbùdōng Hǎiqiū	海丘 Hill	16°46.3'N	114°42.2'E	3
49	漫步北海丘 Manbubei Haiqiu	Mànbùběi Hǎiqiū	海丘 Hill	16°48.0'N	114°44.1'E	3
50	漫步南海丘 Manbunan Haiqiu	Mànbùnán Hǎiqiū	海丘 Hill	16°41.1'N	114°36.0'E	3
51	排洪海丘 Paihong Haiqiu	Páihóng Hǎiqiū	海丘 Hill	16°42.1'N	114°09.5'E	3
52	控湃海脊 Kongpai Haiji	Kòngpài Hǎijǐ	海脊 Ridge	16°32.4'N	114°07.8'E	3
53	控湃北海脊 Kongpaibei Haiji	Kòngpàiběi Hǎijǐ	海脊 Ridge	16°33.6'N	114°05.0'E	3

续表

序号 No.	标准名称 Standard Name	汉语拼音 Chinese Phonetic Alphabet	类别 Generic Term	中心点坐标 Center Coordinates		实体等级 Order
				纬度 Latitude	经度 Longitude	
54	武勇海丘 Wuyong Haiqiu	Wǔyǒng Hǎiqiū	海丘 Hill	16°20.0'N	113°57.6'E	3
55	涛静海丘 Taojing Haiqiu	Tāojìng Hǎiqiū	海丘 Hill	16°32.6'N	114°24.3'E	3
56	涛静北海丘 Taojingbei Haiqiu	Tāojìngběi Hǎiqiū	海丘 Hill	16°38.4'N	114°33.0'E	3
57	屏南海山 Pingnan Haishan	Píngnán Hǎishān	海山 Seamount	16°51.3'N	114°50.1'E	3
58	安定海底峰 Anding Haidifeng	Āndìng Hǎidǐfēng	海底峰 Peak	16°49.2'N	114°56.5'E	4
59	果淀海底峰 Guodian Haidifeng	Guǒdiàn Hǎidǐfēng	海底峰 Peak	16°45.1'N	114°58.5'E	4
60	华夏海底峰 Huaxia Haidifeng	Huáxià Hǎidǐfēng	海底峰 Peak	16°39.4'N	115°04.0'E	4
61	华夏南海底峰 Huaxianan Haidifeng	Huáxiànán Hǎidǐfēng	海底峰 Peak	16°34.8'N	115°10.2'E	4
62	南扉海山 Nanfei Haishan	Nánfēi Hǎishān	海山 Seamount	16°45.9'N	115°11.1'E	3

3.1 中沙南海盆

标准名称 Standard Name	中沙南海盆 Zhongshanan Haipen	类别 Generic Term	海盆 Basin
中心点坐标 Center Coordinates	15°10.0'N, 113°20.0'E	规模（千米 × 千米） Dimension（km × km）	120 × 40
最小水深（米） Min Depth (m)	3380	最大水深（米） Max Depth (m)	3800
地理实体描述 Feature Description	中沙南海盆位于南海西部陆坡的中部，中沙海台和盆西海岭之间。海盆周围被多座海山环绕，形成地形低陷的小型盆地，其内部地形平缓，最大水深 3800 米（图 3–2）。 Zhongshanan Haipen is located in the middle of Nanhaixibu Lupo and is between Zhongsha Haitai and Penxi Hailing. It is a small basin with sinking topography that is surrounded by multiple seamounts. The topography inside the Basin is gentle with the maximum depth of 3800 m (Fig.3–2).		
命名由来 Origin of Name	该海盆位于中沙大环礁以南，因此得名。 The Basin is located on the south of Zhongsha atoll, and “Nan” means south in Chinese, so the word “Zhongshanan” was used to name the Basin.		

3.2 果淀海山

标准名称 Standard Name	果淀海山 Guodian Haishan	类别 Generic Term	海山 Seamount
中心点坐标 Center Coordinates	15°29.1'N, 113°11.8'E	规模（千米 × 千米） Dimension（km × km）	40 × 20
最小水深（米） Min Depth (m)	940	最大水深（米） Max Depth (m)	4000
地理实体描述 Feature Description	果淀海山位于中沙南海盆以北，中沙海台以西，平面形态呈弧形（图 3–2）。 Guodian Haishan is located on the north of Zhongshanan Haipen and is in the west of Zhongsha Haitai with an arc-shaped planform (Fig.3–2).		
命名由来 Origin of Name	该海山位于我国中沙大环礁附近，故以中沙大环礁中的暗沙、浅滩名进行地名的团组化命名。“果淀暗沙”属于中沙大环礁的一个暗沙，取“果淀”两字命名该海山。 The Seamount is located near the Zhongsha atoll in China, so it is named after a group of shoals and banks in Zhongsha atoll. “Guodian Ansha” is one of the shoals belonging to Zhongsha atoll, so the word “Guodian” was used to name the Seamount.		

3.3 波洑海山

标准名称 Standard Name	波洑海山 Bofu Haishan	类别 Generic Term	海山 Seamount
中心点坐标 Center Coordinates	15°24.2'N, 113°41.1'E	规模（千米 × 千米） Dimension（km × km）	42 × 14
最小水深（米） Min Depth (m)	1520	最大水深（米） Max Depth (m)	3600
地理实体描述 Feature Description	波洑海山位于中沙南海盆东北侧，平面形态呈近东—西向的长条形。海山的北坡较缓，南坡较陡（图 3-2）。 Bofu Haishan is located on the northeast of Zhongshanan Haipen, with a planform in the shape of a long strip nearly in the direction of E−W. Its north slope is relatively gentler, while its south slope is relatively steeper (Fig.3−2).		
命名由来 Origin of Name	该海山位于我国中沙大环礁附近，故以中沙大环礁中的暗沙、浅滩名进行地名的团组化命名。“波洑暗沙”属于中沙大环礁的一个暗沙，取“波洑”两字命名该海山。 The Seamount is located near the Zhongsha atoll in China, so it is named after a group of shoals and banks in Zhongsha atoll. “Bofu Ansha” is one of the shoals belonging to Zhongsha atoll, so the word “Bofu” was used to name the Seamount.		

3.4 布德海山

标准名称 Standard Name	布德海山 Bude Haishan	类别 Generic Term	海山 Seamount
中心点坐标 Center Coordinates	15°10.0'N, 113°40.9'E	规模（千米 × 千米） Dimension（km × km）	42 × 18
最小水深（米） Min Depth (m)	2800	最大水深（米） Max Depth (m)	4450
地理实体描述 Feature Description	布德海山位于中沙南海盆东部，平面形态呈东—西走向，由东向西宽度逐渐变窄（图 3-2）。 Bude Haishan is located on the east of Zhongshanan Haipen. It stretches in the direction of E−W and with an increasingly narrower width from east to west (Fig.3−2).		
命名由来 Origin of Name	该海山位于我国中沙大环礁附近，故以中沙大环礁中的暗沙、浅滩名进行地名的团组化命名。“布德暗沙”属于中沙大环礁的一个暗沙，取“布德”两字命名该海山。 The Seamount is located near the Zhongsha atoll in China, so it is named after a group of shoals and banks in Zhongsha atoll. “Bude Ansha” is one of the shoals belonging to Zhongsha atoll, so the word “Bude” was used to name the Seamount.		

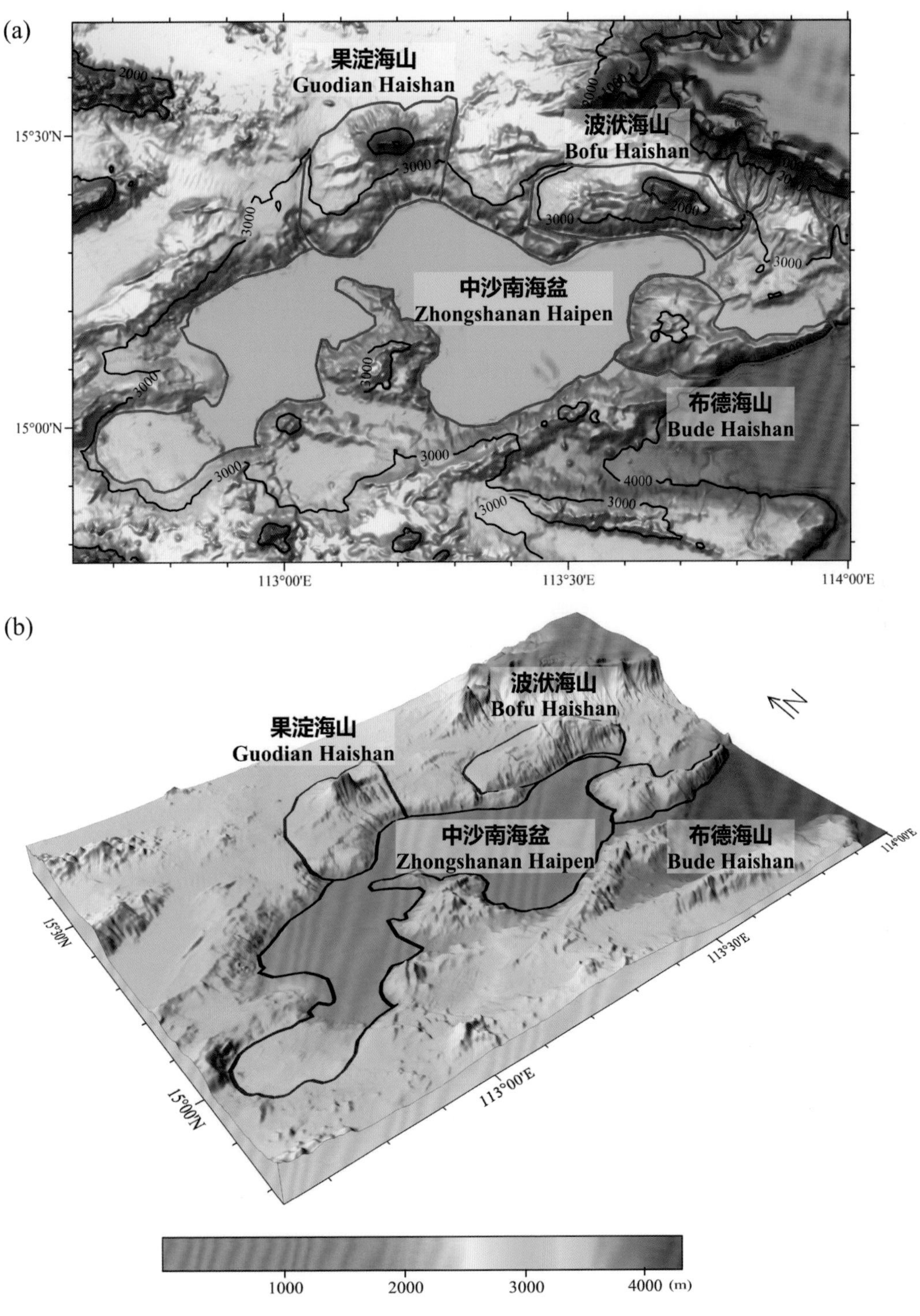

图 3-2　中沙南海盆、果淀海山、波洑海山、布德海山

(a) 海底地形图（等深线间隔 1000 米）；(b) 三维海底地形图

Fig.3-2　Zhongshanan Haipen, Guodian Haishan, Bofu Haishan, Bude Haishan

(a) Seafloor topographic map (with contour interval of 1000 m); (b) 3-D seafloor topographic map

3.5 南安暗沙

标准名称 Standard Name	南安暗沙 Nan'an Ansha	类别 Generic Term	暗沙 Shoal
中心点坐标 Center Coordinates	15°44.2′N, 113°54.3′E	规模（千米 × 千米） Dimension（km × km）	2 × 2
最小水深（米） Min Depth (m)	40	最大水深（米） Max Depth (m)	60
地理实体描述 Feature Description	南安暗沙位于涛静暗沙东北侧，因水深较大，珊瑚礁发育处于衰退时期，可见少量的礁石（图 3–3）。 Nan'an Ansha is located to the northeast of Taojing Ansha. Due to the deep water, the development of the coral reefs is in the recession and a small number of reefs can be seen (Fig.3–3).		
命名由来 Origin of Name	"南安"寓意南海局势始终安定向好，充分展示了我国与东盟国家继续合作，努力维护南海局势向好局面的共同意愿。 "Nan'an" (safety in south) implies the situation in the South China Sea is always stable and promising, fully indicating the common will of China and other ASEAN countries to keep cooperation and make joint efforts to safeguard the improving situation in the South China Sea.		

(a)

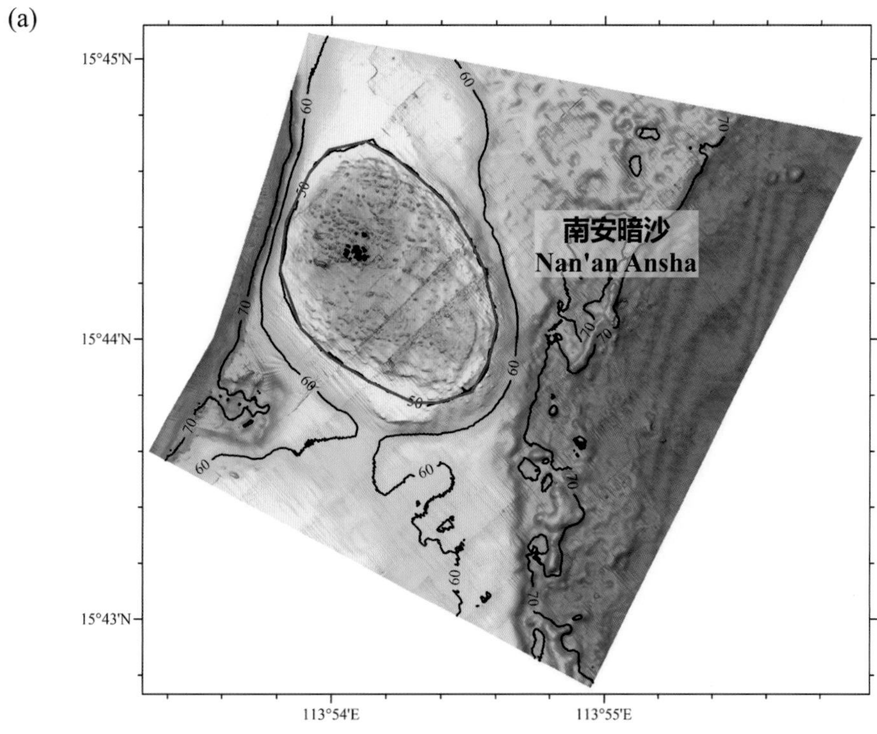

(b)

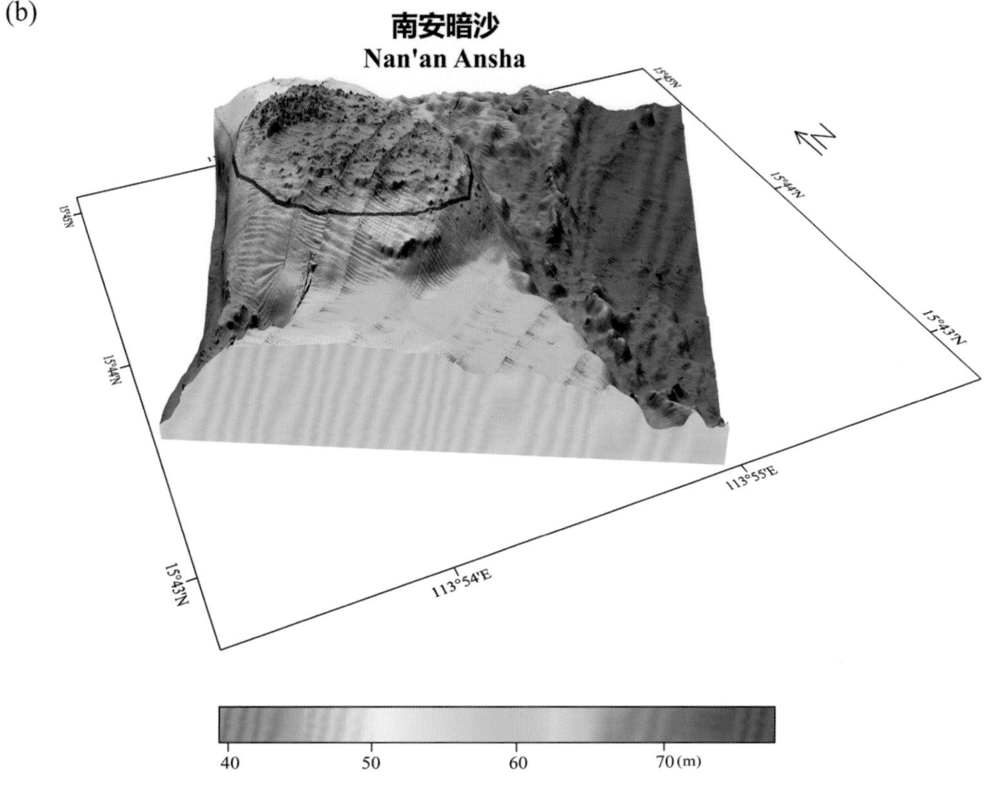

图 3-3 南安暗沙

(a) 海底地形图（等深线间隔 50 米）；(b) 三维海底地形图

Fig.3-3 Nan’an Ansha

(a) Seafloor topographic map (with contour interval of 50 m); (b) 3-D seafloor topographic map

3.6 钟灵暗沙

标准名称 Standard Name	钟灵暗沙 Zhongling Ansha	类别 Generic Term	暗沙 Shoal
中心点坐标 Center Coordinates	15°46.8′N, 113°54.1′E	规模（千米 × 千米） Dimension（km × km）	2 × 1
最小水深（米） Min Depth (m)	20	最大水深（米） Max Depth (m)	50
地理实体描述 Feature Description	钟灵暗沙位于控湃暗沙南部，环状分布。从外环到内环，形状从尖锥状慢慢过渡为谷堆状（图 3–4）。 Zhongling Ansha is located to the south of Kongpai Ansha with circular distribution. The shape of the Shoal gradually changes from pointed cone to grain heap from the outer circle to the inner circle (Fig.3–4).		
命名由来 Origin of Name	取词自清朝著名学者曹雪芹的长篇小说《红楼梦》第三十六回："亦且琼闺绣阁中亦染此风，真真有负天地钟灵毓秀之德了！"。"钟灵"寓意该暗沙是天地之造化，凝聚了天地间灵气。 The Shoal is named after the phrase "Zhongling" (well-endowed spirits of the universe, implying the feature is a well-endowed place) from the quote "and even the boudoirs of ladies were contaminated with such corrupt moral. What an abuse of the well-endowed spirits of the universe!" in chapter 36 of *A Dream in Red Mansion* by Cao Xueqin (1715–1763 A.D.), a famous scholar of the Qing Dynasty (1636–1912 A.D.).		

(a)

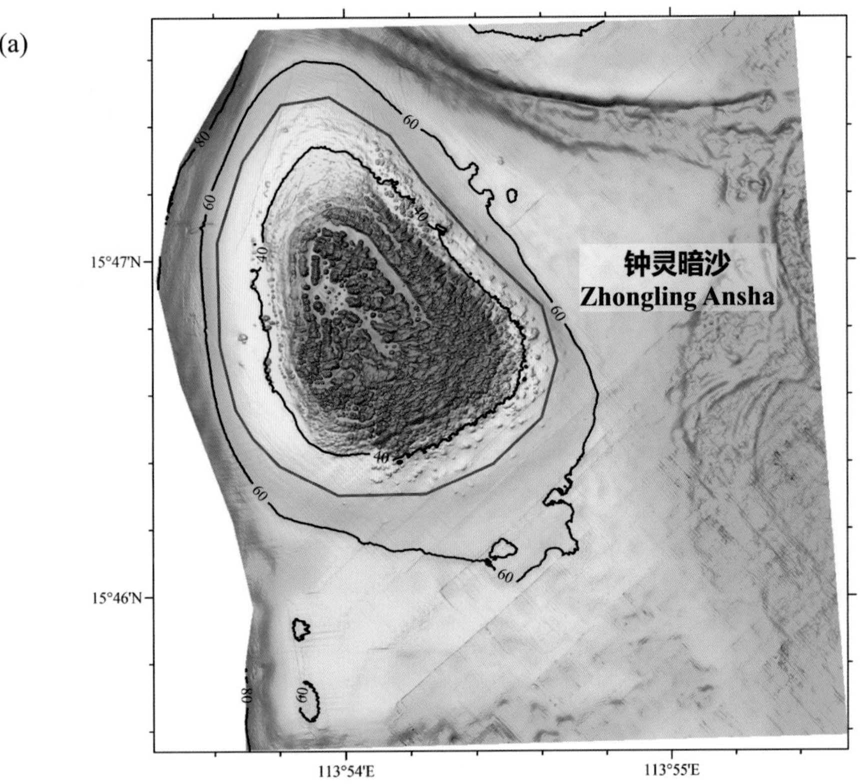

(b)

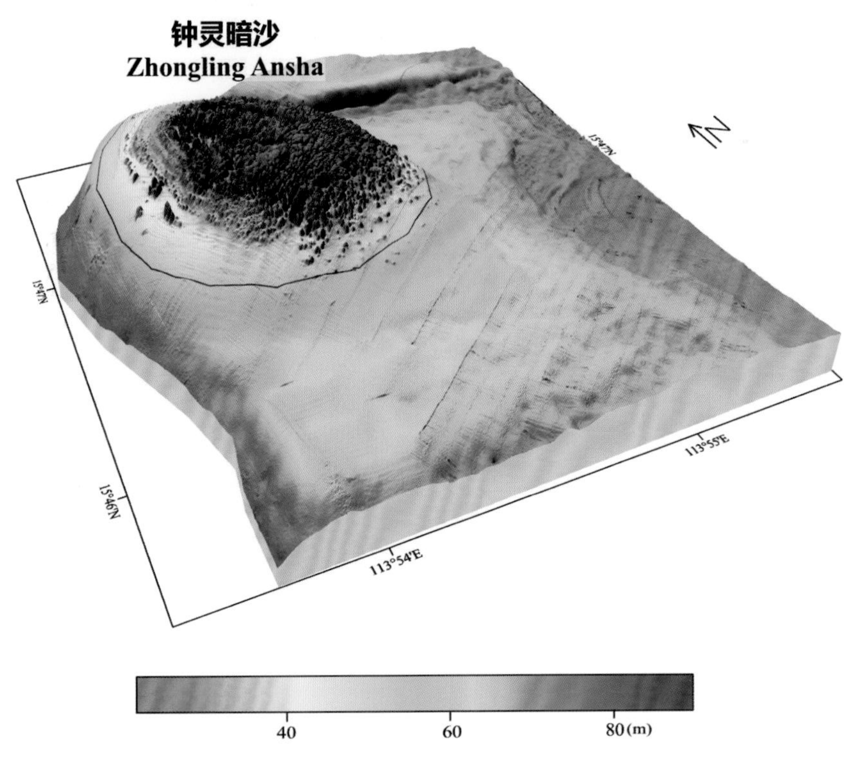

图 3-4 钟灵暗沙

(a) 海底地形图（等深线间隔 50 米）；(b) 三维海底地形图

Fig.3-4 Zhongling Ansha

(a) Seafloor topographic map (with contour interval of 50 m); (b) 3-D seafloor topographic map

3.7 炎黄暗沙

标准名称 Standard Name	炎黄暗沙 Yanhuang Ansha	类别 Generic Term	暗沙 Shoal
中心点坐标 Center Coordinates	15°51.0′N, 113°54.9′E	规模（千米 × 千米） Dimension（km × km）	6 × 3
最小水深（米） Min Depth (m)	20	最大水深（米） Max Depth (m)	50
地理实体描述 Feature Description	炎黄暗沙位于控湃暗沙北部，单体礁石呈尖锥状（图 3–5）。 Yanhuang Ansha is located to the north of Kongpai Ansha, with a cone-shaped single reef (Fig.3–5).		
命名由来 Origin of Name	“炎黄”指的是我国古代帝王炎帝和黄帝，是中华民族的祖先。 The specific term of this undersea feature name “Yanhuang”, refers to the Yan Emperor and the Huang Emperor in ancient China, who are two ancestors of Chinese nation.		

(a)

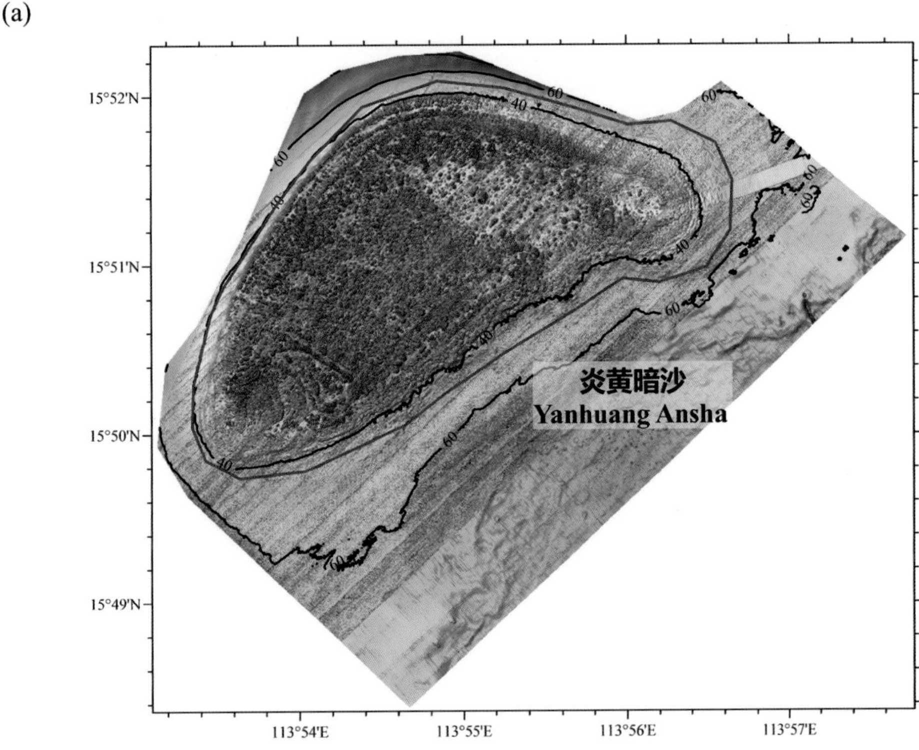

(b)

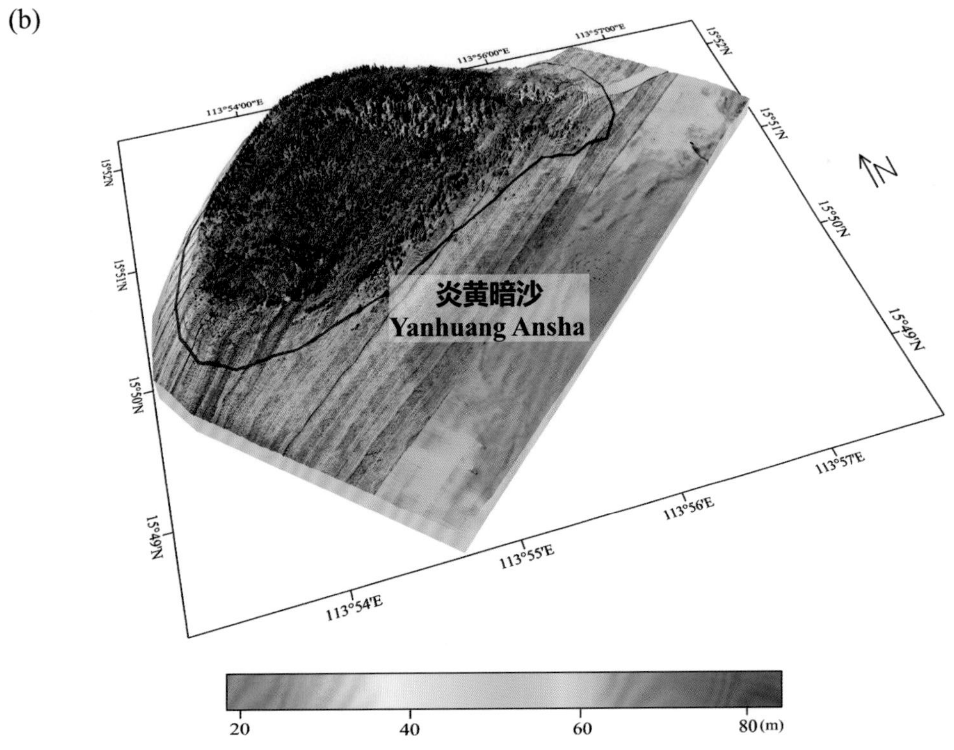

图 3-5　炎黄暗沙

(a) 海底地形图（等深线间隔 50 米）；(b) 三维海底地形图

Fig.3-5　Yanhuang Ansha

(a) Seafloor topographic map (with contour interval of 50 m)；(b) 3-D seafloor topographic map

3.8 明珠暗沙

标准名称 Standard Name	明珠暗沙 Mingzhu Ansha	类别 Generic Term	暗沙 Shoal
中心点坐标 Center Coordinates	15°55.0′N, 114°13.9′E	规模（千米 × 千米） Dimension（km × km）	1 × 1
最小水深（米） Min Depth (m)	20	最大水深（米） Max Depth (m)	50
地理实体描述 Feature Description	明珠暗沙位于本固暗沙东南方向，单体礁石发育的水深范围主要为 20 ~ 28 米。其顶较平，中部礁石连片发育，分布较集中，礁石的高度一般为 1 米左右（图 3-6）。 Mingzhu Ansha is located to the southeast of Bengu Ansha, with the depth range of single reef is mainly from 20 to 28 m. Its top is relatively flat, and the central reefs are contiguously developed and intensively distributed, with reef height of about 1 m in general (Fig.3-6).		
命名由来 Origin of Name	出自东汉文学家班固的《白虎通·封禅》："江出大贝，海出明珠。""明珠"形容该暗沙位于中沙大环礁中部，其形态近圆形，犹如一颗璀璨的明珠。 The Shoal is located in the middle of Zhongsha Dahuanjiao and is round-shaped like a bright pearl. It is named after "Mingzhu" (bright pearl) from the quote "shells are produced by river, while bright pearls are the products of the sea" in *Bai Hu Tong-Offering Sacrifices to Heaven* by Ban Gu (32–92 A.D.), a famous scholar of the Eastern Han Dynasty (25–220 A.D.).		

(a)

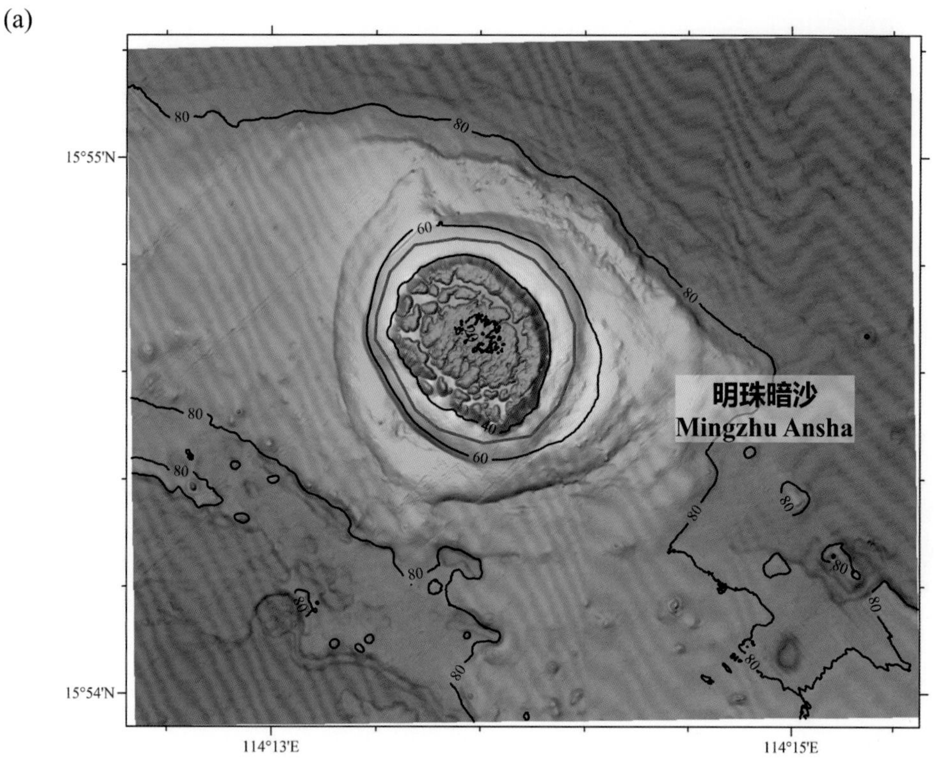

(b)

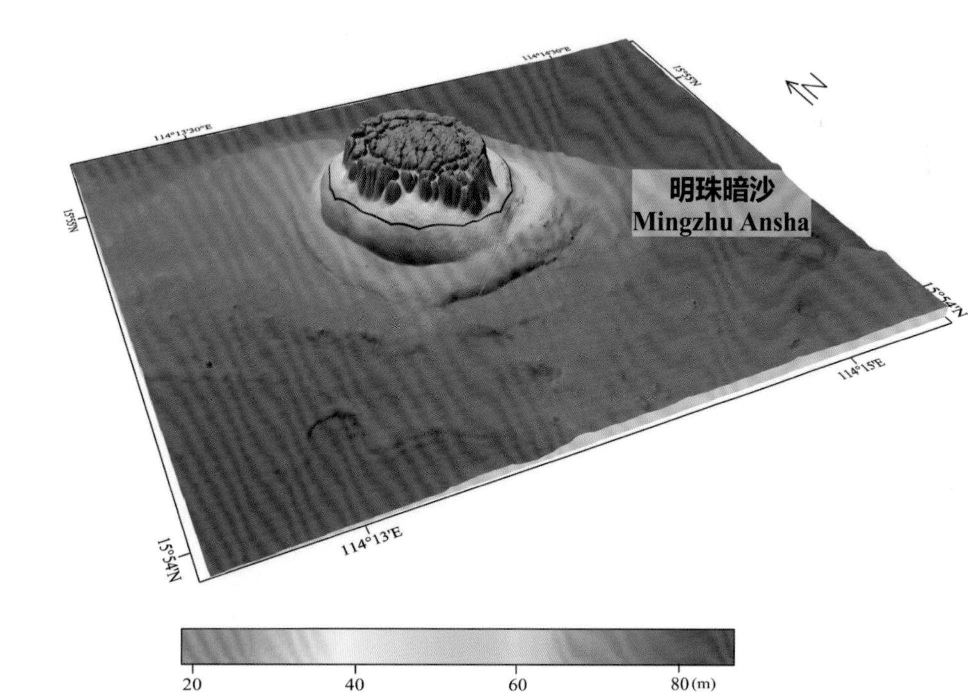

图 3-6 明珠暗沙

(a) 海底地形图（等深线间隔 50 米）；(b) 三维海底地形图

Fig.3-6 Mingzhu Ansha

(a) Seafloor topographic map (with contour interval of 50 m); (b) 3-D seafloor topographic map

3.9 中沙西海底峡谷群

标准名称 Standard Name	中沙西海底峡谷群 Zhongshaxi Haidixiaguqun	类别 Generic Term	海底峡谷群 Canyons
中心点坐标 Center Coordinates	15°57.1'N, 113°49.6'E	规模（千米 × 千米） Dimension（km × km）	86 × 20
最小水深（米） Min Depth (m)	40	最大水深（米） Max Depth (m)	2660
地理实体描述 Feature Description	中沙西海底峡谷群位于中沙海台西部海域，由数十条北西—南东向发育的海底峡谷组成。峡谷群头部起始于 200 ～ 500 米水深的中沙海台顶部边缘处，尾部终止于 1800 ～ 2400 米水深的中沙海槽边缘，峡谷群的长度 5 ～ 25 千米（图 3-7）。 Zhongshaxi Haidixiagu are located in the sea area in the west of Zhongsha Haitai and are composed of dozens canyons developing in the direction of NW-SE. The heads of the Canyons start from the edge of the top of Zhongsha Haitai with depth of 200~500 m, and the tails end at the edge of Zhongsha Haicao with depth of 1800~2400 m. The length of the Canyons are about 5~25 km (Fig.3-7).		
命名由来 Origin of Name	该海底峡谷群位于中沙大环礁以西，因此得名。 These Canyons are located in the west of Zhongsha atoll, and “Xi” means west in Chinese, so the word “Zhongshaxi” was used to name these Canyons.		

3.10 排洪海山

标准名称 Standard Name	排洪海山 Paihong Haishan	类别 Generic Term	海山 Seamount
中心点坐标 Center Coordinates	15°34.4'N, 113°34.6'E	规模（千米 × 千米） Dimension（km × km）	25 × 20
最小水深（米） Min Depth (m)	380	最大水深（米） Max Depth (m)	2680
地理实体描述 Feature Description	排洪海山位于中沙海台西部海域，平面形态呈长轴为东—西向的菱形（图 3-7）。 Paihong Haishan is located in the west sea area of Zhongsha Haitai, with a diamond-shaped planform whose long axis is in the direction of E-W (Fig.3-7).		
命名由来 Origin of Name	该海山位于我国中沙大环礁附近，故以中沙大环礁中的暗沙、浅滩名进行地名的团组化命名。“排洪滩”属于中沙大环礁的浅滩之一，取“排洪”两字命名该海山。 The Seamount is located near the Zhongsha atoll in China, so it is named after a group of shoals and banks in Zhongsha atoll. “Paihong Tan” is one of the banks belonging to Zhongsha atoll, so the word “Paihong” was used to name the Seamount.		

3.11 排波海脊

标准名称 Standard Name	排波海脊 Paibo Haiji	类别 Generic Term	海脊 Ridge
中心点坐标 Center Coordinates	16°01.8'N, 113°53.2'E	规模（千米 × 千米） Dimension（km × km）	32 × 10
最小水深（米） Min Depth (m)	370	最大水深（米） Max Depth (m)	2650
地理实体描述 Feature Description	排波海脊位于中沙海台西部海域，平面形态呈北东—南西向的长条形。海脊的顶部较平坦，平均水深 360 米，向四周水深迅速变大（图 3–7）。 Paibo Haiji is located in the west sea area of Zhongsha Haitai, with a planform in the shape of a long strip in the direction of NE–SW. Its top is relative flat, the average depth is 360 m and the depth of the water increases rapidly from the center to all directions (Fig.3–7).		
命名由来 Origin of Name	该海脊位于我国中沙大环礁附近，故以中沙大环礁中的暗沙、浅滩名进行地名的团组化命名。“排波暗沙”属于中沙大环礁的暗沙之一，取“排波”两字命名该海脊。 The Ridge is located near the Zhongsha atoll in China, so it is named after a group of shoals and banks in Zhongsha atoll. “Paibo Ansha” is one of the shoals belonging to Zhongsha atoll, so the word “Paibo” was used to name the Ridge.		

3.12 中沙南海底峡谷群

标准名称 Standard Name	中沙南海底峡谷群 Zhongshanan Haidixiaguqun	类别 Generic Term	海底峡谷群 Canyons
中心点坐标 Center Coordinates	15°21.2'N, 114°05.2'E	规模（千米 × 千米） Dimension（km × km）	100 × 45
最小水深（米） Min Depth (m)	90	最大水深（米） Max Depth (m)	4180
地理实体描述 Feature Description	中沙南海底峡谷群位于中沙海台南部海域，由数十条南—北向和北东—南西向发育的海底峡谷组成。峡谷群的头部起始于 200 ~ 500 米水深的中沙海台顶部边缘处，尾部终止于 3800 ~ 4100 米水深的南海海盆，峡谷群的长度 5 ~ 55 千米（图 3–7）。 Zhongshanan Haidixiaguqun are located in the south sea area of Zhongsha Haitai and are composed of dozens of canyons developing in the directions of S–N and NE–SW. The heads of Canyons start from the edge of the top of Zhongsha Haitai with depth of 200~500 m, and the tails end at Hainan Haipen with depth of 3800~4100 m. The length of the Canyons are about 5~55 km (Fig.3–7).		
命名由来 Origin of Name	该海底峡谷群位于中沙大环礁以南，因此得名。 These Canyons are located in the south of Zhongsha atoll, and “Nan” means south in Chinese, so the word “Zhongshanan” was used to name these Canyons.		

3.13 美溪海山

标准名称 Standard Name	美溪海山 Meixi Haishan	类别 Generic Term	海山 Seamount
中心点坐标 Center Coordinates	15°18.8'N, 114°36.5'E	规模（千米 × 千米） Dimension（km × km）	42 × 16
最小水深（米） Min Depth (m)	2740	最大水深（米） Max Depth (m)	4500
地理实体描述 Feature Description	美溪海山位于中沙海台南部海域，平面形态呈北东—南西向的三角形。海山的西北坡较缓，东南坡较陡峭（图 3−7）。 Meixi Haishan is located in the south sea area of Zhongsha Haitai, with a planform in the shape of a triangle in the direction of NE−SW. Its northwest slope is relatively gentler while its southeast is relatively steeper (Fig.3−7).		
命名由来 Origin of Name	该海山位于我国中沙大环礁附近，故以中沙大环礁中的暗沙、浅滩名进行地名的团组化命名。“美溪暗沙”属于中沙大环礁的一个暗沙，取“美溪”两字命名该海山。 The Seamount is located near the Zhongsha atoll in China, so it is named after a group of shoals and banks in Zhongsha atoll. “Meixi Ansha” is one of the shoals belonging to Zhongsha atoll, so the word “Meixi” was used to name the Seamount.		

3.14 中沙东海底峡谷群

标准名称 Standard Name	中沙东海底峡谷群 Zhongshadong Haidixiaguqun	类别 Generic Term	海底峡谷群 Canyons
中心点坐标 Center Coordinates	15°51.3'N, 114°51.6'E	规模（千米 × 千米） Dimension（km × km）	110 × 35
最小水深（米） Min Depth (m)	130	最大水深（米） Max Depth (m)	4200
地理实体描述 Feature Description	中沙东海底峡谷群位于中沙海台东部海域，由数十条北西—南东向和东—西向发育的海底峡谷组成。峡谷群的头部起始于 200 ~ 500 米水深的中沙海台顶部边缘处，尾部终止于平均水深 4100 米的南海海盆，峡谷群的长度 5 ~ 80 千米（图 3−7）。 Zhongshadong Haidixiaguqun are located in the east sea area of Zhongsha Haitai and are composed of dozens of canyons developing in the directions of NW−SE and E−W. The heads of the Canyons start from the edge of the top of Zhongsha Haitai with a depth of 200~500 m, and the tails end at Nanhai Haipen with an average depth of 4100 m. The length of the Canyons are about 5~80 km (Fig.3−7).		
命名由来 Origin of Name	该海底峡谷群位于中沙大环礁以东，因此得名。 These Canyons are located in the east of Zhongsha atoll, and “Dong” means east in Chinese, so the word “Zhongshadong” was used to name these Canyons.		

3.15 中沙北海底峡谷群

标准名称 Standard Name	中沙北海底峡谷群 Zhongshabei Haidixiaguqun	类别 Generic Term	海底峡谷群 Canyons
中心点坐标 Center Coordinates	16°24.6'N, 114°51.5'E	规模（千米 × 千米） Dimension（km × km）	145 × 20
最小水深（米） Min Depth (m)	300	最大水深（米） Max Depth (m)	4300
地理实体描述 Feature Description	中沙北海底峡谷群位于中沙海台北部海域，由数十条南—北向和北东—南西向发育的海底峡谷组成。峡谷群的头部起始于 200 ~ 500 米水深的中沙海台顶部边缘处，尾部终止于平均水深 2900 米的中沙北海隆和平均水深 4100 米的南海海盆，峡谷群的长度 5 ~ 50 千米（图 3-7）。 Zhongshabei Haidixiaguqun are located in the north sea area of Zhongsha Haitai and are composed of dozens of canyons developing in the directions of S-N and NE-SW. The head of these Canyons starts from the edge of the top of Zhongsha Haitai with a depth of 200~500 m and the tail ends at Zhongshabei Hailong with an average depth of 2900 m and at Nanhai Haipen with an average depth of 4100 m. The length of the Canyons are about 5~50 km (Fig.3-7).		
命名由来 Origin of Name	该海底峡谷群位于中沙大环礁以北，因此得名。 These Canyons are located in the north of Zhongsha atoll, and ‘Bei’ means north in Chinese, so the word “Zhongshabei” was used to name these Canyons.		

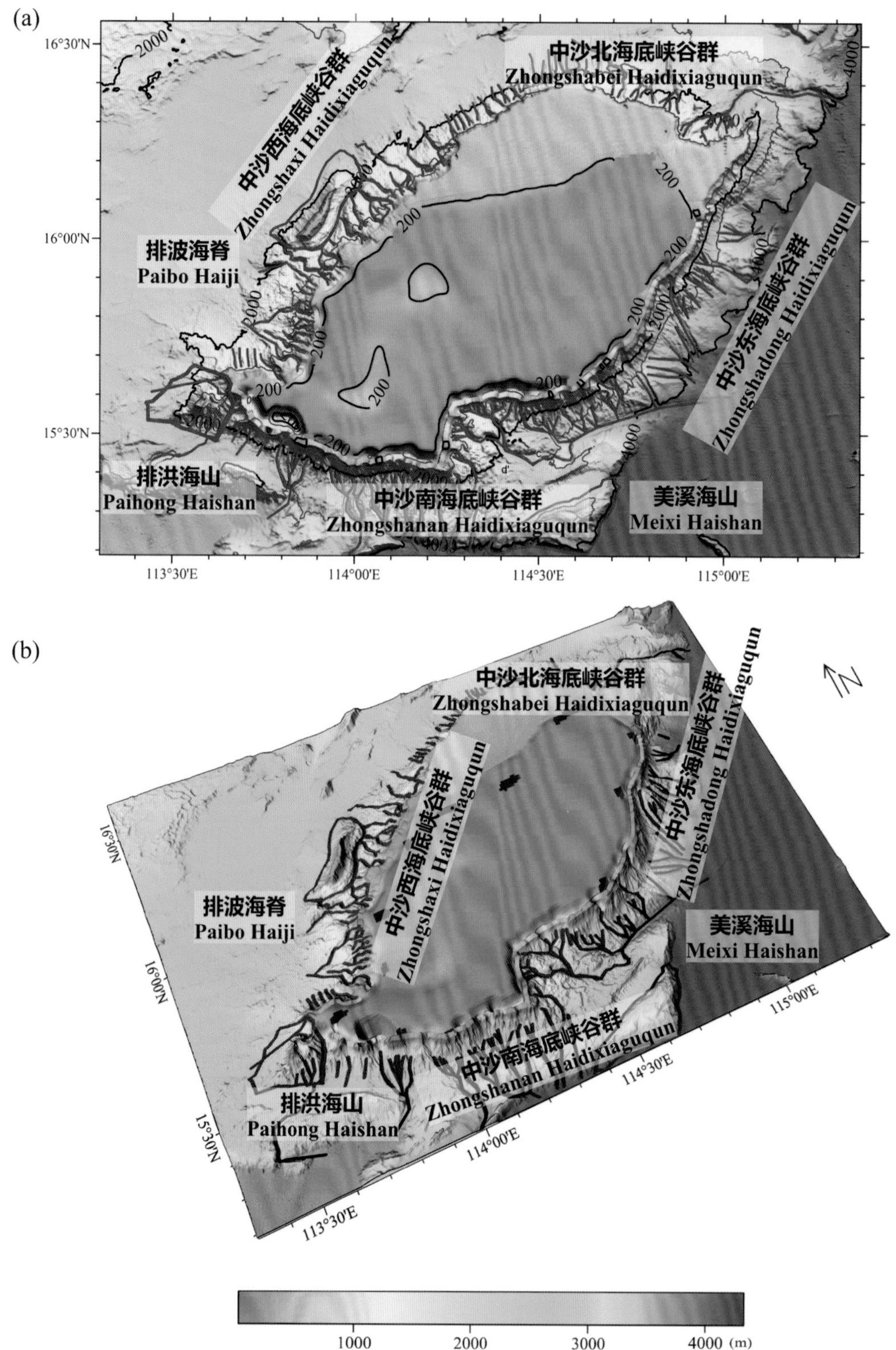

图 3-7　中沙西海底峡谷群、排洪海山、排波海脊、中沙南海底峡谷群、美溪海山、中沙东海底峡谷群、中沙北海底峡谷群

(a) 海底地形图（等深线间隔 1000 米）；(b) 三维海底地形图

Fig.3-7　Zhongshaxi Haidixiaguqun, Paihong Haishan, Paibo Haiji, Zhongshanan Haidixiaguqun, Meixi Haishan, Zhongshadong Haidixiaguqun, Zhongshabei Haidixiaguqun

(a) Seafloor topographic map (with contour interval of 1000 m)；(b) 3-D seafloor topographic map

3.16 中沙海槽

标准名称 Standard Name	中沙海槽 Zhongsha Haicao	类别 Generic Term	海槽 Trough
中心点坐标 Center Coordinates	16°20.0'N, 113°45.0'E	规模（千米 × 千米） Dimension（km × km）	210 × 50
最小水深（米） Min Depth (m)	2350	最大水深（米） Max Depth (m)	3440
地理实体描述 Feature Description	中沙海槽位于西沙海隆和中沙北海隆、中沙海台之间，平面形态呈北东—南西向的长条形。海槽长 210 千米，宽 15 ~ 70 千米。海槽槽底的地形较为平缓，水深从西南往东北逐渐变深，其内发育海丘、海脊等（图 3–8）。 Zhongsha Haicao is located between Xisha Hailong, Zhongshabei Hailong and Zhongsha Haitai with a planform in the shape of a long strip in the direction of NE–SW. The Trough is 210 km long and 15~70 km wide. The bottom of the Trough is relatively flat and gentle and the depth increases gradually from southwest to northeast. Several hills and ridges have developed in it (Fig.3–8).		
命名由来 Origin of Name	1986 年，国务院、外交部和中国地名委员会批准了前地矿部第二海洋地质调查大队（现广州海洋地质调查局）对南海 22 个海底地理实体进行的命名，“中沙海槽”是这 22 个海底地理实体名称之一。 In 1986, the State Council, the Ministry of Foreign Affairs and Chinese Toponymy Committee approved 22 undersea feature names in Nanhai named by former Second Marine Geological Survey Brigade of the Ministry of Geology and Mineral Resources (present Guangzhou Marine Geological Survey of China Geological Survey). Zhongsha Haicao is one of the 22 undersea feature names.		

3.17 中沙北海盆

标准名称 Standard Name	中沙北海盆 Zhongshabei Haipen	类别 Generic Term	海盆 Basin
中心点坐标 Center Coordinates	16°47.3'N, 114°04.4'E	规模（千米 × 千米） Dimension（km × km）	80 × 30
最小水深（米） Min Depth (m)	2920	最大水深（米） Max Depth (m)	3470
地理实体描述 Feature Description	中沙北海盆位于中沙海槽内，西沙海隆和中沙北海隆之间，平面形态呈北东—南西走向的长条形，长 70 千米，宽 20 ~ 40 千米，海盆内地形非常平坦（图 3–8）。 Zhongshabei Haipen is located in Zhongsha Haicao and is between Xisha Hailong and Zhongshabei Hailong, with a planform in the shape of a long strip in the direction of NE–SW. The Basin is 70 km long and 20~40 km width, and is very flat inside (Fig.3–8).		
命名由来 Origin of Name	该海盆位于中沙大环礁以北，因此得名。 The Basin is located to the north of Zhongsha atoll, and “Bei” means north in Chinese, so the word “Zhongshabei” was used to name the Basin.		

3.18 中沙北海隆

标准名称 Standard Name	中沙北海隆 Zhongshabei Hailong	类别 Generic Term	海隆 Rise
中心点坐标 Center Coordinates	17°00.0'N, 114°50.0'E	规模（千米 × 千米） Dimension（km × km）	190 × 135
最小水深（米） Min Depth (m)	500	最大水深（米） Max Depth (m)	4060
地理实体描述 Feature Description	中沙北海隆位于中沙海台以北，中沙海槽与南海海盆之间。平面形态不规则，近北东—南西走向，其上发育众多海山、海丘、海脊等（图 3-8）。 Zhongshabei Hailong is located on the north of Zhongsha Haitai and is between Zhongsha Haicao and Nanhai Haipen, with an irregular planform and nearly stretches in the direction of NE-SW. A great number of seamounts, hills and ridges have developed on it (Fig.3-8).		
命名由来 Origin of Name	1986 年，国务院、外交部和中国地名委员会批准了前地矿部第二海洋地质调查大队（现广州海洋地质调查局）对南海 22 个海底地理实体进行的命名，“中沙北海岭”是这 22 个海底地理实体名称之一。2001 年，经多波束海底地形调查测量确认为海隆，故更名为中沙北海隆。 In 1986, the State Council, the Ministry of Foreign Affairs and Chinese Toponymy Committee approved 22 undersea feature names in Nanhai named by former Second Marine Geological Survey Brigade of the Ministry of Geology and Mineral Resources (present Guangzhou Marine Geological Survey of China Geological Survey). Zhongshabei Hailing is one of the 22 undersea feature names. In 2001, it was confirmed as a rise through multi-beam bathymetric survey and was renamed Zhongshabei Hailong.		

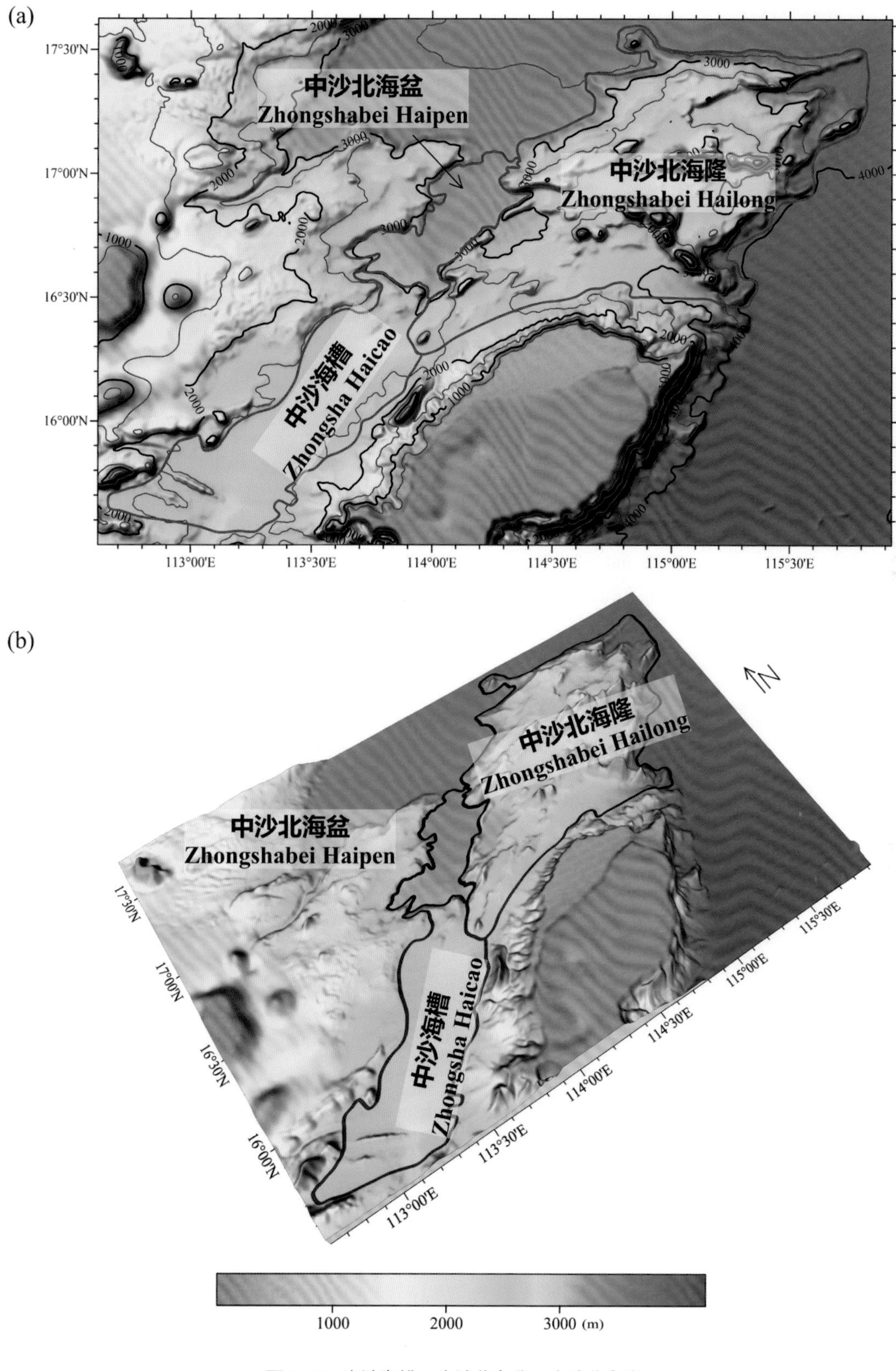

图 3-8　中沙海槽、中沙北海盆、中沙北海隆

(a) 海底地形图（等深线间隔 500 m）；(b) 三维海底地形图

Fig.3-8　Zhongsha Haicao, Zhongshabei Haipen, Zhongshabei Hailong

(a) Seafloor topographic map (with contour interval of 500 m); (b) 3-D seafloor topographic map

3.19 指掌海脊

标准名称 Standard Name	指掌海脊 Zhizhang Haiji	类别 Generic Term	海脊 Ridge
中心点坐标 Center Coordinates	17°31.7'N, 114°49.9'E	规模（千米 × 千米） Dimension（km × km）	30 × 10
最小水深（米） Min Depth (m)	2750	最大水深（米） Max Depth (m)	3850
地理实体描述 Feature Description	指掌海脊发育在中沙北海隆最北端，平面形态呈北西—南东向的长条形（图 3−9）。 Zhizhang Haiji develops on the northernmost end of Zhongshabei Hailong, with a planform in the shape of a long strip in the direction of NW−SE (Fig.3−9).		
命名由来 Origin of Name	该海脊位于我国中沙大环礁附近，故以中沙大环礁中的暗沙、浅滩名进行地名的团组化命名。“指掌暗沙”属于中沙大环礁的一个暗沙，取“指掌”两字命名该海脊。 The Ridge is located near the Zhongsha atoll in China, so it is named after a group of shoals and banks in Zhongsha atoll. “Zhizhang Ansha” is one of the shoals belonging to Zhongsha atoll, so the word “Zhizhang” was used to name the Ridge.		

3.20 天鸡海丘

标准名称 Standard Name	天鸡海丘 Tianji Haiqiu	类别 Generic Term	海丘 Hill
中心点坐标 Center Coordinates	17°23.4'N, 115°03.2'E	规模（千米 × 千米） Dimension（km × km）	15 × 5
最小水深（米） Min Depth (m)	1950	最大水深（米） Max Depth (m)	2870
地理实体描述 Feature Description	天鸡海丘发育在中沙北海隆最北端，平面形态呈东—西向的三角形（图 3−9）。 Tianji Haiqiu develops in the northernmost end of Zhongshabei Hailong, with a planform in the shape of a triangle in the direction of E−W (Fig.3−9).		
命名由来 Origin of Name	以唐朝诗人李白的《梦游天姥吟留别》中的词进行地名的团组化命名。该海底地名的专名取词自该诗作中的：“半壁见海日，空中闻天鸡”。“天鸡”指古代传说中的神鸡。 A group naming of undersea features after the phrases in the poem *A Visit to Mount Tianmu in Dream* by Li Bai (701−762 A.D.), a famous poet in the Tang Dynasty (618−907 A.D.). The specific term of this undersea feature name “Tianji”, means a kind of magic rooster in the ancient Chinese legend, is derived from the poetic lines “I saw the sunrise over the sea at half way up the cliff and heard of magic rooster crowing in the air”.		

3.21 青云海丘

标准名称 Standard Name	青云海丘 Qingyun Haiqiu	类别 Generic Term	海丘 Hill
中心点坐标 Center Coordinates	17°18.3'N, 115°00.2'E	规模（千米 × 千米） Dimension（km × km）	2 × 2
最小水深（米） Min Depth (m)	2020	最大水深（米） Max Depth (m)	2400
地理实体描述 Feature Description	青云海丘发育在中沙北海隆北端，平面形态呈圆形（图 3-9）。 Qingyun Haiqiu develops in the north end of Zhongshabei Hailong, with a circular planform (Fig.3-9).		
命名由来 Origin of Name	以唐朝诗人李白的《梦游天姥吟留别》中的词进行地名的团组化命名。该海底地名的专名取词自该诗作中的："脚著谢公屐，身登青云梯"。"青云"意为很高处的云霄。 A group naming of undersea features after the phrases in the poem *A Visit to Mount Tianmu in Dream* by Li Bai (701-762 A.D.), a famous poet in the Tang Dynasty (618-907 A.D.). The specific term of this undersea feature name "Qingyun", means dark cloud up in the sky, is derived from the poetic lines "Donning the shoes of Xie, I climbed the dark ladder of cloud".		

3.22 谢公海丘

标准名称 Standard Name	谢公海丘 Xiegong Haiqiu	类别 Generic Term	海丘 Hill
中心点坐标 Center Coordinates	17°13.1'N, 115°04.2'E	规模（千米 × 千米） Dimension（km × km）	4 × 3
最小水深（米） Min Depth (m)	1990	最大水深（米） Max Depth (m)	2580
地理实体描述 Feature Description	谢公海丘发育在中沙北海隆北端，平面形态呈圆形（图 3-9）。 Xiegong Haiqiu develops in the north of Zhongshabei Hailong, with a circular planform (Fig.3-9).		
命名由来 Origin of Name	以唐朝诗人李白的《梦游天姥吟留别》中的词进行地名的团组化命名。该海底地名的专名取词自该诗作中的："谢公宿处今尚在，渌水荡漾清猿啼"。"谢公"指的是南朝诗人谢灵运。 A group naming of undersea features after the phrases in the poem *A Visit to Mount Tianmu in Dream* by Li Bai (701-762 A.D.), a famous poet in the Tang Dynasty (618-907 A.D.). The specific term of this undersea feature name "Xiegong", refers to Xie Lingyun (Master Xie), a poet in the Southern Dynasty (420-589 A.D.), is derived from the poetic lines "The hermitage of Master Xie is still there, and the calling of monkeys is cleared heard over the rippling green water".		

3.23 美滨海脊

标准名称 Standard Name	美滨海脊 Meibin Haiji	类别 Generic Term	海脊 Ridge
中心点坐标 Center Coordinates	17°22.7′N, 115°21.5′E	规模（千米 × 千米） Dimension（km × km）	20 × 10
最小水深（米） Min Depth (m)	1940	最大水深（米） Max Depth (m)	4090
地理实体描述 Feature Description	美滨海脊发育在中沙北海隆北端，平面形态呈北西西—南东东向的长条形（图 3-9）。 Meibin Haiji develops in the north of Zhongshabei Hailong, with a planform in the shape of a long strip in the direction of NWW−SEE (Fig.3−9).		
命名由来 Origin of Name	该海脊位于我国中沙大环礁附近，故以中沙大环礁中的暗沙、浅滩名进行地名的团组化命名。"美滨暗沙"属于中沙大环礁的一个暗沙，取"美滨"两字命名该海脊。 The Ridge is located near the Zhongsha atoll in China, so it is named after a group of shoals and banks in Zhongsha atoll. "Meibin Ansha" is one of the shoals belonging to Zhongsha atoll, so the word "Meibin" was used to name the Ridge.		

3.24 美滨海山

标准名称 Standard Name	美滨海山 Meibin Haishan	类别 Generic Term	海山 Seamount
中心点坐标 Center Coordinates	17°17.9′N, 115°09.3′E	规模（千米 × 千米） Dimension（km × km）	20 × 25
最小水深（米） Min Depth (m)	1800	最大水深（米） Max Depth (m)	3050
地理实体描述 Feature Description	美滨海脊发育在中沙北海隆北端，平面形态呈不规则多边形（图 3-9）。 Meibin Haiji develops in the north of Zhongshabei Hailong, with a planform in the shape of an irregular polygon (Fig.3−9).		
命名由来 Origin of Name	该海山位于美滨海脊附近，因此得名。 The Seamount is located near Meibin Haiji, so the word "Meibin" was used to name the Seamount.		

(a)

(b)

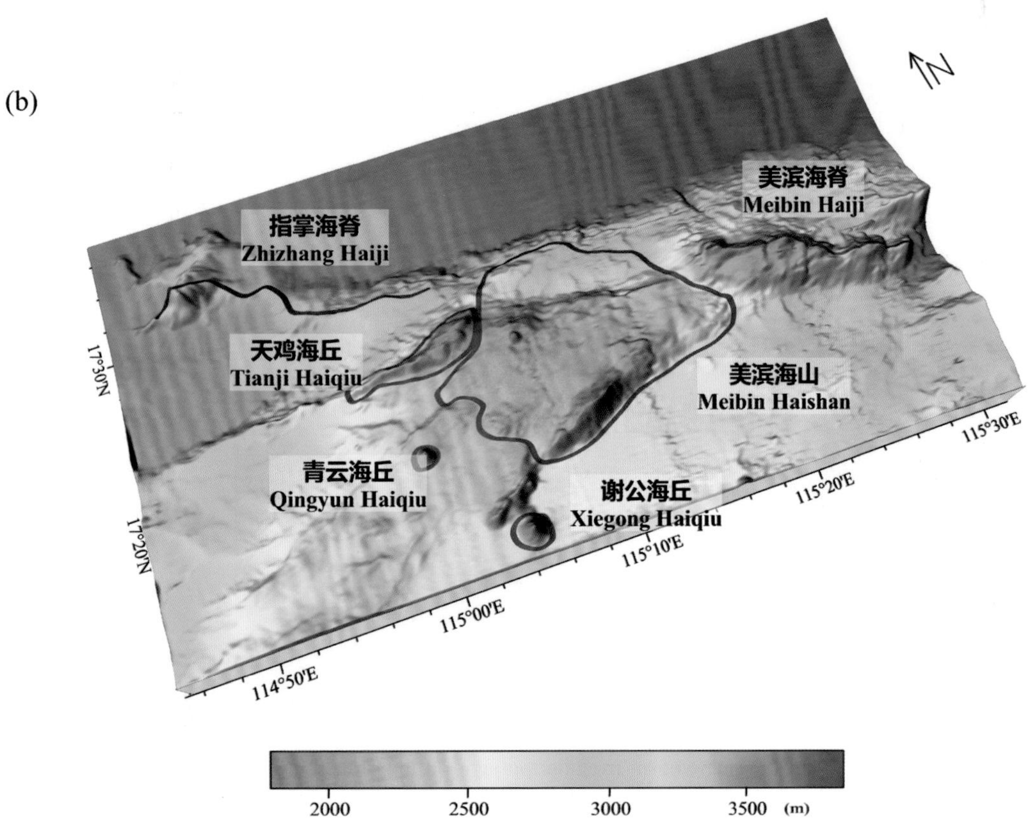

图 3-9 指掌海脊、天鸡海丘、青云海丘、谢公海丘、美滨海脊、美滨海山

(a) 海底地形图（等深线间隔 500 米）；(b) 三维海底地形图

Fig.3-9 Zhizhang Haiji, Tianji Haiqiu, Qingyun Haiqiu, Xiegong Haiqiu, Meibin Haiji, Meibin Haishan

(a) Seafloor topographic map (with contour interval of 500 m)；(b) 3-D seafloor topographic map

3.25 中北海脊

标准名称 Standard Name	中北海脊 Zhongbei Haiji	类别 Generic Term	海脊 Ridge
中心点坐标 Center Coordinates	17°19.8'N, 114°45.2'E	规模（千米 × 千米） Dimension（km × km）	28 × 7
最小水深（米） Min Depth (m)	2580	最大水深（米） Max Depth (m)	3660
地理实体描述 Feature Description	中北海脊发育在中沙北海隆北端，平面形态呈北西—南东向的长条形。海脊的西北侧水深较深，东南侧水深较浅（图 3-10）。 Zhongbei Haiji develops in the north of Zhongshabei Hailong, with a planform in the shape of a long strip in the direction of NW−SE. Its northwest side of the Ridge is deeper, while its southeast side is shallower (Fig.3−10).		
命名由来 Origin of Name	该海脊位于我国中沙大环礁附近，故以中沙大环礁中的暗沙、浅滩名进行地名的团组化命名。“中北暗沙”属于中沙大环礁的一个暗沙，取“中北”两字命名该海脊。 The Ridge is located near the Zhongsha atoll in China, so it is named after a group of shoals and banks in Zhongsha atoll. “Zhongbei Ansha” is one of the shoals belonging to Zhongsha atoll, so the word “Zhongbei” was used to name the Ridge.		

3.26 鲁班海丘

标准名称 Standard Name	鲁班海丘 Luban Haiqiu	类别 Generic Term	海丘 Hill
中心点坐标 Center Coordinates	17°06.8'N, 114°48.3'E	规模（千米 × 千米） Dimension（km × km）	22 × 10
最小水深（米） Min Depth (m)	1800	最大水深（米） Max Depth (m)	2700
地理实体描述 Feature Description	鲁班海丘发育在中沙北海隆西北侧，平面形态呈北东东—南西西向的不规则多边形。海丘的北坡较陡，南坡较缓（图 3-10）。 Luban Haiqiu develops in the northwest side of Zhongshabei Hailong, with a planform in the shape of an irregular polygon in the direction of NEE−SWW. Its north slope is relatively steeper while its south slope is relatively gentler (Fig.3−10).		
命名由来 Origin of Name	该海丘位于我国中沙大环礁附近，故以中沙大环礁中的暗沙、浅滩名进行地名的团组化命名。“鲁班暗沙”属于中沙大环礁的一个暗沙，取“鲁班”两字命名该海丘。 The Hill is located near the Zhongsha atoll in China, so it is named after a group of shoals and banks in Zhongsha atoll. “Luban Ansha” is one of the shoals belonging to Zhongsha atoll, so the word “Luban” was used to name the Hill.		

3.27 湖月圆丘

标准名称 Standard Name	湖月圆丘 Huyue Yuanqiu	类别 Generic Term	圆丘 Knoll
中心点坐标 Center Coordinates	17°05.8'N, 114°40.8'E	规模（千米 × 千米） Dimension（km × km）	3 × 3
最小水深（米） Min Depth (m)	1980	最大水深（米） Max Depth (m)	2870
地理实体描述 Feature Description	湖月圆丘发育在中沙北海隆西北侧，鲁班海丘和西门海山之间，平面形态呈圆形（图 3-10）。 Huyue Yuanqiu develops in the northwest side of Zhongshabei Hailong and is between Luban Haiqiu and Ximen Haishan, with a circular planform (Fig.3-10).		
命名由来 Origin of Name	以唐朝诗人李白的《梦游天姥吟留别》中的词进行地名的团组化命名。该海底地名的专名取词自该诗作中的："湖月照我影，送我至剡溪"。"湖月"指湖中的月光。 A group naming of undersea features after the phrases in the poem *A Visit to Mount Tianmu in Dream* by Li Bai (701–762 A.D.), a famous poet in the Tang Dynasty (618–907 A.D.). The specific term of this undersea feature name "Huyue", means moon over lake, is derived from the poetic lines "The moon over the lake is shedding light on my shadow and accompanying to Shan River".		

(a)

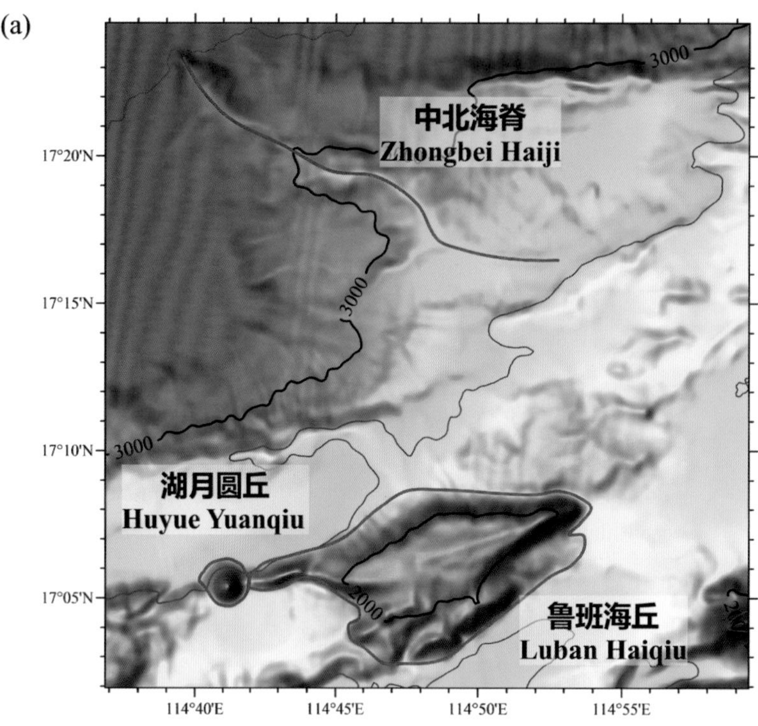

(b)

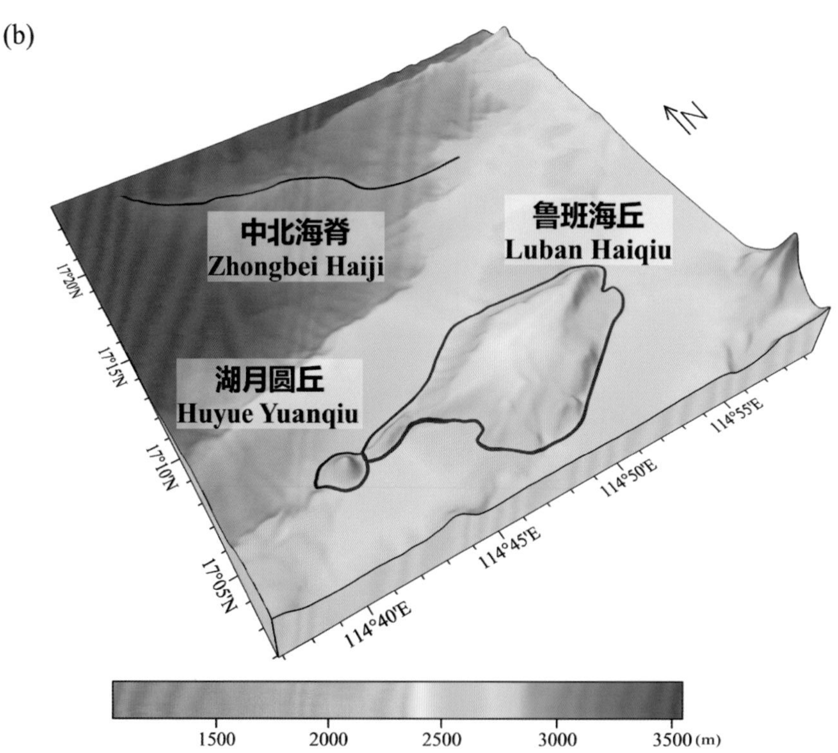

图 3-10　中北海脊、鲁班海丘、湖月圆丘

(a) 海底地形图（等深线间隔 500 米）；(b) 三维海底地形图

Fig.3-10　Zhongbei Haiji, Luban Haiqiu, Huyue Yuanqiu

(a) Seafloor topographic map (with contour interval of 500 m)；(b) 3-D seafloor topographic map

3.28 隐矶海山

标准名称 Standard Name	隐矶海山 Yinji Haishan	类别 Generic Term	海山 Seamount
中心点坐标 Center Coordinates	17°04.8'N, 115°01.4'E	规模（千米 × 千米） Dimension（km × km）	54 × 10
最小水深（米） Min Depth (m)	650	最大水深（米） Max Depth (m)	4160
地理实体描述 Feature Description	隐矶海山是一座发育在中沙北海隆上、呈东—西走向的大型链状海山，其顶部发育 4 座峰（图 3-11）。 Yinji Haishan develops in Zhongshabei Hailong and is a large-scale seamount chain in the direction of E-W, with four peaks developed on its top (Fig.3-11).		
命名由来 Origin of Name	该海山位于我国中沙大环礁附近，故以中沙大环礁中的暗沙、浅滩名进行地名的团组化命名。“隐矶滩”属于中沙大环礁的一个浅滩，取“隐矶”两字命名该海山。 The Seamount is located near the Zhongsha atoll in China, so it is named after a group of shoals and banks in Zhongsha atoll. “Yinji Tan” is one of the banks belonging to Zhongsha atoll, so the word “Yinji” was used to name the Seamount.		

3.29 烟霞海底峰

标准名称 Standard Name	烟霞海底峰 Yanxia Haidifeng	类别 Generic Term	海底峰 Peak
中心点坐标 Center Coordinates	17°04.7'N, 115°10.1'E	规模（千米 × 千米） Dimension（km × km）	15 × 6
最小水深（米） Min Depth (m)	670	最大水深（米） Max Depth (m)	2640
地理实体描述 Feature Description	烟霞海底峰是属于隐矶海山的一个山峰，平面形态呈东—西走向的长条形（图 3-11）。 Yanxia Haidifeng is a peak of Yinji Haishan, with a planform in the shape of a long strip in the direction of E-W (Fig.3-11).		
命名由来 Origin of Name	以唐朝诗人李白的《梦游天姥吟留别》中的词进行地名的团组化命名。该海底地名的专名取词自该诗作中的：“惟觉时之枕席，失向来之烟霞”。“烟霞”指烟雾袅绕的意思。 A group naming of undersea features after the phrases in the poem *A Visit to Mount Tianmu in Dream* by Li Bai (701-762 A.D.), a famous poet in the Tang Dynasty (618-907 A.D.). The specific term of this undersea feature name “Yanxia”, means misty cloud, is derived from the poetic lines “I woke up only to find my pillow and mat, without any trace of the misty cloud I perceived in my dream”.		

3.30 济猛海底峰

标准名称 Standard Name	济猛海底峰 Jimeng Haidifeng	类别 Generic Term	海底峰 Peak
中心点坐标 Center Coordinates	17°03.9'N, 115°15.4'E	规模（千米 × 千米） Dimension（km × km）	10 × 8
最小水深（米） Min Depth (m)	710	最大水深（米） Max Depth (m)	2980
地理实体描述 Feature Description	济猛海底峰是属于隐矶海山的一个山峰，平面形态呈东—西走向（图 3−11）。 Jimeng Haidifeng is peak of Yinji Haishan, with an planform in the direction of E−W (Fig.3−11).		
命名由来 Origin of Name	该海底峰位于我国中沙大环礁附近，故以中沙大环礁中的暗沙、浅滩名进行地名的团组化命名。“济猛暗沙”属于中沙大环礁的一个暗沙，取“济猛”两字命名该海山。 The Peak is located near the Zhongsha atoll in China, so it is named after a group of shoals and banks in Zhongsha atoll. “Jimeng Ansha” is one of the shoals belonging to Zhongsha atoll, so the word “Jimeng” was used to name the Peak.		

3.31 天姥海底峰

标准名称 Standard Name	天姥海底峰 Tianmu Haidifeng	类别 Generic Term	海底峰 Peak
中心点坐标 Center Coordinates	17°02.8'N, 115°21.9'E	规模（千米 × 千米） Dimension（km × km）	18 × 8
最小水深（米） Min Depth (m)	630	最大水深（米） Max Depth (m)	3100
地理实体描述 Feature Description	天姥海底峰是属于隐矶海山的一个山峰，平面形态呈东—西走向，其顶部较为平坦（图 3−11）。 Tianmu Haidifeng is a peak of Yinji Haishan, with a planform in the direction of E−W and a relatively flat top (Fig.3−11).		
命名由来 Origin of Name	以唐朝诗人李白的《梦游天姥吟留别》中的词进行地名的团组化命名。该海底地名的专名取词自该诗作的题名，“天姥”指浙江省绍兴市新昌县境内的天姥山。 A group naming of undersea features after the phrases in the poem *A Visit to Mount Tianmu in Dream* by Li Bai (701−762 A.D.), a famous poet in the Tang Dynasty (618−907 A.D.). The specific term of this undersea feature name “Tianmu”, means Mount Tianmu in Xinchang county, Shaoxing, Zhejiang province.		

3.32 海鸠海丘

标准名称 Standard Name	海鸠海丘 Haijiu Haiqiu	类别 Generic Term	海丘 Hill
中心点坐标 Center Coordinates	17°11.8'N, 115°15.1'E	规模（千米 × 千米） Dimension（km × km）	2 × 2
最小水深（米） Min Depth (m)	2210	最大水深（米） Max Depth (m)	2540
地理实体描述 Feature Description	海鸠海丘发育在中沙北海隆中北部，平面形态呈圆形（图 3−11）。 Haijiu Haiqiu is development in the north central of Zhongshabei Hailong with a circular planform (Fig.3−11).		
命名由来 Origin of Name	该海丘位于我国中沙大环礁附近，故以中沙大环礁中的暗沙、浅滩名进行地名的团组化命名。“海鸠暗沙”属于中沙大环礁的一个暗沙，取“海鸠”两字命名该海丘。 The Hill is located near the Zhongsha atoll in China, so it is named after a group of shoals and banks in Zhongsha atoll. “Haijiu Ansha” is one of the shoals belonging to Zhongsha atoll, so the word “Haijiu” was used to name the Hill.		

3.33 本固海丘

标准名称 Standard Name	本固海丘 Bengu Haiqiu	类别 Generic Term	海丘 Hill
中心点坐标 Center Coordinates	17°07.1'N, 115°09.9'E	规模（千米 × 千米） Dimension（km × km）	8 × 5
最小水深（米） Min Depth (m)	1400	最大水深（米） Max Depth (m)	2210
地理实体描述 Feature Description	本固海丘发育在中沙北海隆中北部，紧邻隐矶海山，平面形态近似圆形（图 3−11）。 Bengu Haiqiu develops in the north central of Zhongshabei Hailong and is adjacent to Yinji Haishan with a nearly circular planform (Fig.3−11).		
命名由来 Origin of Name	该海丘位于我国中沙大环礁附近，故以中沙大环礁中的暗沙、浅滩名进行地名的团组化命名。“本固暗沙”属于中沙大环礁的一个暗沙，取“本固”两字命名该海丘。 The Hill is located near the Zhongsha atoll in China, so it is named after a group of shoals and banks in Zhongsha atoll. “Bengu Ansha” is one of the shoals belonging to Zhongsha atoll, so the word “Bengu” was used to name the Hill.		

3.34 隐矶南海山

标准名称 Standard Name	隐矶南海山 Yinjinan Haishan	类别 Generic Term	海山 Seamount
中心点坐标 Center Coordinates	17°01.8'N, 115°01.0'E	规模（千米 × 千米） Dimension（km × km）	18 × 5
最小水深（米） Min Depth (m)	1610	最大水深（米） Max Depth (m)	2710
地理实体描述 Feature Description	隐矶南海山发育在中沙北海隆中部，紧邻隐矶海山，平面形态呈东—西向的长条形（图 3-11）。 Yinjinan Haishan develops in the middle of Zhongshabei Hailong and is adjacent to Yinji Haishan, with a planform in the shape of a long strip in the direction of E–W (Fig.3-11).		
命名由来 Origin of Name	该海山位于隐矶海山以南，因此得名。 The Seamount is located on the south of Yinji Haishan, and "Nan" means south in Chinese, so the word "Yinjinan" was used to name the Seamount.		

3.35 越人海山

标准名称 Standard Name	越人海山 Yueren Haishan	类别 Generic Term	海山 Seamount
中心点坐标 Center Coordinates	17°00.6'N, 115°08.3'E	规模（千米 × 千米） Dimension（km × km）	6 × 5
最小水深（米） Min Depth (m)	1580	最大水深（米） Max Depth (m)	2800
地理实体描述 Feature Description	越人海山发育在中沙北海隆中部，紧邻隐矶海山，平面形态呈圆形（图 3-11）。 Yueren Haishan develops in the middle of Zhongshabei Hailong and is adjacent to Yinji Haishan, with a circular planform (Fig.3-11).		
命名由来 Origin of Name	以唐朝诗人李白的《梦游天姥吟留别》中的词进行地名的团组化命名。该海底地名的专名取词自该诗作中的："越人语天姥，云霞明灭或可睹"。"越人"指浙江绍兴一带的人。 A group naming of undersea features after the phrases in the poem *A Visit to Mount Tianmu in Dream* by Li Bai (701–762 A.D.), a famous poet in the Tang Dynasty (618–907 A.D.). The specific term of this undersea feature name "Yueren", means people of Shaoxing, is derived from the poetic lines "Local people of Shaoxing mentioned of Mount Tianmu, saying it was always clouded in mist and only occasionally revealed itself".		

3.36 天台海丘

标准名称 Standard Name	天台海丘 Tiantai Haiqiu	类别 Generic Term	海丘 Hill
中心点坐标 Center Coordinates	16°57.6'N, 114°57.4'E	规模（千米 × 千米） Dimension（km × km）	2 × 2
最小水深（米） Min Depth (m)	2550	最大水深（米） Max Depth (m)	2770
地理实体描述 Feature Description	天台海丘发育在中沙北海隆中部，平面形态呈圆形（图 3-11）。 Tiantai Haiqiu develops in the middle of Zhongshabei Hailong with a circular planform (Fig.3-11).		
命名由来 Origin of Name	以唐朝诗人李白的《梦游天姥吟留别》中的词进行地名的团组化命名。该海底地名的专名取词自该诗作中的：“天台四万八千丈，对此欲倒东南倾”。“天台”指天台山。 A group naming of undersea features after the phrases in the poem *A Visit to Mount Tianmu in Dream* by Li Bai (701–762 A.D.), a famous poet in the Tang Dynasty (618–907 A.D.). The specific term of this undersea feature name “Tiantai”, means Mount Tiantai, is derived from the poetic lines “Lofty as mountain Tiantai is in the southeast, it is dwarfed before Mount Tianmu”.		

3.37 吴越海丘

标准名称 Standard Name	吴越海丘 Wuyue Haiqiu	类别 Generic Term	海丘 Hill
中心点坐标 Center Coordinates	16°56.1'N, 115°06.6'E	规模（千米 × 千米） Dimension（km × km）	6 × 5
最小水深（米） Min Depth (m)	2580	最大水深（米） Max Depth (m)	2950
地理实体描述 Feature Description	吴越海丘发育在中沙北海隆中部，平面形态近似圆形（图 3-11）。 Wuyue Haiqiu develops in the middle of Zhongshabei Hailong with a nearly circular planform (Fig.3-11).		
命名由来 Origin of Name	以唐朝诗人李白的《梦游天姥吟留别》中的词进行地名的团组化命名。该海底地名的专名取词自该诗作中的：“我欲因之梦吴越，一夜飞度镜湖月”。“吴越”指古代吴国、越国之地，如今的江浙一带。 A group naming of undersea features after the phrases in the poem *A Visit to Mount Tianmu in Dream* by Li Bai (701–762 A.D.), a famous poet in the Tang Dynasty (618–907 A.D.). The specific term of this undersea feature name “Wuyue”, means the land of Wu and Yue kingdom in ancient China (present Zhejiang province and Jiangsu province), is derived from the poetic lines “I thus dreamed of visiting land of Wu and Yue, and flying across the Mirror Lake in the moonlit night”.		

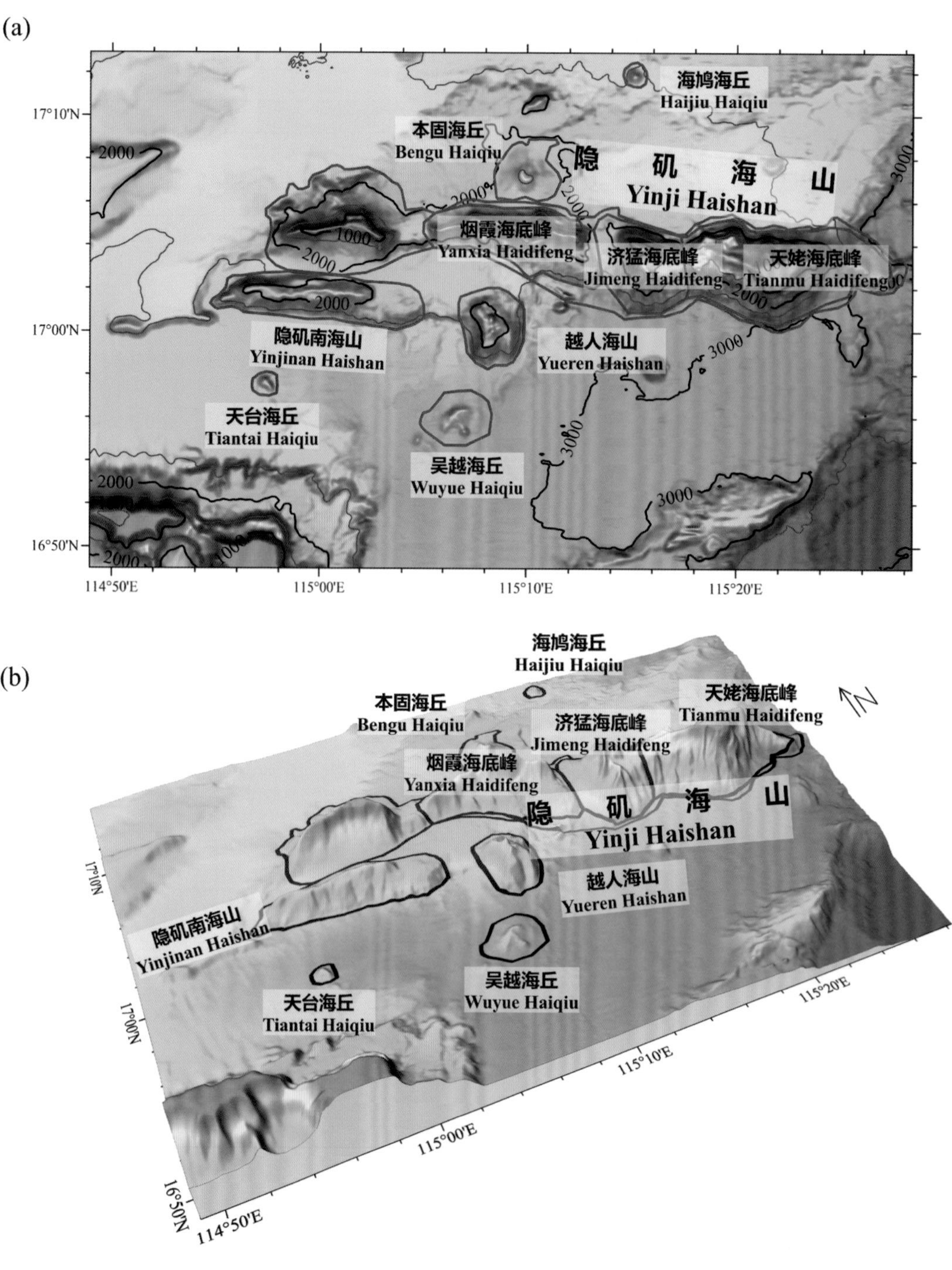

图 3-11 隐矶海山、烟霞海底峰、济猛海底峰、天姥海底峰、海鸠海丘、本固海丘、隐矶南海山、越人海山、天台海丘、吴越海丘

(a) 海底地形图（等深线间隔 500 米）；(b) 三维海底地形图

Fig.3-11 Yinji Haishan, Yanxia Haidifeng, Jimeng Haidifeng, Tianmu Haidifeng, Haijiu Haiqiu, Bengu Haiqiu, Yinjinan Haishan, Yueren Haishan, Tiantai Haiqiu, Wuyue Haiqiu

(a) Seafloor topographic map (with contour interval of 500 m); (b) 3-D seafloor topographic map

3.38 天横圆丘

标准名称 Standard Name	天横圆丘 Tianheng Yuanqiu	类别 Generic Term	圆丘 Knoll
中心点坐标 Center Coordinates	16°58.3'N, 115°16.1'E	规模（千米 × 千米） Dimension（km × km）	3 × 3
最小水深（米） Min Depth (m)	2460	最大水深（米） Max Depth (m)	3000
地理实体描述 Feature Description	天横圆丘发育在中沙北海隆中部，平面形态呈圆形（图 3−12）。 Tianheng Yuanqiu develops in the middle of Zhongshabei Hailong with a circular planform (Fig.3−12).		
命名由来 Origin of Name	以唐朝诗人李白的《梦游天姥吟留别》中的词进行地名的团组化命名。该海底地名的专名取词自该诗作中的："天姥连天向天横，势拔五岳掩赤城"。"天横"即天潢，星宿名。 A group naming of undersea features after the phrases in the poem *A Visit to Mount Tianmu in Dream* by Li Bai (701−762 A.D.), a famous poet in the Tang Dynasty (618−907 A.D.). The specific term of this undersea feature name "Tianheng" means the name of a constellation, is derived from the poetic lines "Mount Tianmu is soaring towards Tianheng constellation, surpassing the five major mountains and Mount Chicheng."		

3.39 指掌海丘

标准名称 Standard Name	指掌海丘 Zhizhang Haiqiu	类别 Generic Term	海丘 Hill
中心点坐标 Center Coordinates	16°51.2'N, 115°19.7'E	规模（千米 × 千米） Dimension（km × km）	22 × 10
最小水深（米） Min Depth (m)	2520	最大水深（米） Max Depth (m)	3650
地理实体描述 Feature Description	指掌海丘发育在中沙北海隆东端，平面形态呈北东—南西向的长条形。海丘的西北坡较缓，东南坡较陡（图 3−12）。 Zhizhang Haiqiu develops in the east of Zhongshabei Hailong, with a planform in the shape of a long strip in the direction of NE−SW. Its northwest slope is relatively gentler while its southeast slope is relatively steeper (Fig.3−12).		
命名由来 Origin of Name	该海丘位于我国中沙大环礁附近，故以中沙大环礁中的暗沙、浅滩名进行地名的团组化命名。"指掌暗沙"属于中沙大环礁的一个暗沙，取"指掌"两字命名该海丘。 The Hill is located near the Zhongsha atoll in China, so it is named after a group of shoals and banks in Zhongsha atoll. "Zhizhang Ansha" is one of the shoals belonging to Zhongsha atoll, so the word "Zhizhang" was used to name the Hill.		

3.40 南扉海脊

标准名称 Standard Name	南扉海脊 Nanfei Haiji	类别 Generic Term	海脊 Ridge
中心点坐标 Center Coordinates	16°56.1′N, 115°26.1′E	规模（千米 × 千米） Dimension（km × km）	13 × 7
最小水深（米） Min Depth (m)	2800	最大水深（米） Max Depth (m)	4000
地理实体描述 Feature Description	南扉海脊发育在中沙北海隆东端，平面形态呈北东—南西向的长条形。海脊的西北坡较缓，东南坡较陡（图 3-12）。 Nanfei Haiji develops in on the east of Zhongshabei Hailong with a planform in the shape of a long strip in the direction of NE-SW. Its northwest slope is relatively gentler while its southeast slope is relatively steeper (Fig.3-12).		
命名由来 Origin of Name	该海脊位于我国中沙大环礁附近，故以中沙大环礁中的暗沙、浅滩名进行地名的团组化命名。“南扉暗沙”属于中沙大环礁的一个暗沙，取“南扉”两字命名该海脊。 The Ridge is located near the Zhongsha atoll in China, so it is named after a group of shoals and banks in Zhongsha atoll. “Nanfei Ansha” is one of the shoals belonging to Zhongsha atoll, so the word “Nanfei” was used to name the Ridge.		

3.41 东流海丘

标准名称 Standard Name	东流海丘 Dongliu Haiqiu	类别 Generic Term	海丘 Hill
中心点坐标 Center Coordinates	16°56.5′N, 115°33.9′E	规模（千米 × 千米） Dimension（km × km）	13 × 4
最小水深（米） Min Depth (m)	3500	最大水深（米） Max Depth (m)	4000
地理实体描述 Feature Description	东流海丘发育在中沙北海隆东端，平面形态呈北东东—南西西向的长条形（图 3-12）。 Zhizhang Haiqiu develops in on the east of Zhongshabei Hailong with a planform in the shape of a long strip in the direction of NEE-SWW (Fig.3-12).		
命名由来 Origin of Name	以唐朝诗人李白的《梦游天姥吟留别》中的词进行地名的团组化命名。该海底地名的专名取词自该诗作中的：“世间行乐亦如此，古来万事东流水”。“东流”形容事物像东流的水一样一去不复返。 A group naming of undersea features after the phrases in the poem *A Visit to Mount Tianmu in Dream* by Li Bai (701-762 A.D.), a famous poet in the Tang Dynasty (618-907 A.D.). The specific term of this undersea feature name “Dongliu”, means things that do not return like the water flows eastward, is derived from the poetic lines “And so with the pleasures of this life, all pass, as water flows eastward”.		

(a)

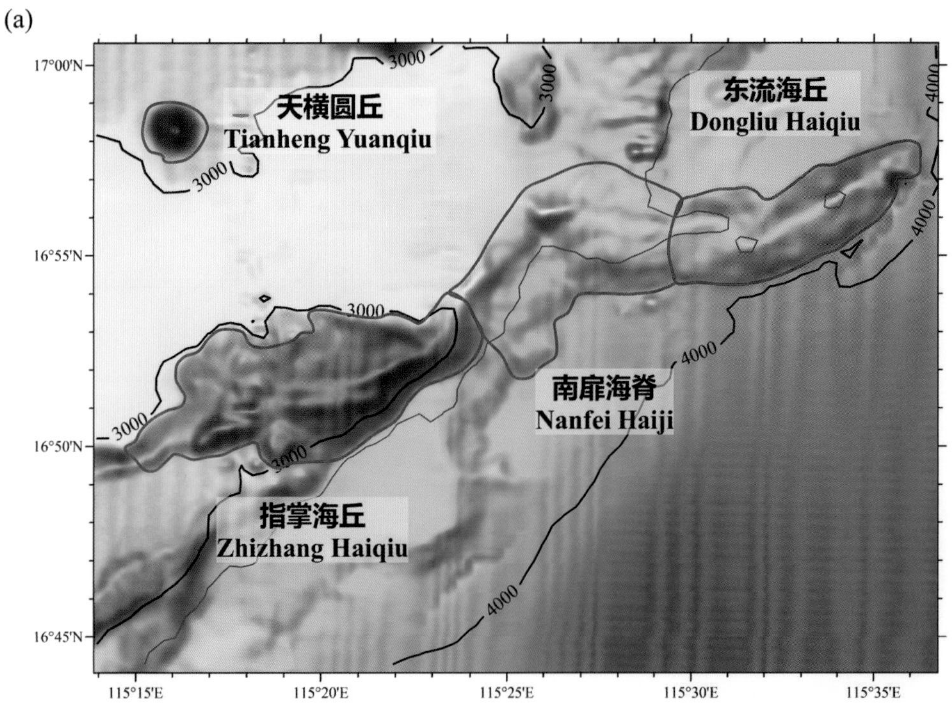

(b)

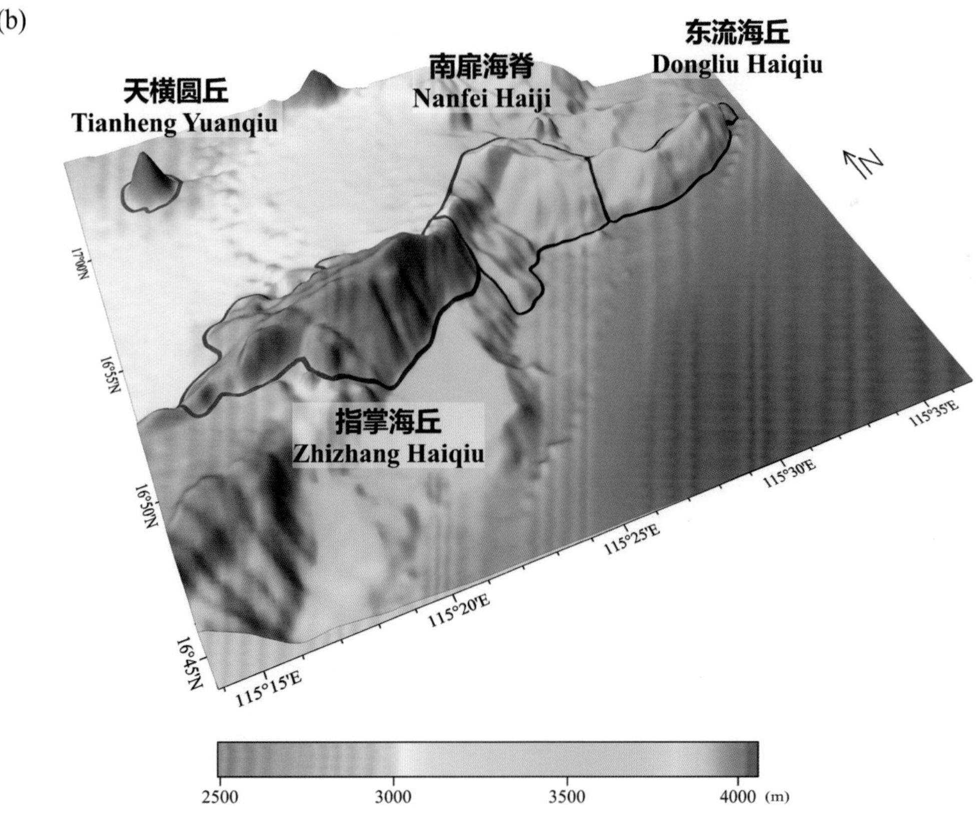

图 3-12 天横圆丘、指掌海丘、南扉海脊、东流海丘

(a) 海底地形图（等深线间隔 500 米）；(b) 三维海底地形图

Fig.3-12 Tianheng Yuanqiu, Zhizhang Haiqiu, Nanfei Haiji, Dongliu Haiqiu

(a) Seafloor topographic map (with contour interval of 500 m); (b) 3-D seafloor topographic map

3.42 比微海山

标准名称 Standard Name	比微海山 Biwei Haishan	类别 Generic Term	海山 Seamount
中心点坐标 Center Coordinates	17°11.5'N, 115°43.5'E	规模（千米 × 千米） Dimension（km × km）	28 × 10
最小水深（米） Min Depth (m)	2870	最大水深（米） Max Depth (m)	4160
地理实体描述 Feature Description	比微海山发育在中沙北海隆东北端，平面形态呈北东—南西向的菱形。海山的西北坡较缓，东南坡较陡，其顶部发育多座峰（图 3-13）。 Biwei Haishan develops in the northeast of Zhongshabei Hailong, with a planform in the shape of a diamond in the direction of NE−SW. Its northwest slope is relatively gentler while its southeast slope is relatively steeper, with multiple peaks developed on its top (Fig.3−13).		
命名由来 Origin of Name	该海山位于我国中沙大环礁附近，故以中沙大环礁中的暗沙、浅滩名进行地名的团组化命名。“比微暗沙”属于中沙大环礁的一个暗沙，取“比微”两字命名该海山。 The Seamount is located near the Zhongsha atoll in China, so it is named after a group of shoals and banks in Zhongsha atoll. “Biwei Ansha” is one of the shoals belonging to Zhongsha atoll, so the word “Biwei” was used to name the Seamount.		

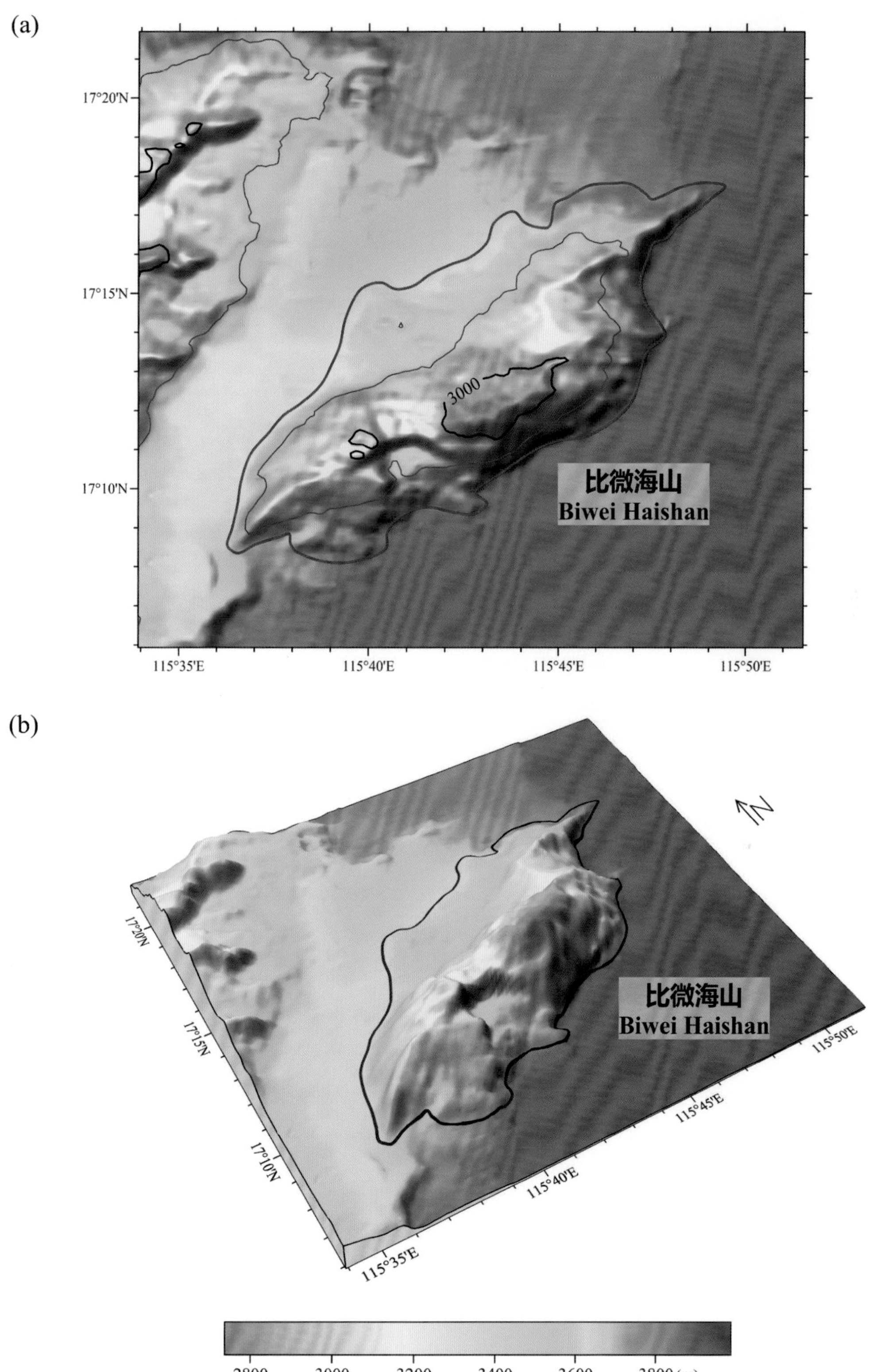

图 3-13 比微海山

(a) 海底地形图（等深线间隔 500 米）；(b) 三维海底地形图

Fig.3-13 Biwei Haishan

(a) Seafloor topographic map (with contour interval of 500 m); (b) 3-D seafloor topographic map

3.43 西门海山

标准名称 Standard Name	西门海山 Ximen Haishan	类别 Generic Term	海山 Seamount
中心点坐标 Center Coordinates	16°57.2'N, 114°32.9'E	规模（千米 × 千米） Dimension（km × km）	43 × 25
最小水深（米） Min Depth (m)	2020	最大水深（米） Max Depth (m)	3460
地理实体描述 Feature Description	西门海山发育在中沙北海隆西端，平面形态呈北东—南西走向。海山的西北坡较缓，东南坡较陡（图 3-14）。 Ximen Haishan develops in the west of Zhongshabei Hailong and stretches in the direction of NE-SW. Its northwest slope is relatively gentler, while its southeast slope is relatively steeper (Fig.3-14).		
命名由来 Origin of Name	该海山位于我国中沙大环礁附近，故以中沙大环礁中的暗沙、浅滩名进行地名的团组化命名。“西门暗沙”属于中沙大环礁的一个暗沙，取“西门”两字命名该海山。 The Seamount is located near the Zhongsha atoll in China, so it is named after a group of shoals and banks in Zhongsha atoll. “Ximen Ansha” is one of the shoals belonging to Zhongsha atoll, so the word “Ximen” was used to name the Seamount.		

3.44 乐西海脊

标准名称 Standard Name	乐西海脊 Lexi Haiji	类别 Generic Term	海脊 Ridge
中心点坐标 Center Coordinates	16°50.9'N, 114°19.7'E	规模（千米 × 千米） Dimension（km × km）	20 × 3
最小水深（米） Min Depth (m)	2680	最大水深（米） Max Depth (m)	3420
地理实体描述 Feature Description	乐西海脊发育在中沙北海隆西端，其脊线呈北东—南西走向（图 3-14）。 Lexi Haiji develops in the west of Zhongshabei Hailong, with a ridge line in the direction of NE-SW (Fig.3-14).		
命名由来 Origin of Name	该海脊位于我国中沙大环礁附近，故以中沙大环礁中的暗沙、浅滩名进行地名的团组化命名。“乐西暗沙”属于中沙大环礁的一个暗沙，取“乐西”两字命名该海脊。 The Ridge is located near the Zhongsha atoll in China, so it is named after a group of shoals and banks in Zhongsha atoll. “Lexi Ansha” is one of the shoals belonging to Zhongsha atoll, so the word “Lexi” was used to name the Ridge.		

3.45 霹雳海丘

标准名称 Standard Name	霹雳海丘 Pili Haiqiu	类别 Generic Term	海丘 Hill
中心点坐标 Center Coordinates	16°51.2'N, 114°29.1'E	规模（千米 × 千米） Dimension（km × km）	20 × 7
最小水深（米） Min Depth (m)	2140	最大水深（米） Max Depth (m)	3010
地理实体描述 Feature Description	霹雳海丘发育在中沙北海隆西端，平面形态呈东—西向的长条形，其顶部发育多座峰（图 3–14）。 Pili Haiqiu develops in the west of Zhongshabei Hailong, with a planform in the shape of a long strip in the direction of E–W and has multiple peaks developed on the top (Fig.3–14).		
命名由来 Origin of Name	以唐朝诗人李白的《梦游天姥吟留别》中的词进行地名的团组化命名。该海底地名的专名取词自该诗作中的："列缺霹雳，丘峦崩摧"。"霹雳"形容雷声轰鸣。 A group naming of undersea features after the phrases in the poem *A Visit to Mount Tianmu in Dream* by Li Bai (701–762 A.D.), a famous poet in the Tang Dynasty (618–907 A.D.). The specific term of this undersea feature name "Pili", means roaring thunder, is derived from the poetic lines "Among roaring thunder and swift lighting, the hills and peaks are broken asunder".		

3.46 青崖海丘

标准名称 Standard Name	青崖海丘 Qingya Haiqiu	类别 Generic Term	海丘 Hill
中心点坐标 Center Coordinates	16°52.9'N, 114°38.7'E	规模（千米 × 千米） Dimension（km × km）	12 × 5
最小水深（米） Min Depth (m)	2050	最大水深（米） Max Depth (m)	2570
地理实体描述 Feature Description	青崖海丘发育在中沙北海隆西端，霹雳海丘和屏南海山之间，平面形态呈东—西向的长条形（图 3–14）。 Qingya Haiqiu develops in the west of Zhongshabei Hailong and is between Pili Haiqiu and Pingnan Haishan, with a planform in the shape of a long strip in the direction of E–W (Fig.3–14).		
命名由来 Origin of Name	以唐朝诗人李白的《梦游天姥吟留别》中的词进行地名的团组化命名。该海底地名的专名取词自该诗作中的："且放白鹿青崖间，须行即骑访名山"。"青崖"指青青的山崖。 A group naming of undersea features after the phrases in the poem *A Visit to Mount Tianmu in Dream* by Li Bai (701–762 A.D.), a famous poet in the Tang Dynasty (618–907 A.D.). The specific term of this undersea feature name "Qingya", means verdant cliffs, is derived from the poetic lines "I would let the white deer graze among the verdant cliffs for the moment, and would visit famous mountains on it whenever I feel like to".		

(a)

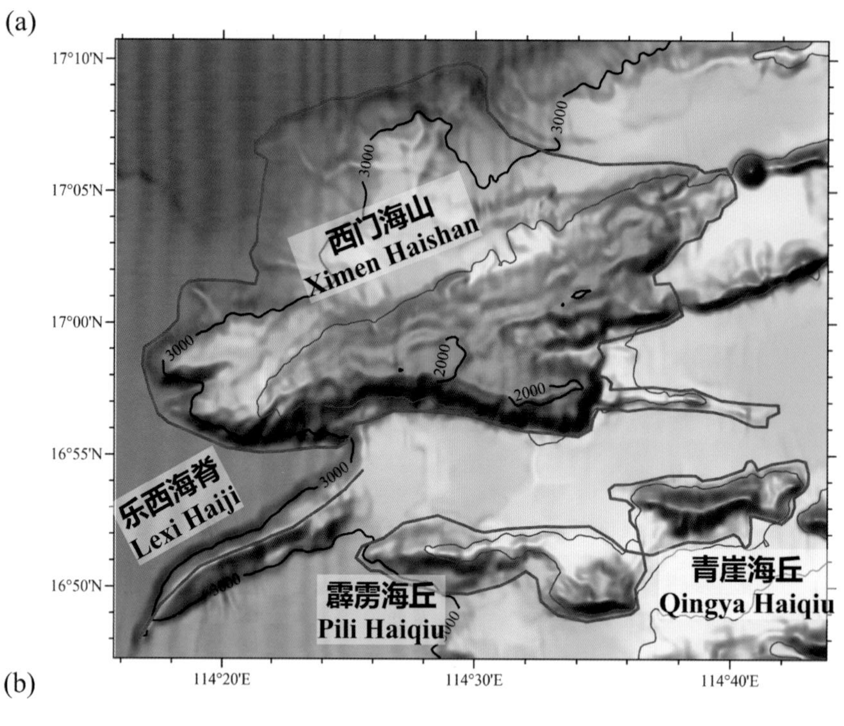

(b)

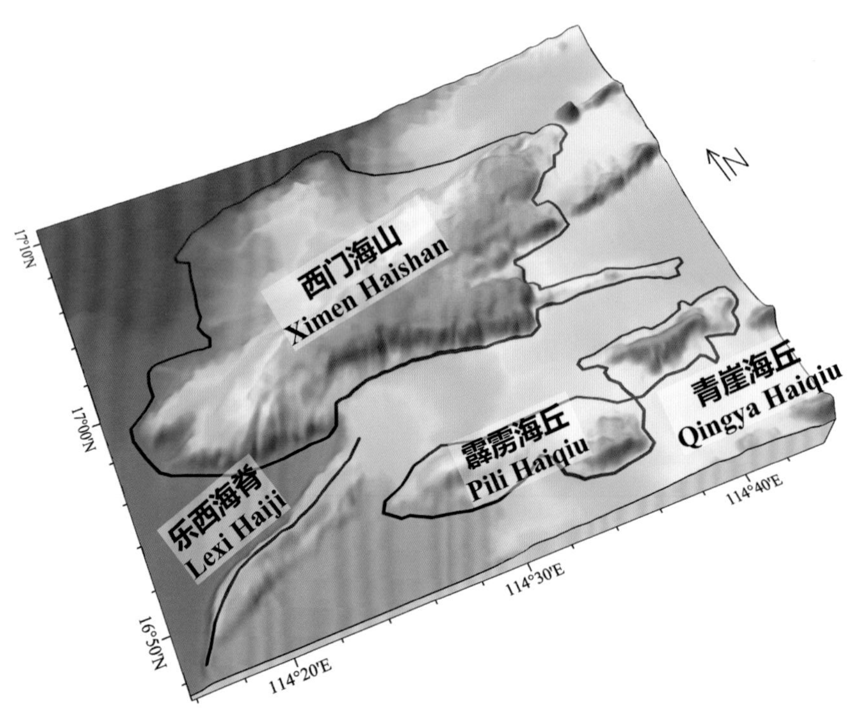

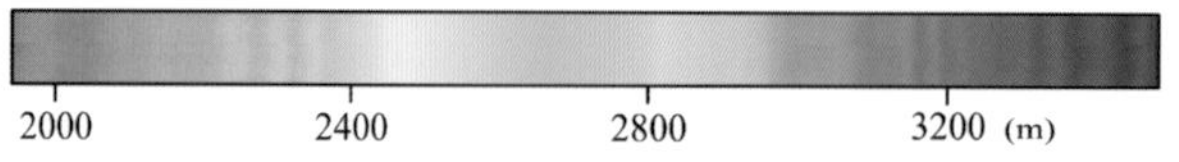

图 3-14 西门海山、乐西海脊、霹雳海丘、青崖海丘

(a) 海底地形图（等深线间隔 500 米）；(b) 三维海底地形图

Fig.3-14 Ximen Haishan, Lexi Haiji, Pili Haiqiu, Qingya Haiqiu

(a) Seafloor topographic map (with contour interval of 500 m)；(b) 3-D seafloor topographic map

3.47 漫步海山

标准名称 Standard Name	漫步海山 Manbu Haishan	类别 Generic Term	海山 Seamount
中心点坐标 Center Coordinates	16°45.7'N, 114°37.0'E	规模（千米 × 千米） Dimension（km × km）	10 × 7
最小水深（米） Min Depth (m)	710	最大水深（米） Max Depth (m)	2720
地理实体描述 Feature Description	漫步海山发育在中沙北海隆中部，紧邻漫步东海丘，平面形态近似圆形（图 3–15）。 Manbu Haishan develops in the middle of Zhongshabei Hailong and is adjacent to Manbudong Haiqiu with a nearly circular planform (Fig.3–15).		
命名由来 Origin of Name	该海山位于我国中沙大环礁附近，故以中沙大环礁中的暗沙、浅滩名进行地名的团组化命名。“漫步暗沙”属于中沙大环礁的一个暗沙，取“漫步”两字命名该海山。 The Seamount is located near the Zhongsha atoll in China, so it is named after a group of shoals and banks in Zhongsha atoll. “Manbu Ansha” is one of the shoals belonging to Zhongsha atoll, so the word “Manbu” was used to name the Seamount.		

3.48 漫步东海丘

标准名称 Standard Name	漫步东海丘 Manbudong Haiqiu	类别 Generic Term	海丘 Hill
中心点坐标 Center Coordinates	16°46.3'N, 114°42.2'E	规模（千米 × 千米） Dimension（km × km）	8 × 8
最小水深（米） Min Depth (m)	1850	最大水深（米） Max Depth (m)	2780
地理实体描述 Feature Description	漫步东海丘发育在中沙北海隆中部，紧邻漫步海山，平面形态近似圆形（图 3–15）。 Manbudong Haiqiu develops in the middle of Zhongshabei Hailong and is adjacent to Manbu Haishan with a nearly circular planform (Fig.3–15).		
命名由来 Origin of Name	该海丘位于漫步海山以东，因此得名。 The Hill is located on the east of Manbu Haishan, and “Dong” means east in Chinese, so the word “Manbudong” was used to name the Hill.		

3.49 漫步北海丘

标准名称 Standard Name	漫步北海丘 Manbubei Haiqiu	类别 Generic Term	海丘 Hill
中心点坐标 Center Coordinates	16°48.0'N, 114°44.1'E	规模（千米 × 千米） Dimension（km × km）	12 × 2
最小水深（米） Min Depth (m)	2080	最大水深（米） Max Depth (m)	2580
地理实体描述 Feature Description	漫步北海丘发育在中沙北海隆中部，漫步东海丘北部，平面形态呈东—西向的长条形，其顶部发育两座峰（图 3–15）。 Manbubei Haiqiu develops in the middle of Zhongshabei Hailong and on the north of Manbudong Haiqiu, with a planform in the shape of a long strip in the direction of E–W, and has two peaks developed on its top (Fig.3–15).		
命名由来 Origin of Name	该海丘位于漫步海山以北，因此得名。 The Hill is located to the north of Manbu Haishan, and “Bei” means north in Chinese, so the word “Manbubei” was used to name the Hill.		

3.50 漫步南海丘

标准名称 Standard Name	漫步南海丘 Manbunan Haiqiu	类别 Generic Term	海丘 Hill
中心点坐标 Center Coordinates	16°41.1'N, 114°36.0'E	规模（千米 × 千米） Dimension（km × km）	4 × 3
最小水深（米） Min Depth (m)	2360	最大水深（米） Max Depth (m)	2790
地理实体描述 Feature Description	漫步南海丘发育在中沙北海隆中部，漫步海山南部，平面形态呈椭圆形（图 3–15）。 Manbunan Haiqiu develops in the middle of Zhongshabei Hailong and to the south of Manbu Haishan with an oval-shaped planform (Fig.3–15).		
命名由来 Origin of Name	该海丘位于漫步海山以南，因此得名。 The Hill is located to the south of Manbu Haishan, and “Nan” means south in Chinese, so the word “Manbunan” was used to name the Hill.		

(a)

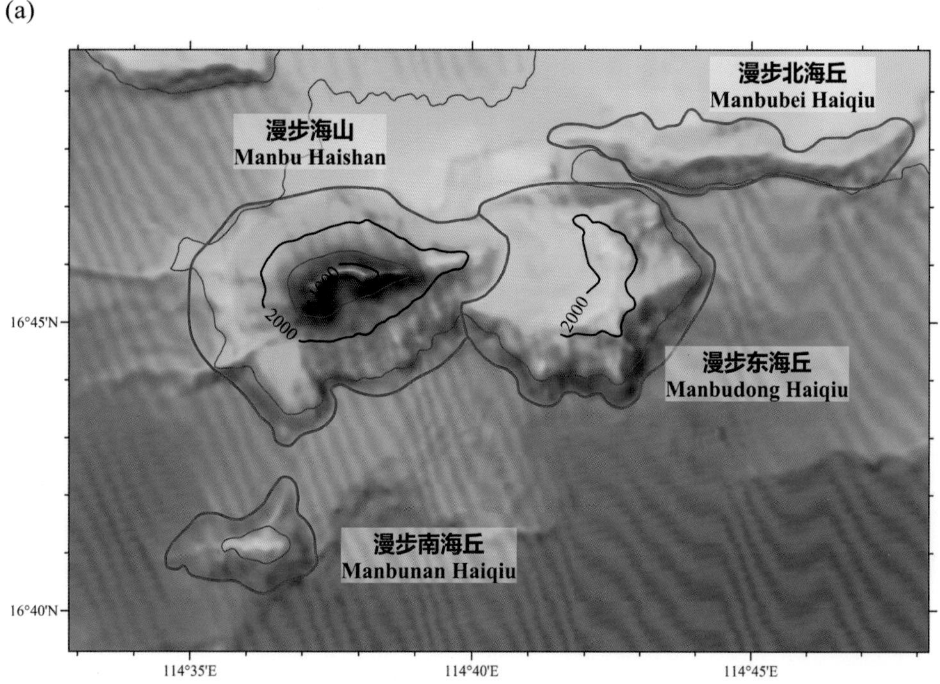

(b)

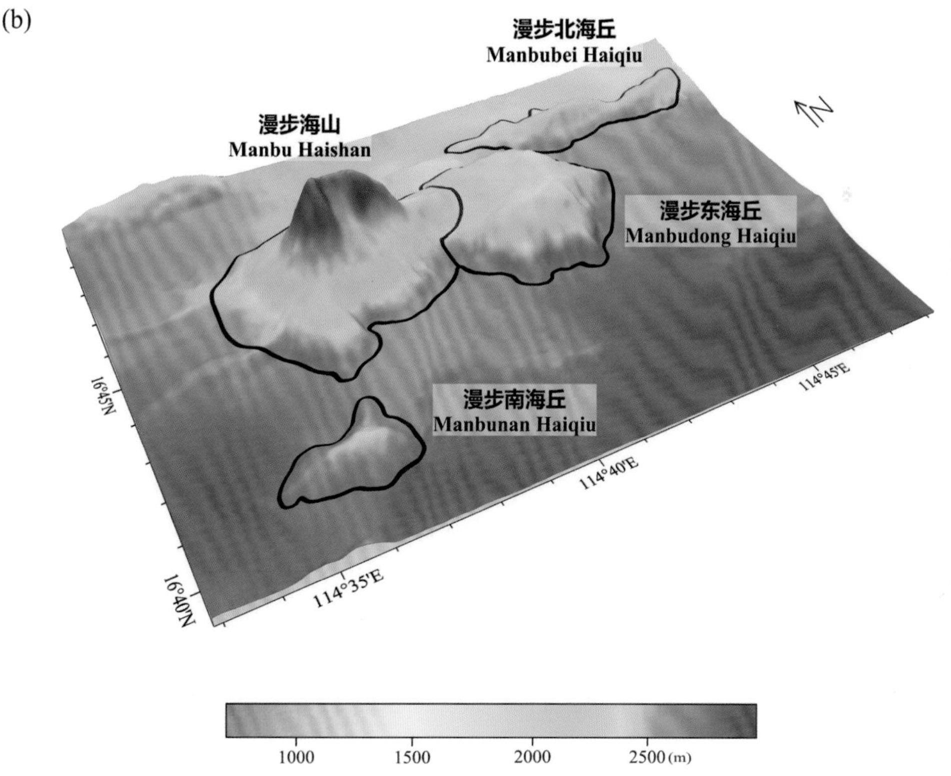

图 3-15 漫步海山、漫步东海丘、漫步北海丘、漫步南海丘

(a) 海底地形图（等深线间隔 500 米）；(b) 三维海底地形图

Fig.3-15 Manbu Haishan, Manbudong Haiqiu, Manbubei Haiqiu, Manbunan Haiqiu

(a) Seafloor topographic map (with contour interval of 500 m); (b) 3-D seafloor topographic map

3.51 排洪海丘

标准名称 Standard Name	排洪海丘 Paihong Haiqiu	类别 Generic Term	海丘 Hill
中心点坐标 Center Coordinates	16°42.1'N, 114°09.5'E	规模（千米 × 千米） Dimension（km × km）	17 × 5
最小水深（米） Min Depth (m)	2540	最大水深（米） Max Depth (m)	3360
地理实体描述 Feature Description	排洪海丘发育在中沙北海隆西端，平面形态呈北东—南西向的长条形（图 3−16）。 Paihong Haiqiu develops in the west of Zhongshabei Hailong, with a planform in the shape of a long strip in the direction of NE−SW (Fig.3−16).		
命名由来 Origin of Name	该海丘位于我国中沙大环礁附近，故以中沙大环礁中的暗沙、浅滩名进行地名的团组化命名。“排洪滩”属于中沙大环礁的一个浅滩，取“排洪”两字命名该海丘。 The Hill is located near the Zhongsha atoll in China, so it is named after a group of shoals and banks in Zhongsha atoll. “Paihong Ansha” is one of the banks belonging to Zhongsha atoll, so the word “Paihong” was used to name the Hill.		

3.52 控湃海脊

标准名称 Standard Name	控湃海脊 Kongpai Haiji	类别 Generic Term	海脊 Ridge
中心点坐标 Center Coordinates	16°32.4'N, 114°07.8'E	规模（千米 × 千米） Dimension（km × km）	20 × 4
最小水深（米） Min Depth (m)	1790	最大水深（米） Max Depth (m)	3170
地理实体描述 Feature Description	控湃海脊发育在中沙北海隆西端，平面形态呈北东—南西向的长条形（图 3−16）。 Kongpai Haiji develops in the west of Zhongshabei Hailong, with a planform in the shape of a long strip in the direction of NE−SW (Fig.3−16).		
命名由来 Origin of Name	该海脊位于我国中沙大环礁附近，故以中沙大环礁中的暗沙、浅滩名进行地名的团组化命名。“控湃暗沙”属于中沙大环礁的一个暗沙，取“控湃”两字命名该海丘。 The Ridge is located near the Zhongsha atoll in China, so it is named after a group of shoals and banks in Zhongsha atoll. “Kongpai Ansha” is one of the shoals belonging to Zhongsha atoll, so the word “Kongpai” was used to name the Ridge.		

3.53 控湃北海脊

标准名称 Standard Name	控湃北海脊 Kongpaibei Haiji	类别 Generic Term	海脊 Ridge
中心点坐标 Center Coordinates	16°33.6'N, 114°05.0'E	规模（千米 × 千米） Dimension（km × km）	10 × 4
最小水深（米） Min Depth (m)	2050	最大水深（米） Max Depth (m)	3260
地理实体描述 Feature Description	控湃北海脊发育在中沙北海隆西端，平面形态近东—西向的长条形（图 3−16）。 Kongpaibei Haiji develops in the west of Zhongshabei Hailong, with a planform in the shape of a long strip in the direction of E−W (Fig.3−16).		
命名由来 Origin of Name	该海脊位于控湃海脊以北，因此得名。 The Ridge is located to the north of Kongpai Haiji, and "Bei" means north in Chinese, so the word "Kongpaibei" was used to name the Ridge.		

3.54 武勇海丘

标准名称 Standard Name	武勇海丘 Wuyong Haiqiu	类别 Generic Term	海丘 Hill
中心点坐标 Center Coordinates	16°20.0'N, 113°57.6'E	规模（千米 × 千米） Dimension（km × km）	9 × 4
最小水深（米） Min Depth (m)	1950	最大水深（米） Max Depth (m)	2890
地理实体描述 Feature Description	武勇海丘发育在中沙北海隆西南端，平面形态呈北东—南西向的菱形（图 3−16）。 Wuyong Haiqiu develops in the southwest of Zhongshabei Hailong, with a planform in the shape of a diamond in the direction of NE−SW (Fig.3−16).		
命名由来 Origin of Name	该海丘位于我国中沙大环礁附近，故以中沙大环礁中的暗沙、浅滩名进行地名的团组化命名。"武勇暗沙"属于中沙大环礁的一个暗沙，取"武勇"两字命名该海丘。 The Hill is located near the Zhongsha atoll in China, so it is named after a group of shoals and banks in Zhongsha atoll. "Wuyong Ansha" is one of the shoals belonging to Zhongsha atoll, so the word "Wuyong" was used to name the Hill.		

(a)

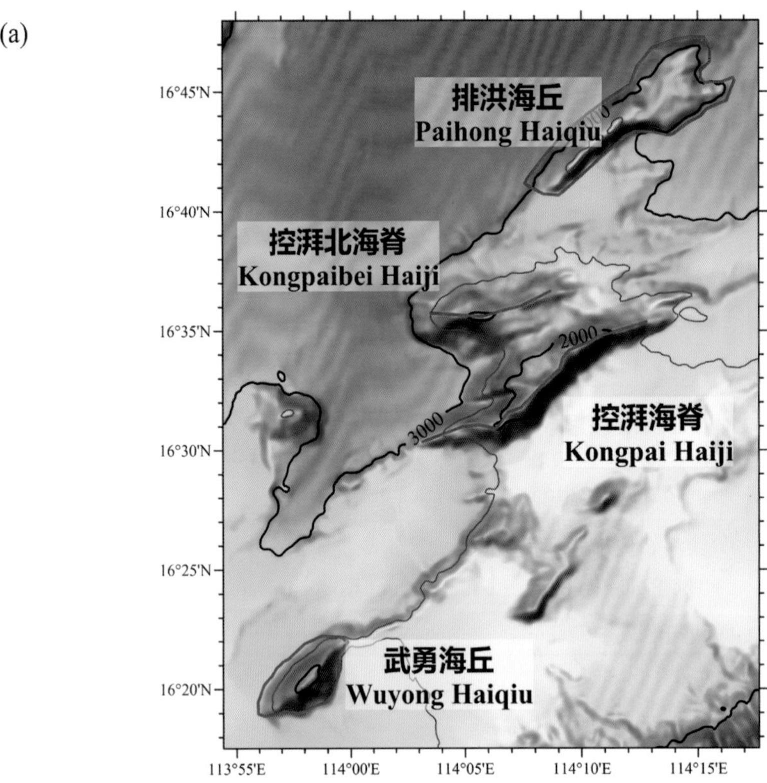

(b)

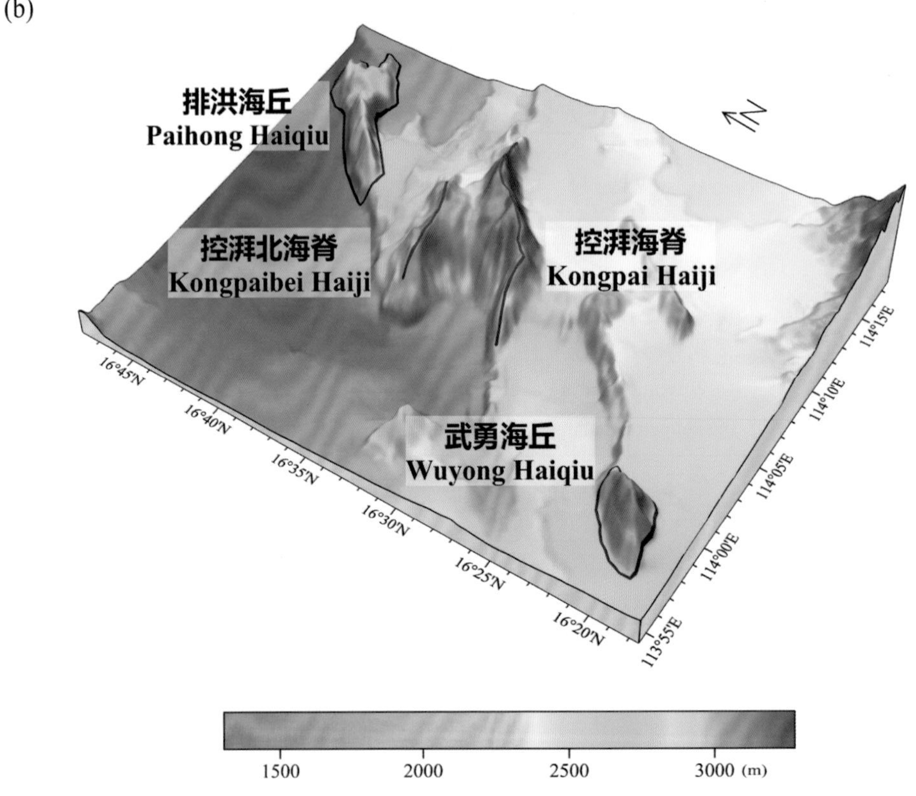

图 3-16 排洪海丘、控湃海脊、控湃北海脊、武勇海丘

(a) 海底地形图（等深线间隔 500 米）；(b) 三维海底地形图

Fig.3-16 Paihong Haiqiu, Kongpai Haiji, Kongpaibei Haiji, Wuyong Haiqiu

(a) Seafloor topographic map (with contour interval of 500 m); (b) 3-D seafloor topographic map

3.55 涛静海丘

标准名称 Standard Name	涛静海丘 Taojing Haiqiu	类别 Generic Term	海丘 Hill
中心点坐标 Center Coordinates	16°32.6'N, 114°24.3'E	规模（千米 × 千米） Dimension（km × km）	25 × 9
最小水深（米） Min Depth (m)	1990	最大水深（米） Max Depth (m)	2880
地理实体描述 Feature Description	涛静海丘发育在中沙北海隆南端，平面形态呈北东东—南西西向的菱形，其顶部发育多座峰（图 3–17）。 Taojing Haiqiu develops in the south of Zhongshaobei Hailong, with a planform in the shape of a diamond in the direction of NEE–SWW. Multiple peaks developed on the top (Fig.3–17).		
命名由来 Origin of Name	该海丘位于我国中沙大环礁附近，故以中沙大环礁中的暗沙、浅滩名进行地名的团组化命名。“涛静暗沙”属于中沙大环礁的一个暗沙，取“涛静”两字命名该海丘。 The Hill is located near the Zhongsha atoll in China, so it is named after a group of shoals and banks in Zhongsha atoll. “Taojing Ansha” is one of the shoals belonging to Zhongsha atoll, so the word “Taojing” was used to name the Hill.		

3.56 涛静北海丘

标准名称 Standard Name	涛静北海丘 Taojingbei Haiqiu	类别 Generic Term	海丘 Hill
中心点坐标 Center Coordinates	16°38.4'N, 114°33.0'E	规模（千米 × 千米） Dimension（km × km）	12 × 5
最小水深（米） Min Depth (m)	2530	最大水深（米） Max Depth (m)	2870
地理实体描述 Feature Description	涛静北海丘发育在中沙北海隆南端，平面形态呈北东东—南西西向的菱形（图 3–17）。 Taojingbei Haiqiu develops in the south of Zhongshaobei Hailong, with a planform in the shape of a diamond in the direction of NEE–SWW (Fig.3–17).		
命名由来 Origin of Name	该海丘位于涛静海丘以北，因此得名。 The Hill is located to the north of Taojing Haiqiu, and “Bei” means north in Chinese, so the word “Taojingbei” was used to name the Hill.		

(a)

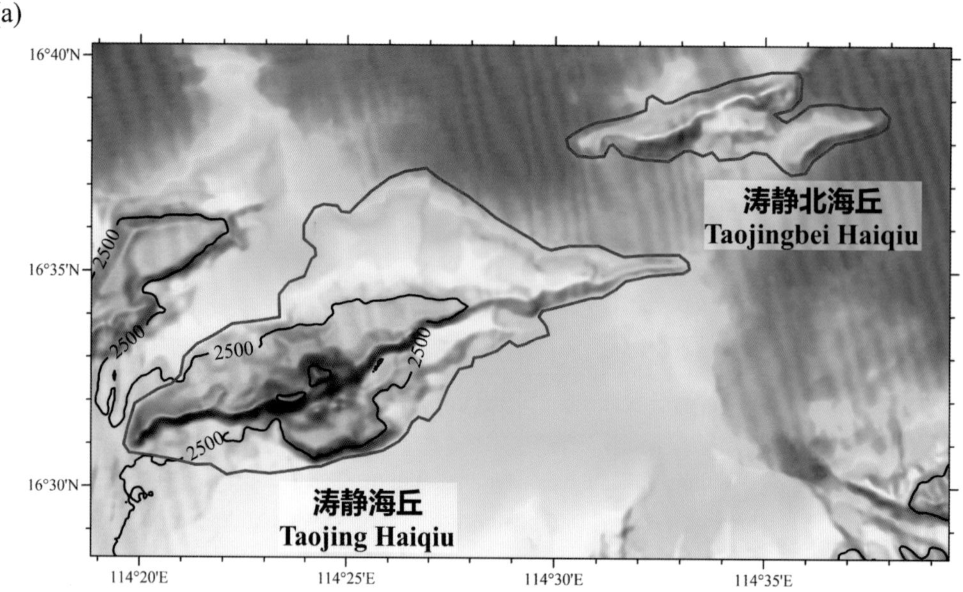

(b)

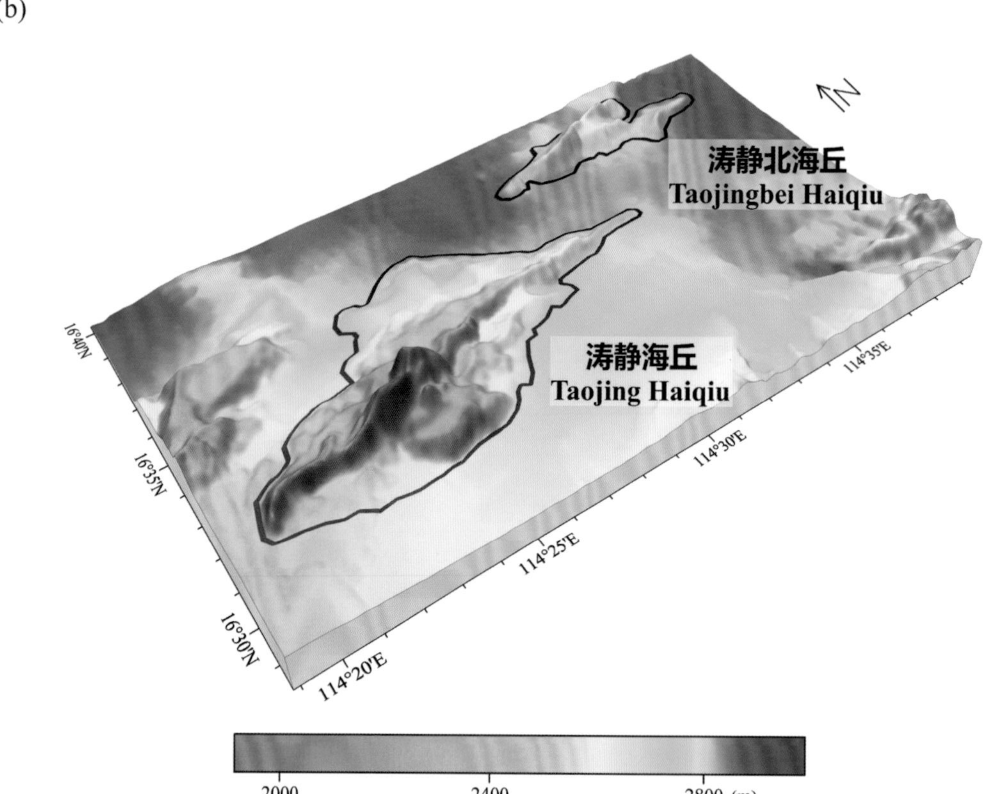

图 3-17　涛静海丘、涛静北海丘

(a) 海底地形图（等深线间隔 500 米）；(b) 三维海底地形图

Fig.3-17　Taojing Haiqiu, Taojingbei Haiqiu

(a) Seafloor topographic map (with contour interval of 500 m)；(b) 3-D seafloor topographic map

3.57 屏南海山

标准名称 Standard Name	屏南海山 Pingnan Haishan	类别 Generic Term	海山 Seamount
中心点坐标 Center Coordinates	16°51.3'N, 114°50.1'E	规模（千米 × 千米） Dimension（km × km）	68 × 18
最小水深（米） Min Depth (m)	490	最大水深（米） Max Depth (m)	4160
地理实体描述 Feature Description	屏南海山是一座发育在中沙北海隆中西部、呈北西—南东向的大型链状海山，其顶部发育 5 座峰（图 3–18）。 Pingnan Haishan develops in the mid-west of Zhongshabei Hailong and is a large-scale seamount chain in the direction of NW–SE, with five peaks developed on its top (Fig.3–18).		
命名由来 Origin of Name	该海山位于我国中沙大环礁附近，故以中沙大环礁中的暗沙、浅滩名进行地名的团组化命名。“屏南暗沙”属于中沙大环礁的一个暗沙，取“屏南”两字命名该海山。 The Seamount is located near the Zhongsha atoll in China, so it is named after a group of shoals and banks in Zhongsha atoll. “Pingnan Ansha” is one of the shoals belonging to Zhongsha atoll, so the word “Pingnan” was used to name the Seamount.		

3.58 安定海底峰

标准名称 Standard Name	安定海底峰 Anding Haidifeng	类别 Generic Term	海底峰 Peak
中心点坐标 Center Coordinates	16°49.2'N, 114°56.5'E	规模（千米 × 千米） Dimension（km × km）	15 × 12
最小水深（米） Min Depth (m)	500	最大水深（米） Max Depth (m)	2890
地理实体描述 Feature Description	安定海底峰属于屏南海山的一个山峰，平面形态呈北西西—南东东走向，其顶部较为平坦（图 3–18）。 Anding Haidifeng is a peak of Pingnan Haishan and stretches in the direction of NWW–SEE with a flat top (Fig.3–18).		
命名由来 Origin of Name	该海底峰位于我国中沙大环礁附近，故以中沙大环礁中的暗沙、浅滩名进行地名的团组化命名。“安定连礁”属于中沙大环礁的一个暗沙，取“安定”两字命名该海底峰。 The Peak is located near the Zhongsha atoll in China, so it is named after a group of shoals and banks in Zhongsha atoll. “Anding Lianjiao” is one of the shoals belonging to Zhongsha atoll, so the word “Anding” was used to name the Peak.		

3.59 果淀海底峰

标准名称 Standard Name	果淀海底峰 Guodian Haidifeng	类别 Generic Term	海底峰 Peak
中心点坐标 Center Coordinates	16°45.1'N, 114°58.5'E	规模（千米 × 千米） Dimension（km × km）	10 × 8
最小水深（米） Min Depth (m)	570	最大水深（米） Max Depth (m)	2640
地理实体描述 Feature Description	果淀海底峰属于屏南海山的一个山峰，平面形态呈北北西—南南东走向，其顶部较为平坦（图 3−18）。 Guodian Haidifeng is a peak of Pingnan Haishan and stretches in the direction of NNW−SSE with a flat top (Fig.3−18).		
命名由来 Origin of Name	该海底峰位于我国中沙大环礁附近，故以中沙大环礁中的暗沙、浅滩名进行地名的团组化命名。“果淀暗沙”属于中沙大环礁的一个暗沙，取“果淀”两字命名该海底峰。 The Peak is located near the Zhongsha atoll in China, so it is named after a group of shoals and banks in Zhongsha atoll. “Guoding Ansha” is one of the shoals belonging to Zhongsha atoll, so the word “Guoding” was used to name the Peak.		

3.60 华夏海底峰

标准名称 Standard Name	华夏海底峰 Huaxia Haidifeng	类别 Generic Term	海底峰 Peak
中心点坐标 Center Coordinates	16°39.4'N, 115°04.0'E	规模（千米 × 千米） Dimension（km × km）	13 × 6
最小水深（米） Min Depth (m)	590	最大水深（米） Max Depth (m)	3060
地理实体描述 Feature Description	华夏海底峰属于屏南海山的一个山峰，平面形态呈北西—南东走向，其顶部较为平坦（图 3−18）。 Huaxia Haidifeng is a peak of Pingnan Haishan and stretches in the direction of NW−SE with a flat top (Fig.3−18).		
命名由来 Origin of Name	该海底峰位于我国中沙大环礁附近，故以中沙大环礁中的暗沙、浅滩名进行地名的团组化命名。“华夏暗沙”属于中沙大环礁的一个暗沙，取“华夏”两字命名该海底峰。 The Peak is located near the Zhongsha atoll in China, so it is named after a group of shoals and banks in Zhongsha atoll. “Huaxia Ansha” is one of the shoals belonging to Zhongsha atoll, so the word “Huaxia” was used to name the Peak.		

3.61 华夏南海底峰

标准名称 Standard Name	华夏南海底峰 Huaxianan Haidifeng	类别 Generic Term	海底峰 Peak
中心点坐标 Center Coordinates	16°34.8'N, 115°10.2'E	规模（千米 × 千米） Dimension（km × km）	10 × 6
最小水深（米） Min Depth (m)	1410	最大水深（米） Max Depth (m)	3180
地理实体描述 Feature Description	华夏南海底峰是属于屏南海山的一个山峰，平面形态呈东—西走向（图 3-18）。 Huaxianan Haidifeng is a peak of Pingnan Haishan and stretches in the direction of E−W (Fig.3−18).		
命名由来 Origin of Name	该海底峰位于华夏海底峰以南，因此得名。 The Peak is located to the south of Huaxia Haidifeng, and "Nan" means south in Chinese, so the word "Huaxianan" was used to name the Peak.		

3.62 南扉海山

标准名称 Standard Name	南扉海山 Nanfei Haishan	类别 Generic Term	海山 Seamount
中心点坐标 Center Coordinates	16°45.9'N, 115°11.1'E	规模（千米 × 千米） Dimension（km × km）	21 × 16
最小水深（米） Min Depth (m)	2480	最大水深（米） Max Depth (m)	3780
地理实体描述 Feature Description	南扉海山发育在中沙北海隆东部，其西北坡较缓，东南坡较陡（图 3-18）。 Nanfei Haishan develops in the east of Zhongshabei Hailong, and its northwest slope is relatively gentler, while its southeast slope is relatively steeper (Fig.3−18).		
命名由来 Origin of Name	该海山位于我国中沙大环礁附近，故以中沙大环礁中的暗沙、浅滩名进行地名的团组化命名。"南扉暗沙"属于中沙大环礁的一个暗沙，取"南扉"两字命名该海山。 The Seamount is located near the Zhongsha atoll in China, so it is named after a group of shoals and banks in Zhongsha atoll. "Nanfei Ansha" is one of the shoals belonging to Zhongsha atoll, so the word "Nanfei" was used to name the Seamount.		

(a)

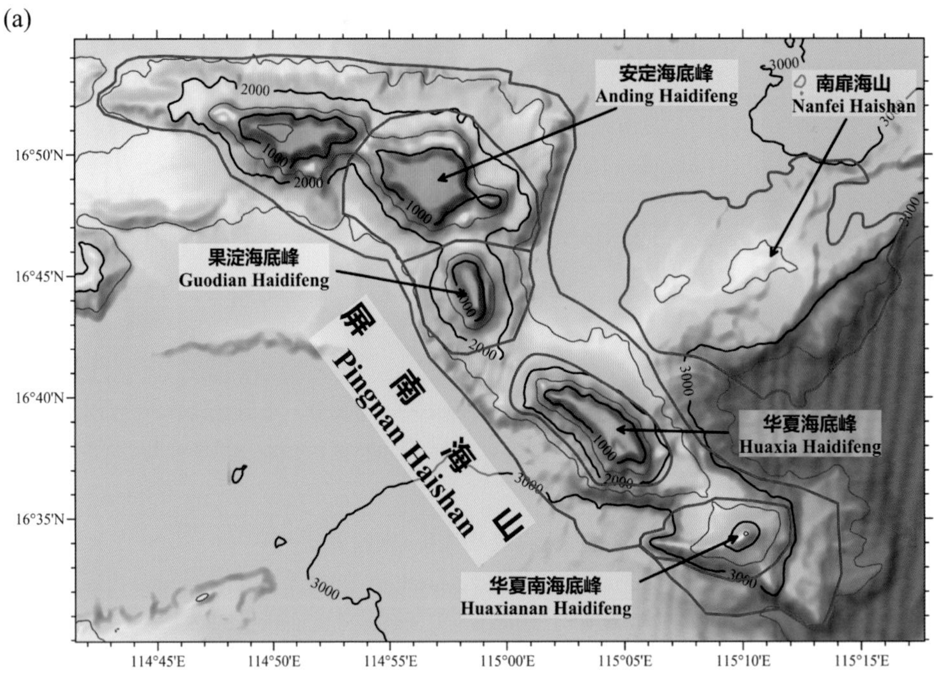

(b)

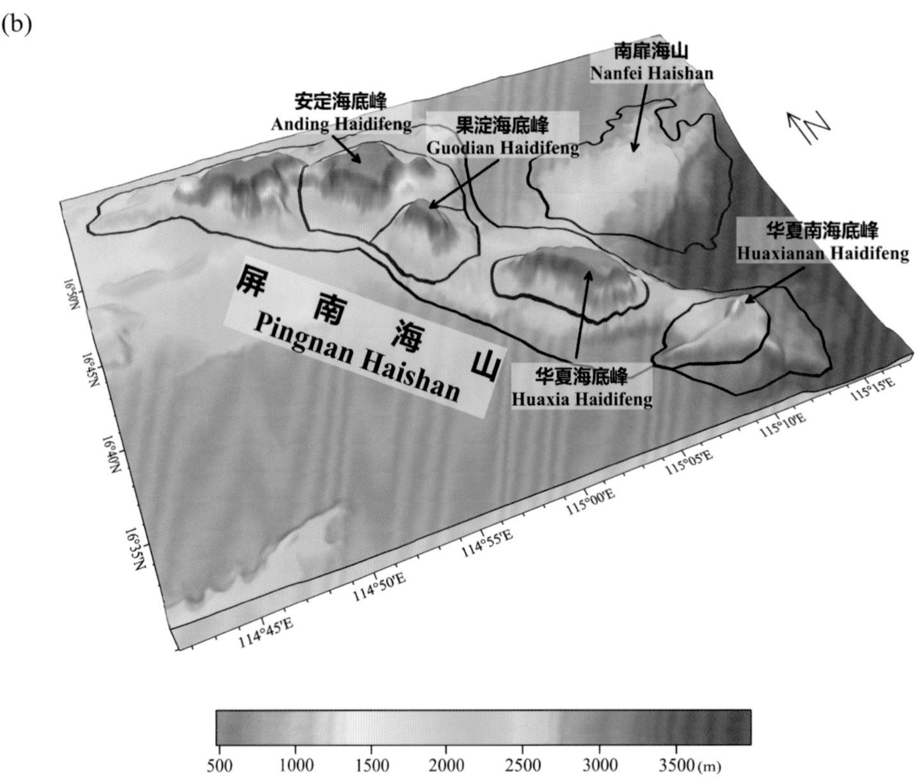

图 3-18 屏南海山、安定海底峰、果淀海底峰、华夏海底峰、华夏南海底峰、南扉海山

(a) 海底地形图（等深线间隔 500 米）；(b) 三维海底地形图

Fig.3-18 Pingnan Haishan, Anding Haidifeng, Guodian Haidifeng, Huaxia Haidifeng, Huaxianan Haidifeng, Nanfei Haishan

(a) Seafloor topographic map (with contour interval of 500 m); (b) 3-D seafloor topographic map

4

盆西海岭海区海底地理实体

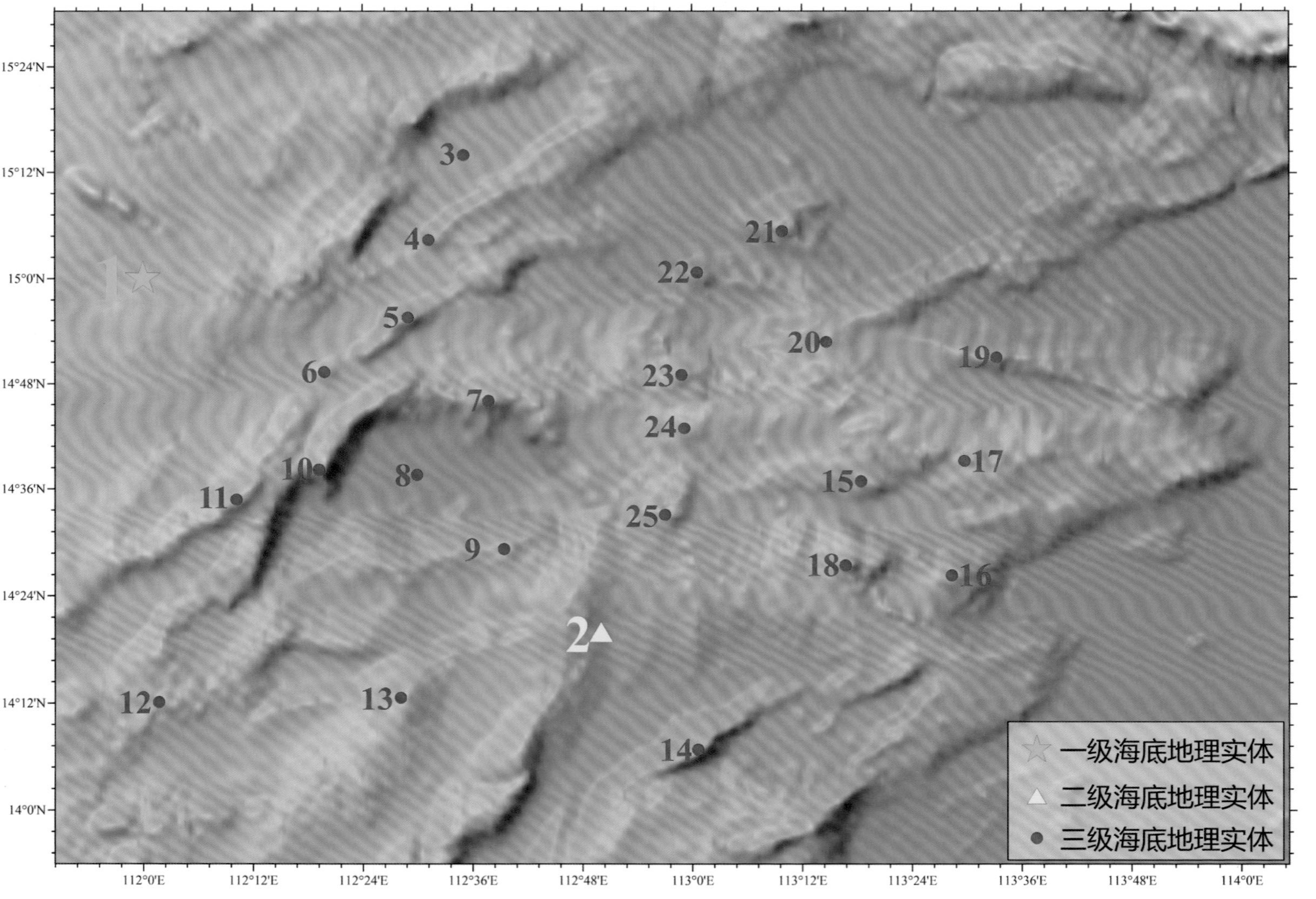

图 4-1　盆西海岭海区海底地理实体中心点位置示意图，序号含义见表 4-1

Fig.4-1　Location of center coordinates of undersea features in the Penxi Hailing, with the meanings of the serial numbers shown in Tab. 4-1

表 4-1 盆西海岭海区海底地理实体列表

Tab.4-1 List of undersea features in the Penxi Hailing

序号 No.	标准名称 Standard Name	汉语拼音 Chinese Phonetic Alphabet	类别 Generic Term	中心点坐标 Center Coordinates 纬度 Latitude	 经度 Longitude	实体等级 Order
1	南海西部陆坡 Nanhaixibu Lupo	Nánhǎixībù Lùpō	大陆坡 Slope	15°00.0'N	112°00.0'E	1
2	盆西海岭 Penxi Hailing	Pénxīnán Hǎilǐng	海岭 Ridge	14°20.0'N	112°50.0'E	2
3	兰宫洼地 Langong Wadi	Lángōng Wādì	海底洼地 Depression	15°14.0'N	112°35.0'E	3
4	羌笛海山 Qiangdi Haishan	Qiāngdí Hǎishān	海山 Seamount	15°04.4'N	112°31.2'E	3
5	桂殿海山 Guidian Haishan	Guìdiàn Hǎishān	海山 Seamount	14°55.6'N	112°28.9'E	3
6	萦回海丘 Yinghui Haiqiu	Yínghuí Hǎiqiū	海丘 Hill	14°49.3'N	112°19.8'E	3
7	玉门海山 Yumen Haishan	Yùmén Hǎishān	海山 Seamount	14°46.1'N	112°37.7'E	3
8	玉门海盆 Yumen Haipen	Yùmén Hǎipén	海盆 Basin	14°37.6'N	112°29.9'E	3
9	玉门海脊 Yumen Haiji	Yùmén Hǎijǐ	海脊 Ridge	14°29.3'N	112°39.4'E	3
10	万仞海脊 Wanren Haiji	Wànrèn Hǎijǐ	海脊 Ridge	14°38.2'N	112°19.2'E	3
11	杨柳海山 Yangliu Haishan	Yángliǔ Hǎishān	海山 Seamount	14°34.8'N	112°10.2'E	3
12	春风海脊 Chunfeng Haiji	Chūnfēng Hǎijǐ	海脊 Ridge	14°12.1'N	112°01.7'E	3
13	春风海山 Chunfeng Haishan	Chūnfēng Hǎishān	海山 Seamount	14°12.6'N	112°28.1'E	3
14	紫电海山 Zidian Haishan	Zǐdiàn Hǎishān	海山 Seamount	14°06.8'N	113°00.6'E	3
15	流丹海丘 Liudan Haiqiu	Liúdān Hǎiqiū	海丘 Hill	14°36.9'N	113°18.4'E	3

续表

序号 No.	标准名称 Standard Name	汉语拼音 Chinese Phonetic Alphabet	类别 Generic Term	中心点坐标 Center Coordinates		实体等级 Order
				纬度 Latitude	经度 Longitude	
16	鹤汀海山 Heting Haishan	Hètīng Hǎishān	海山 Seamount	14°26.4'N	113°28.3'E	3
17	青海海山 Qinghai Haishan	Qīnghǎi Hǎishān	海山 Seamount	14°39.2'N	113°29.7'E	3
18	青海南海山群 Qinghainan Haishanqun	Qīnghǎinán Hǎishānqún	海山群 Seamounts	14°27.5'N	113°16.7'E	3
19	青海北海脊 Qinghaibei Haiji	Qīnghǎiběi Hǎijǐ	海脊 Ridge	14°51.0'N	113°33.2'E	3
20	边色海脊 Bianse Haiji	Biānsè Hǎijǐ	海脊 Ridge	14°52.8'N	113°14.6'E	3
21	孤城海山 Gucheng Haishan	Gūchéng Hǎishān	海山 Seamount	15°05.4'N	113°09.8'E	3
22	飞阁海山 Feige Haishan	Fēigé Hǎishān	海山 Seamount	15°00.7'N	113°00.5'E	3
23	重霄海山 Chongxiao Haishan	Chóngxiāo Hǎishān	海山 Seamount	14°49.0'N	112°58.8'E	3
24	耸翠海丘 Songcui Haiqiu	Sǒngcuì Hǎiqiū	海丘 Hill	14°42.9'N	112°59.1'E	3
25	云海海山 Yunhai Haishan	Yúnhǎi Hǎishān	海山 Seamount	14°33.1'N	112°57.0'E	3

4.1 南海西部陆坡

标准名称 Standard Name	南海西部陆坡 Nanhaixibu Lupo	类别 Generic Term	大陆坡 Slope
中心点坐标 Center Coordinates	15°00.0'N, 120°00.0'E	规模（千米 × 千米） Dimension（km × km）	1000 × 477
最小水深（米） Min Depth (m)	200	最大水深（米） Max Depth (m)	4310
地理实体描述 Feature Description	南海西部陆坡北以西沙海槽为界、南至广雅斜坡，向东延伸至 3500 ～ 4000 米水深与南海海盆接壤。陆坡的南北长约 1000 千米，呈现为北宽南窄趋势。总体上看，西部陆坡 1500 ～ 2000 米以浅地区地形比较平缓，而 1500 ～ 2000 米以深地区至深海平原之间，北东—南西向延伸的盆西海岭和盆西南海岭（图 4–2）。 Nanhaixibu Lupo borders on Xisha Haicao on the north, stretches to Guangya Xiepo on the south and borders on Nanhai Haipen at depth of 3500~4000 m on the east. The Slope is 1000 km long from north to south and is wider in the north and narrower in the south. Generally, the topography of the Slope with depth of less than 1500~2000 m is relatively flatter and gentler, while the topographies between the abyssal plain and the region with depth of more than 1500~2000 m are Penxi Hailing and Penxinan Hailing stretching in the direction of NE–SW (Fig.4–2).			
命名由来 Origin of Name	该大陆坡位于南海西部，因此得名。 The Slope is located in the western part of the South China Sea, and “Xibu” means western part in Chinese, so the word “Nanhaixibu” was used to name the Slope.			

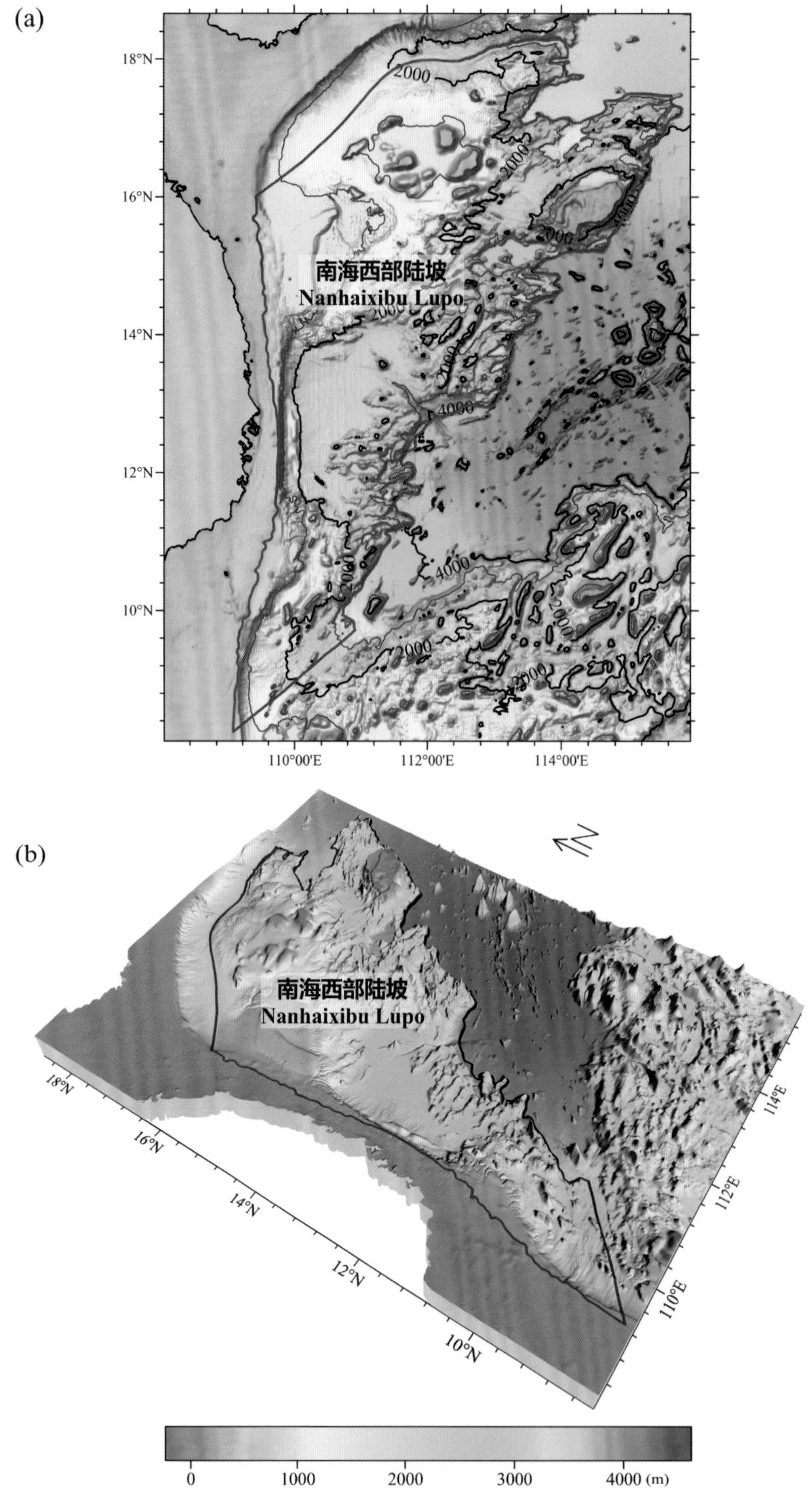

图 4-2　南海西部陆坡

(a) 海底地形图（等深线间隔 1000 米）；(b) 三维海底地形图

Fig.4-2　Nanhaixibu Lupo

(a) Seafloor topographic map (with contour interval of 1000 m)；(b) 3-D seafloor topographic map

4.2 盆西海岭

标准名称 Standard Name	盆西海岭 Penxi Hailing	类别 Generic Term	海岭 Ridge
中心点坐标 Center Coordinates	14°20.0'N, 112°50.0'E	规模（千米 × 千米） Dimension（km × km）	270 × 180
最小水深（米） Min Depth (m)	300	最大水深（米） Max Depth (m)	4330
地理实体描述 Feature Description	盆西海岭位于南海西部陆坡，其西界为中建南斜坡和中建南海盆，北界为中沙海槽与中沙海台，南部与盆西南海岭以盆西海谷为界，东部与南海海盆相接。海岭上密布众多的线性海山和山间盆地、海谷，这些海山多以北东—南西向平行排列，在线性海山之间分布众多的山间盆地（图 4–3）。 Penxi Hailing is located in the Nanhaixibu Lupo, with Zhongjiannan Xiepo and Zhongjiannan Haipen on the west, Zhongsha Haicao and Zhongsha Haitai on the north, Penxinan Hailing and Penxi Haigu on the south, and Nanhai Haipen on the east. It is densely distributed with multiple linear seamounts and inter-mountain basins and valleys. Most of these seamounts are arranged in parallel in the direction of NE–SW. A great number of inter-mountain basins are distributed among linear seamounts (Fig.4–3).		
命名由来 Origin of Name	1986 年，国务院、外交部和中国地名委员会批准了前地矿部第二海洋地质调查大队（现广州海洋地质调查局）对南海 22 个海底地理实体进行的命名，“盆西海岭”是这 22 个海底地理实体名称之一。该地理实体位于南海海盆以西，因此得名。 In 1986, the State Council, the Ministry of Foreign Affairs and Chinese Toponymy Committee approved 22 undersea feature names in Nanhai named by former Second Marine Geological Survey Brigade of the Ministry of Geology and Mineral Resources (present Guangzhou Marine Geological Survey of China Geological Survey). Penxi Hailing is one of the 22 undersea feature names. The feature is located on the west of Nanhai Haipen, and “Xi” means west in Chinese, so the word “Penxi” was useol to named the Ridge.		

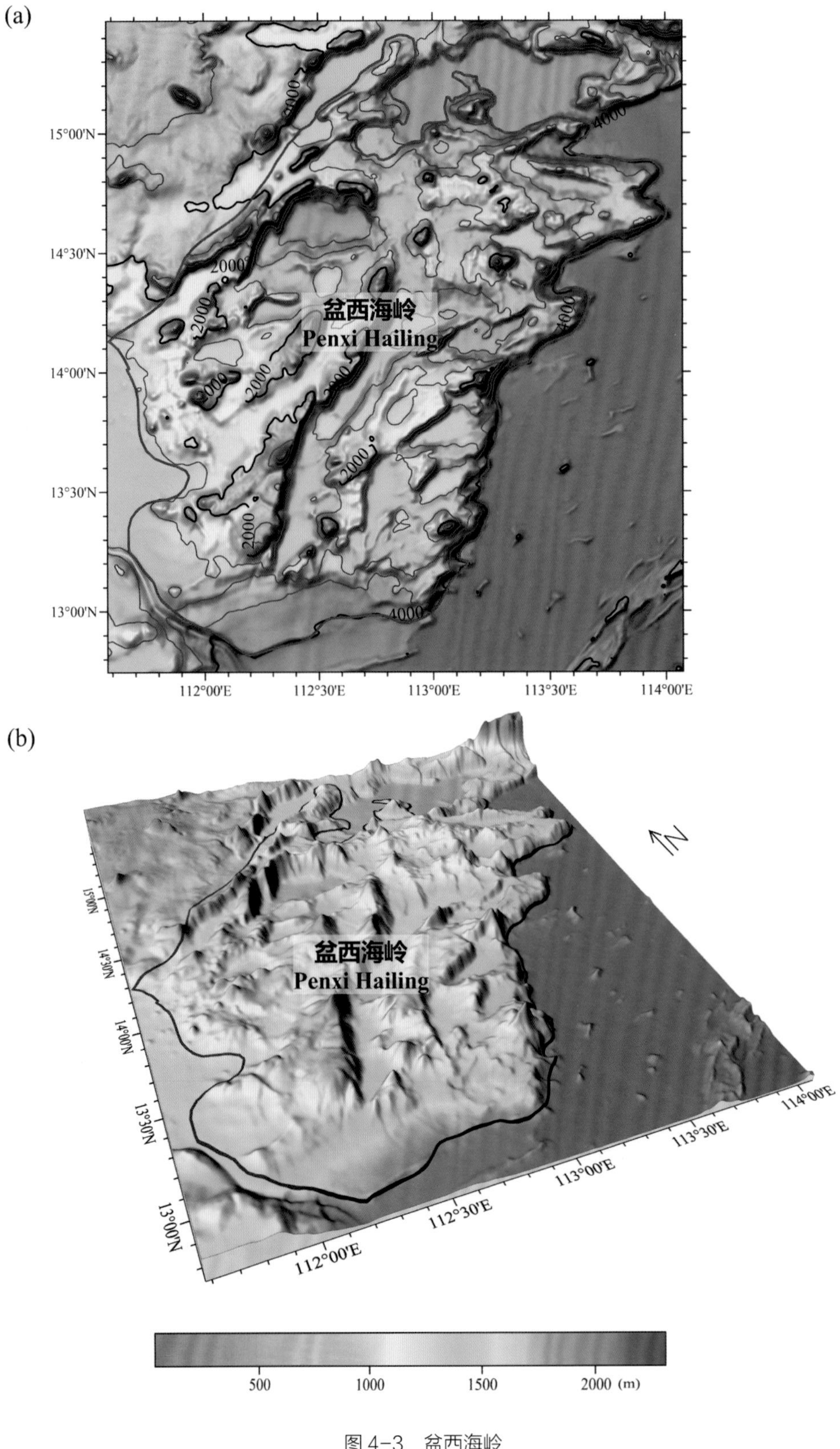

图 4-3　盆西海岭

(a) 海底地形图（等深线间隔 500 米）；(b) 三维海底地形图

Fig.4-3　Penxi Hailing

(a) Seafloor topographic map (with contour interval of 500 m)；(b) 3-D seafloor topographic map

4.3 兰宫洼地

标准名称 Standard Name	兰宫洼地 Langong Wadi	类别 Generic Term	海底洼地 Depression
中心点坐标 Center Coordinates	15°14.0'N, 112°35.0'E	规模（千米 × 千米） Dimension（km × km）	69 × 16
最小水深（米） Min Depth (m)	2675	最大水深（米） Max Depth (m)	2930
地理实体描述 Feature Description	兰宫洼地发育在盆西海岭北部，平面形态呈北东—南西向的长条形（图 4–4）。 Langong Wadi develops in the north of Penxi Hailing, with a planform in the shape of a long strip in the direction of NE–SW (Fig.4–4).		
命名由来 Origin of Name	以唐朝诗人王勃的《滕王阁序》中的词进行地名的团组化命名。该海底地名的专名取词自该诗作中的："桂殿兰宫，即冈峦之体势"。"兰宫"指建筑气派华美的宫殿。 A group naming of undersea features after the phrases in the poem *A Tribute to King Teng's Tower* by Wang Bo (650–676 A.D.), a famous poet in the Tang Dynasty (618–907 A.D.). The specific term of this undersea feature name "Langong", means magnificent building and splendid palaces, is derived from the poetic lines "The magnificent buildings in the splendid palace rise high and low continuously like undulating hills".		

4.4 羌笛海山

标准名称 Standard Name	羌笛海山 Qiangdi Haishan	类别 Generic Term	海山 Seamount
中心点坐标 Center Coordinates	15°04.4'N, 112°31.2'E	规模（千米 × 千米） Dimension（km × km）	58 × 18
最小水深（米） Min Depth (m)	2000	最大水深（米） Max Depth (m)	3710
地理实体描述 Feature Description	羌笛海山发育在盆西海岭北部，中沙南海盆西侧，平面形态呈北东—南西向的长条形（图 4–4）。 Qiangdi Haishan develops in the north of Penxi Hailing and is on the west of Zhongshanan Haipen, with a planform in the shape of a long strip in the direction of NE–SW (Fig.4–4).		
命名由来 Origin of Name	以唐朝诗人王之涣的《凉州词》中的词进行地名的团组化命名。该海底地名的专名取词自该诗作中的："羌笛何须怨杨柳，春风不度玉门关"。"羌笛"指的是一种古羌族乐器。 A group naming of undersea features after the phrases in the poem *Out of the Great Wall* by Wang Zhihuan (688–742 A.D.), a famous poet in the Tang Dynasty (618–907 A.D.). The specific term of this undersea feature name "Qiangdi" means a kind of music instrument of ancient Qiang nationality, is derived from the poetic lines "Why should the Mongol flute (Qiangdi) complain no willows grow? Beyond the Gate of Jade no vernal wind will blow".		

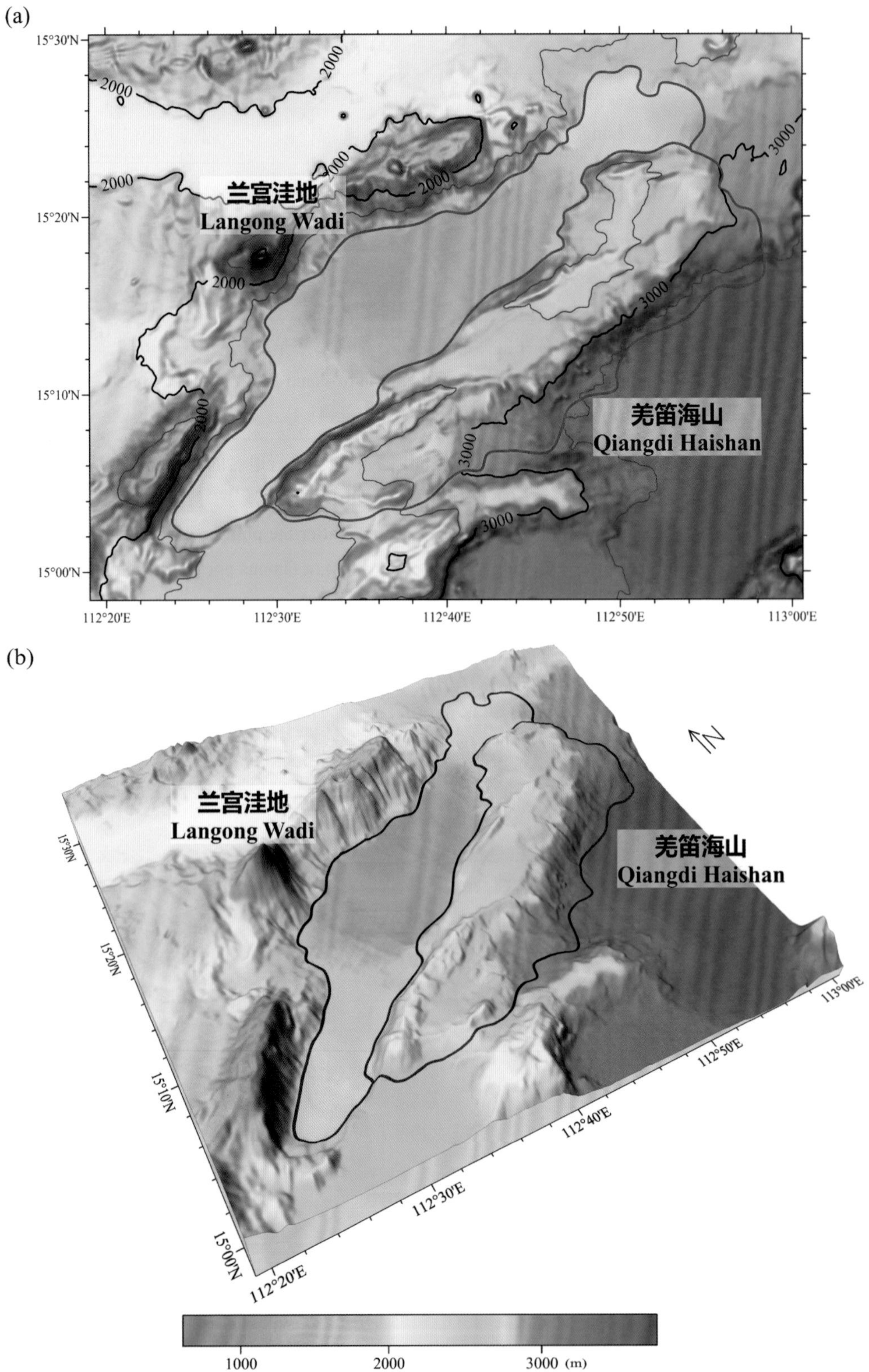

图 4-4　兰宫洼地、羌笛海山

(a) 海底地形图（等深线间隔 500 米）；(b) 三维海底地形图

Fig.4-4　Langong Wadi, Qiangdi Haishan

(a) Seafloor topographic map (with contour interval of 500 m)；(b) 3-D seafloor topographic map

4.5 桂殿海山

标准名称 Standard Name	桂殿海山 Guidian Haishan	类别 Generic Term	海山 Seamount
中心点坐标 Center Coordinates	14°55.6'N, 112°28.9'E	规模（千米 × 千米） Dimension（km × km）	55 × 6
最小水深（米） Min Depth (m)	1259	最大水深（米） Max Depth (m)	2541
地理实体描述 Feature Description	桂殿海山发育在盆西海岭北部，平面形态呈北东—南西向的长条形（图 4–5）。 Guidian Haishan is located in the north of Penxi Hailing, with a planform in the shape of a long strip in the direction of NE–SW (Fig.4–5).		
命名由来 Origin of Name	以唐朝诗人王勃的《滕王阁序》中的词进行地名的团组化命名。该海底地名的专名取词自该诗作中的：“桂殿兰宫，即冈峦之体势”。“桂殿”指建筑气派，设备华美的宫殿。 A group naming of undersea features after the phrases in the poem *A Tribute to King Teng's Tower* by Wang Bo (650–676 A.D.), a famous poet in the Tang Dynasty (618–907 A.D.). The specific term of this undersea feature name “Guidian”, means magnificent building and splendid palaces, is derived from the poetic lines “The magnificent buildings in the splendid palace rise high and low continuously like undulating hills”.		

4.6 萦回海丘

标准名称 Standard Name	萦回海丘 Yinghui Haiqiu	类别 Generic Term	海丘 Hill
中心点坐标 Center Coordinates	14°49.3'N, 112°19.8'E	规模（千米 × 千米） Dimension（km × km）	4 × 4
最小水深（米） Min Depth (m)	1940	最大水深（米） Max Depth (m)	2470
地理实体描述 Feature Description	萦回海丘发育在盆西海岭北部，平面形态呈三角形（图 4–5）。 Yinghui Haiqiu develops in the north of Penxi Hailing with a planform in the shape of a triangle (Fig.4–5).		
命名由来 Origin of Name	以唐朝诗人王勃的《滕王阁序》中的词进行地名的团组化命名。该海底地名的专名取词自该诗作中的：“鹤汀凫渚，穷岛屿之萦回”。“萦回”意为回旋环绕。 A group naming of undersea features after the phrases in the poem *A Tribute to King Teng's Tower* by Wang Bo (650–676 A.D.), a famous poet in the Tang Dynasty (618–907 A.D.). The specific term of this undersea feature name “Yinghui”, means circling around, is derived from the poetic lines “Crane and wild ducks are perching on the islet, or circling around the twisting island”.		

(a)

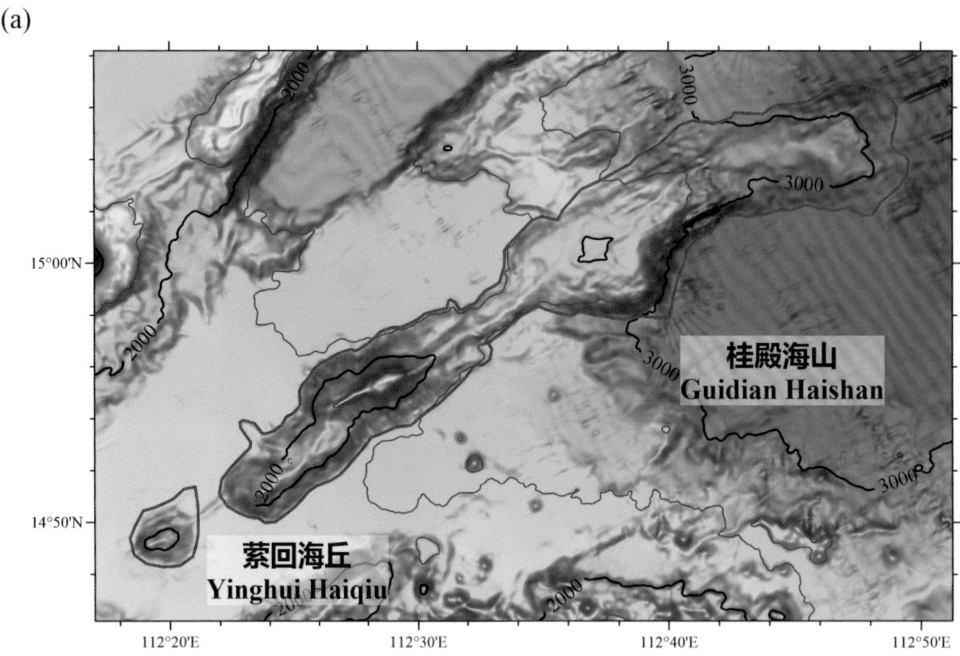

(b)

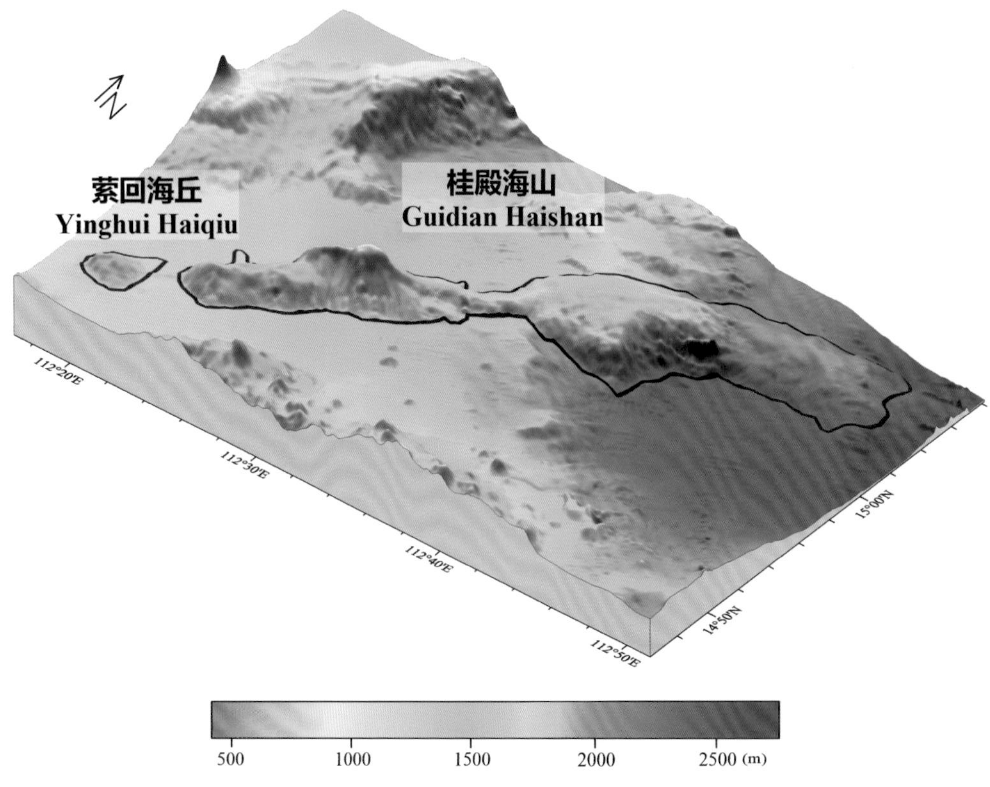

图 4-5　桂殿海山、萦回海丘

(a) 海底地形图（等深线间隔 500 米）；(b) 三维海底地形图

Fig.4-5　Guidian Haishan, Yinghui Haiqiu

(a) Seafloor topographic map (with contour interval of 500 m)；(b) 3-D seafloor topographic map

4.7 玉门海山

标准名称 Standard Name	玉门海山 Yumen Haishan	类别 Generic Term	海山 Seamount
中心点坐标 Center Coordinates	14°46.1'N, 112°37.7'E	规模（千米 × 千米） Dimension（km × km）	28 × 20
最小水深（米） Min Depth (m)	1330	最大水深（米） Max Depth (m)	3740
地理实体描述 Feature Description	玉门海山发育在盆西海岭北部，玉门海盆北部，平面形态呈东—西向的长条形（图 4–6）。 Yumen Haishan develops in the north of Penxi Hailing and on the north of Yumen Haipen, with a planform in the shape of a long strip in the direction of E–W (Fig.4–6).		
命名由来 Origin of Name	以唐朝诗人王之涣的《凉州词》中的词进行地名的团组化命名。该海底地名的专名取词自该诗作中的："羌笛何须怨杨柳，春风不度玉门关"。"玉门关"为我国古代丝绸之路上的一个地名，故址在今甘肃敦煌西北小方盘城，是古代通往西域的要道。 A group naming of undersea features after the phrases in the poem *Out of the Great Wall*, by Wang Zhihuan (688–742 A.D.), a famous poet in the Tang Dynasty (618–907 A.D.). The specific term of this undersea feature name "Yumen", means Gate of Jade, an important pass on the ancient silk road located in today's Xiaofangpan in the northwest of Dunhuang, Gansu province, is derived from the poetic lines "Why should the Mongol flute (Qiangdi) complain no willows grow? Beyond the Gate of Jade no vernal wind will blow".		

4.8 玉门海盆

标准名称 Standard Name	玉门海盆 Yumen Haipen	类别 Generic Term	海盆 Basin
中心点坐标 Center Coordinates	14°37.6'N, 112°29.9'E	规模（千米 × 千米） Dimension（km × km）	38 × 16
最小水深（米） Min Depth (m)	3010	最大水深（米） Max Depth (m)	3910
地理实体描述 Feature Description	玉门海盆发育在盆西海岭西北部，平面形态呈不规则多边形，东西向长约 38 千米，南北向宽约 16 千米。海盆的西侧水深较浅，东侧水深较深（图 4–6）。 Yumen Haipen develops in the northwest of Penxi Hailing, with a planform in the shape of an irregular polygon. The Basin is 38 km long from west to east, and 16 km wide from north to south, with shallower water on the west side and deeper water on the east side (Fig.4–6).		
命名由来 Origin of Name	该海盆位于玉门海山附近，因此得名。 The Basin is located near Yumen Haishan, so the word "Yumen" was used to name the Basin.		

4.9 玉门海脊

标准名称 Standard Name	玉门海脊 Yumen Haiji	类别 Generic Term	海脊 Ridge
中心点坐标 Center Coordinates	14°29.3'N, 112°39.4'E	规模（千米 × 千米） Dimension（km × km）	40 × 6
最小水深（米） Min Depth (m)	2510	最大水深（米） Max Depth (m)	3850
地理实体描述 Feature Description	玉门海脊发育在盆西海岭西北部，玉门海盆南部，平面形态呈东—西向的长条形。海脊的北侧水深较深，南侧水深较浅（图 4–6）。 Yumen Haiji develops in the northwest of Penxi Hailing and on the south of Yumen Haipen, with a planform in the shape of a long strip in the direction of E–W. Its north side is deeper while its south side is shallower (Fig.4–6).		
命名由来 Origin of Name	该海脊位于玉门海山附近，因此得名。 The Ridge is located near Yumen Haishan, so the word "Yumen" was used to name the Ridge.		

4.10 万仞海脊

标准名称 Standard Name	万仞海脊 Wanren Haiji	类别 Generic Term	海脊 Ridge
中心点坐标 Center Coordinates	14°38.2'N, 112°19.2'E	规模（千米 × 千米） Dimension（km × km）	25 × 7
最小水深（米） Min Depth (m)	1160	最大水深（米） Max Depth (m)	3800
地理实体描述 Feature Description	万仞海脊发育在盆西海岭西北部，玉门海盆北部，平面形态呈北东—南西走向（图 4–6）。 Wanren Haiji develops in the northwest of Penxi Hailing and on the north of Yumen Haipen, stretching in the direction of NE–SW (Fig.4–6).		
命名由来 Origin of Name	以唐朝诗人王之涣的《凉州词》中的词进行地名的团组化命名。该海底地名的专名取词自该诗作中的："黄河远上白云间，一片孤城万仞山"。"万仞"用以形容山非常高耸。 A group naming of undersea features after the phrases in the poem Out of the Great Wall, by Wang Zhihuan (688–742 A.D.), a famous poet in the Tang Dynasty (618–907 A.D.). The specific term of this undersea feature name "Wanren", means towering mountain, is derived from the poetic lines "The yellow river uprises as high as cloud, the lonely town is lost among towering mountains".		

4.11 杨柳海山

标准名称 Standard Name	杨柳海山 Yangliu Haishan	类别 Generic Term	海山 Seamount
中心点坐标 Center Coordinates	14°34.8'N, 112°10.2'E	规模（千米 × 千米） Dimension（km × km）	56 × 22
最小水深（米） Min Depth (m)	310	最大水深（米） Max Depth (m)	2450
地理实体描述 Feature Description	杨柳海山发育在盆西海岭西北部，玉门海盆西部，平面形态呈北东—南西向的倒“V”型（图 4–6）。 Yangliu Haishan develops in the northwest of Penxi Hailing and on the west of Yumen Haipen, with a planform in the shape of reverted “V” in the direction of NE–SW (Fig.4–6).		
命名由来 Origin of Name	以唐朝诗人王之涣的《凉州词》中的词进行地名的团组化命名。该海底地名的专名取词自该诗作中的：“羌笛何须怨杨柳，春风不度玉门关”。“杨柳”指一种叫《折杨柳》的歌曲。 A group naming of undersea features after the phrases in the poem *Out of the Great Wall*, by Wang Zhihuan (688–742 A.D.), a famous poet in the Tang Dynasty (618–907 A.D.). The specific term of this undersea feature name “Yangliu” literally meaning willow, implying to a song titled “Plucking a Willow Twig”, is derived from the poetic lines “Why should the Mongol flute (Qiangdi) complain no willows grow? Beyond the Gate of Jade no vernal wind will blow”.		

(a)

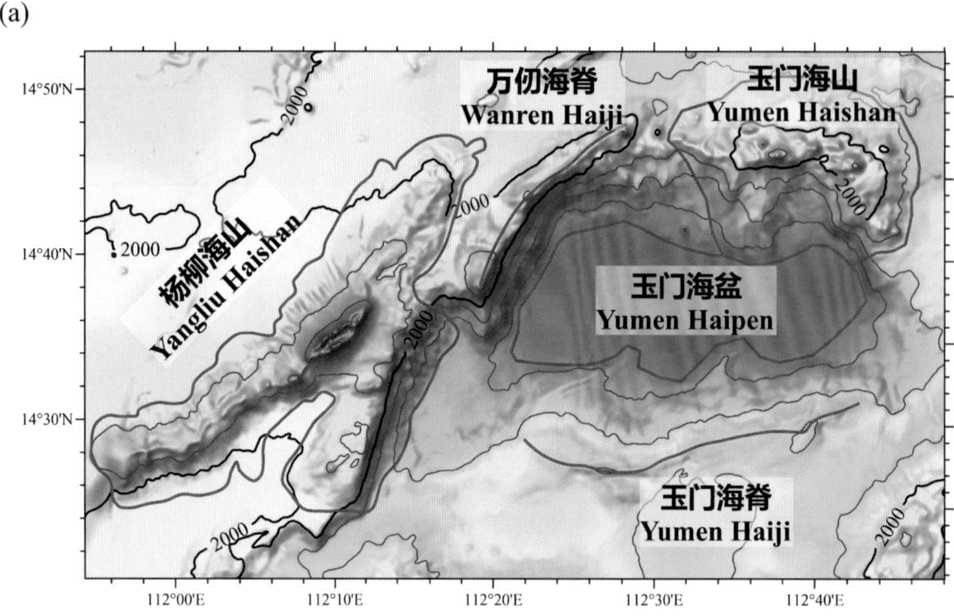

(b)

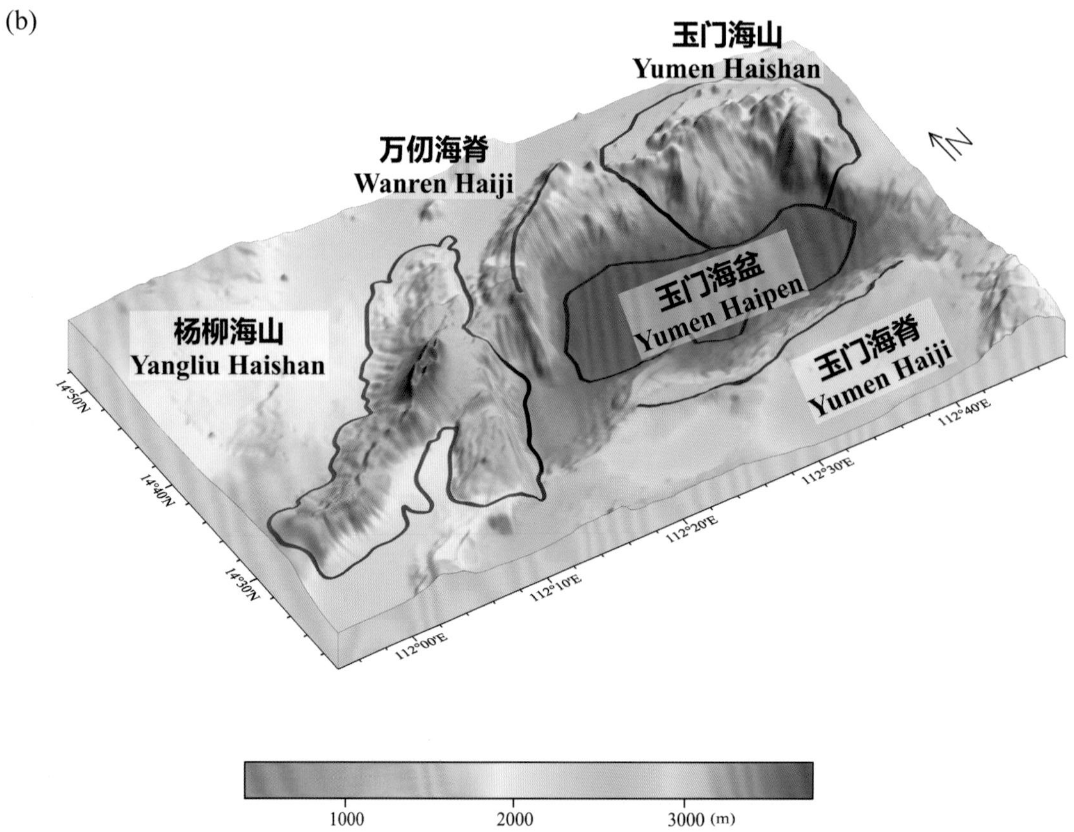

图 4-6 玉门海山、玉门海盆、玉门海脊、万仞海脊、杨柳海山

(a) 海底地形图（等深线间隔 500 米）；(b) 三维海底地形图

Fig.4-6 Yumen Haishan, Yumen Haipen, Yumen Haiji, Wanren Haiji, Yangliu Haishan

(a) Seafloor topographic map (with contour interval of 500 m)；(b) 3-D seafloor topographic map

4.12 春风海脊

标准名称 Standard Name	春风海脊 Chunfeng Haiji	类别 Generic Term	海脊 Ridge
中心点坐标 Center Coordinates	14°12.1'N, 112°01.7'E	规模（千米 × 千米） Dimension（km × km）	44 × 11
最小水深（米） Min Depth (m)	1580	最大水深（米） Max Depth (m)	2870
地理实体描述 Feature Description	春风海脊发育在盆西海岭西部，平面形态呈北东—南西走向（图 4-7）。 Chunfeng Haiji develops in the west of Penxi Hailing, with a planform in the direction of NE−SW (Fig.4−7).		
命名由来 Origin of Name	以唐朝诗人王之涣的《凉州词》中的词进行地名的团组化命名。该海底地名的专名取词自该诗作中的："羌笛何须怨杨柳，春风不度玉门关"。"春风"指春天的风。 A group naming of undersea features after the phrases in the poem *Out of the Great Wall*, by Wang Zhihuan (688−742 A.D.), a famous poet in the Tang Dynasty (618−907 A.D.). The specific term of this undersea feature name "Chunfeng", means vernal wind, is derived from the poetic lines "Why should the Mongol flute (Qiangdi) complain no willows grow? Beyond the Gate of Jade no vernal wind will blow".		

4.13 春风海山

标准名称 Standard Name	春风海山 Chunfeng Haishan	类别 Generic Term	海山 Seamount
中心点坐标 Center Coordinates	14°12.6'N, 112°28.1'E	规模（千米 × 千米） Dimension（km × km）	72 × 22
最小水深（米） Min Depth (m)	1260	最大水深（米） Max Depth (m)	3170
地理实体描述 Feature Description	春风海山发育在盆西海岭中部，平面形态呈北东—南西向的纺锤形。海山的东北部水深较浅，西南部水深较深（图 4-7）。 Chunfeng Haishan develops in the middle of Penxi Hailing, with a spindle-shaped planform in the direction of NE−SW. The depth of the Seamount is shallower on the northeast side and is deeper on the southwest side (Fig.4−7).		
命名由来 Origin of Name	以唐朝诗人王之涣的《凉州词》中的词进行地名的团组化命名。该海底地名的专名取词自该诗作中的："羌笛何须怨杨柳，春风不度玉门关"。"春风"指春天的风。 A group naming of undersea features after the phrases in the poem *Out of the Great Wall*, by Wang Zhihuan (688−742 A.D.), a famous poet in the Tang Dynasty (618−907 A.D.). The specific term of this undersea feature name "Chunfeng", means vernal wind, is derived from the poetic lines "Why should the Mongol flute (Qiangdi) complain no willows grow? Beyond the Gate of Jade no vernal wind will blow".		

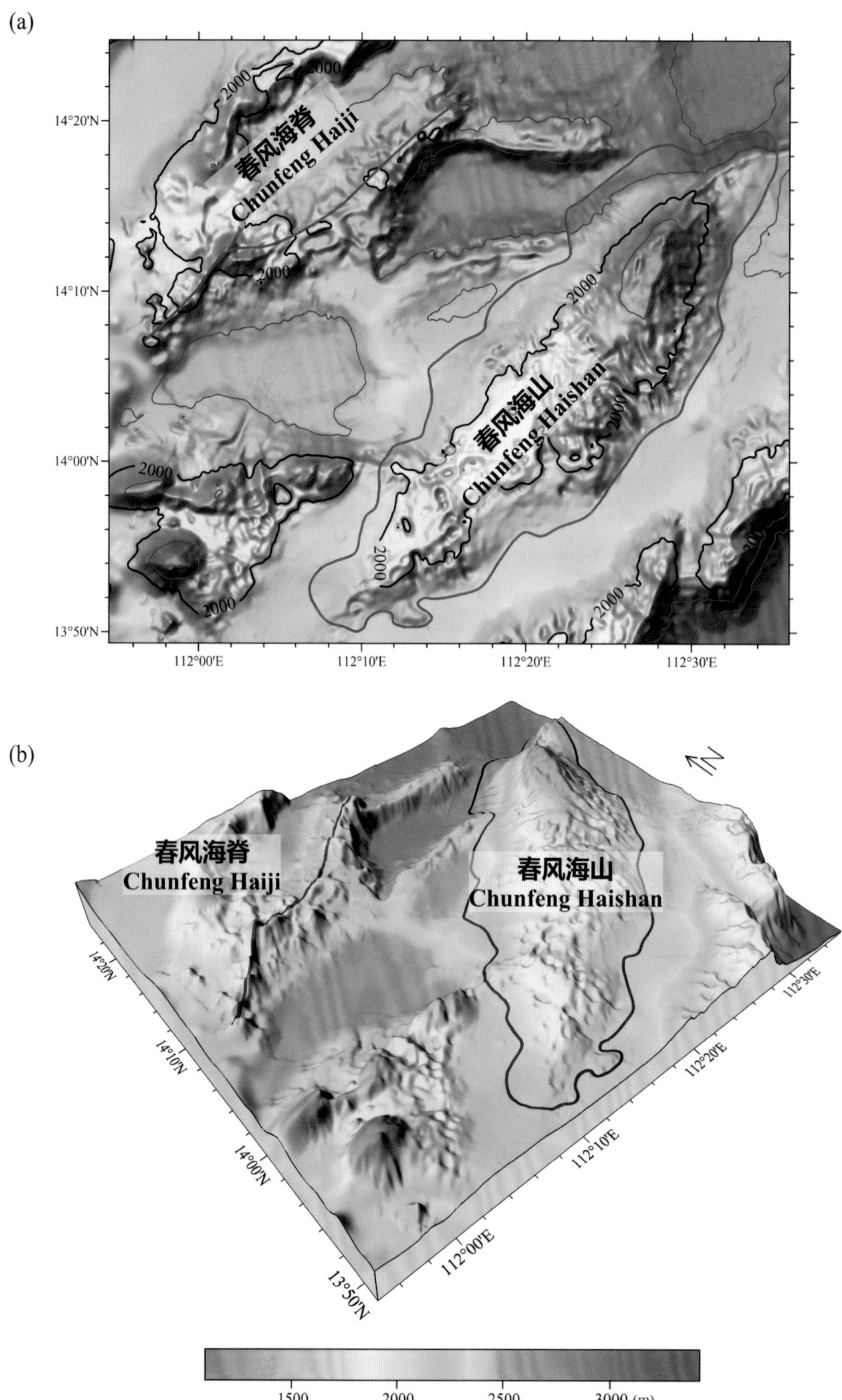

图 4-7　春风海脊、春风海山

(a) 海底地形图（等深线间隔 500 米）；(b) 三维海底地形图

Fig.4-7　Chunfeng Haiji, Chunfeng Haishan

(a) Seafloor topographic map (with contour interval of 500 m)；(b) 3-D seafloor topographic map

4.14 紫电海山

标准名称 Standard Name	紫电海山 Zidian Haishan	类别 Generic Term	海山 Seamount
中心点坐标 Center Coordinates	14°06.8'N, 113°00.6'E	规模（千米 × 千米） Dimension（km × km）	39 × 14
最小水深（米） Min Depth (m)	1580	最大水深（米） Max Depth (m)	3570
地理实体描述 Feature Description	紫电海山发育在盆西海岭中部，平面形态呈北东—南西向的长条形（图 4–8）。 Zidian Haishan develops in the middle of Penxi Hailing, with a planform in the shape of a long strip in the direction of NE–SW (Fig.4–8).		
命名由来 Origin of Name	以唐朝诗人王勃的《滕王阁序》中的词进行地名的团组化命名。该海底地名的专名取词自该诗作中的："紫电青霜，王将军之武库"。"紫电"指紫电剑，形容宝剑十分锋利。 A group naming of undersea features after the phrases in the poem *A Tribute to King Teng's Tower* by Wang Bo (650–676 A.D.), a famous poet in the Tang Dynasty (618–907 A.D.). The specific term of this undersea feature name "Zidian", means purple lightening, referring to a sharp sword, is derived from the poetic lines "General Wang has weapons as sharp as the famous sword 'Purple Lightening' and 'Blue Frost' in his armory".		

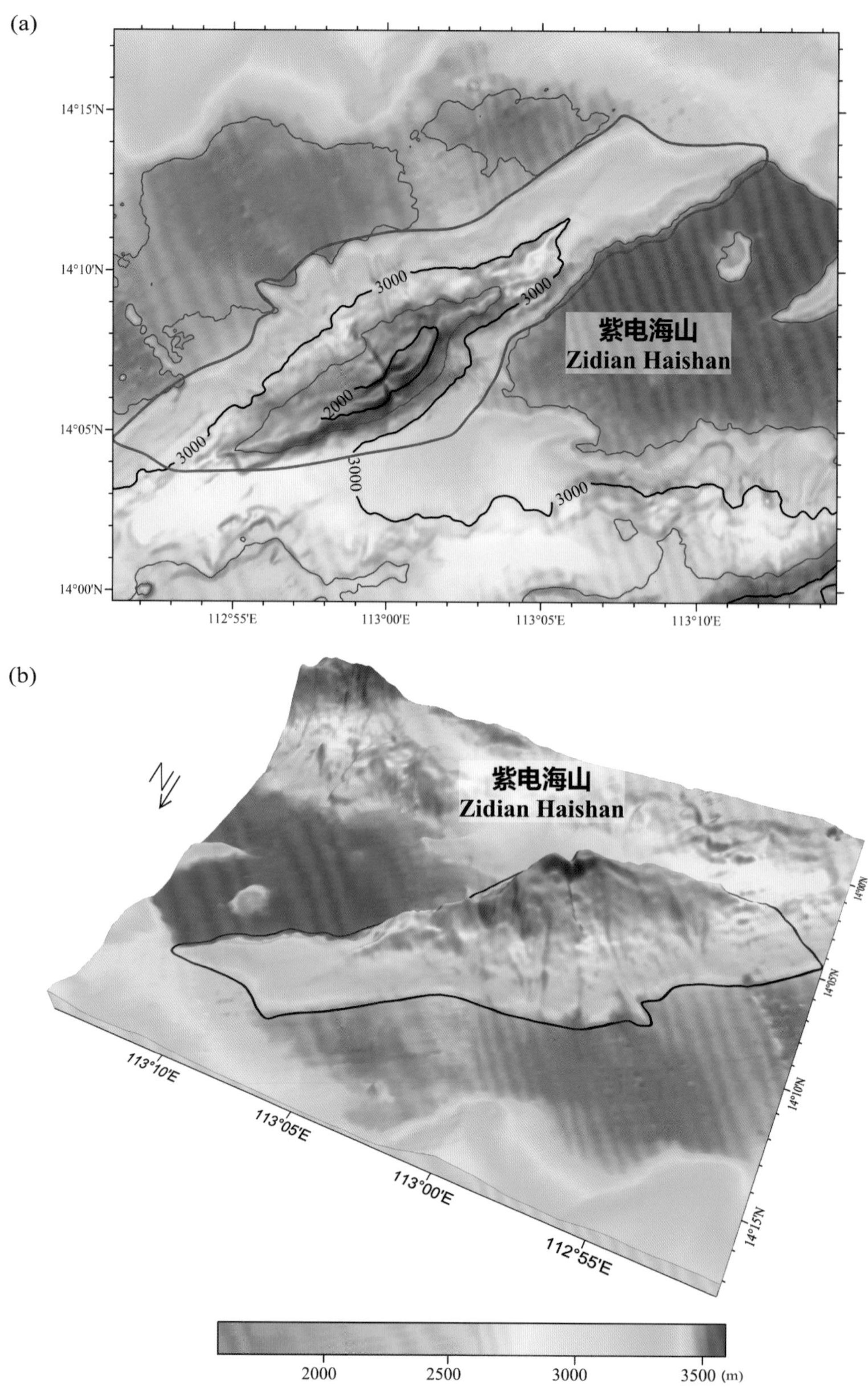

图 4-8　紫电海山

(a) 海底地形图（等深线间隔 500 米）；(b) 三维海底地形图

Fig.4-8　Zidian Haishan

(a) Seafloor topographic map (with contour interval of 500 m)；(b) 3-D seafloor topographic map

4.15 流丹海丘

标准名称 Standard Name	流丹海丘 Liudan Haiqiu	类别 Generic Term	海丘 Hill
中心点坐标 Center Coordinates	14°36.9'N, 113°18.4'E	规模（千米 × 千米） Dimension（km × km）	13 × 7
最小水深（米） Min Depth (m)	1810	最大水深（米） Max Depth (m)	3150
地理实体描述 Feature Description	流丹海丘发育在盆西海岭东北部，平面形态呈东—西向的菱形（图 4–9）。 Liudan Haiqiu develops in the northeast of Penxi Hailing, with a planform in the shape of a diamond in the direction of E–W (Fig.4–9).		
命名由来 Origin of Name	以唐朝诗人王勃的《滕王阁序》中的词进行地名的团组化命名。该海底地名的专名取词自该诗作中的："飞阁流丹，下临无地"。"流丹"意为流动着红色，形容色彩飞动。 A group naming of undersea features after the phrases in the poem *A Tribute to King Teng's Tower* by Wang Bo (650–676 A.D.), a famous poet in the Tang Dynasty (618–907 A.D.). The specific term of this undersea feature name "Liudan", means glowing red, is derived from the poetic lines "The lofty pavilion is glowing in red, soaring so high that ground is too faraway to see".		

4.16 鹤汀海山

标准名称 Standard Name	鹤汀海山 Heting Haishan	类别 Generic Term	海山 Seamount
中心点坐标 Center Coordinates	14°26.4'N, 113°28.3'E	规模（千米 × 千米） Dimension（km × km）	13 × 12
最小水深（米） Min Depth (m)	1990	最大水深（米） Max Depth (m)	2980
地理实体描述 Feature Description	鹤汀海山发育在盆西海岭东北部，平面形态近似圆形（图 4–9）。 Heting Haishan develops in the northeast of Penxi Hailing, with a nearly circular planform (Fig.4–9).		
命名由来 Origin of Name	以唐朝诗人王勃的《滕王阁序》中的词进行地名的团组化命名。该海底地名的专名取词自该诗作中的："鹤汀凫渚，穷岛屿之萦回"。"鹤汀"指野鹤栖居的水中小洲。 A group naming of undersea features after the phrases in the poem *A Tribute to King Teng's Tower* by Wang Bo (650–676 A.D.), a famous poet in the Tang Dynasty (618–907 A.D.). The specific term of this undersea feature name "Heting", means a small islet dwelt by cranes, is derived from the poetic lines "Cranes and wild ducks are perching on the islet, or circling around the island".		

4.17 青海海山

标准名称 Standard Name	青海海山 Qinghai Haishan	类别 Generic Term	海山 Seamount
中心点坐标 Center Coordinates	14°39.2'N, 113°29.7'E	规模（千米 × 千米） Dimension（km × km）	38 × 17
最小水深（米） Min Depth (m)	1390	最大水深（米） Max Depth (m)	4520
地理实体描述 Feature Description	青海海山发育在盆西海岭东北部，平面形态呈不规则多边形，其顶部发育多座峰（图 4–9）。 Qinghai Haishan develops in the northeast of Penxi Hailing, with a planform in the shape of an irregular polygon. Multiple peaks have developed on its top (Fig.4–9).		
命名由来 Origin of Name	以唐朝诗人李白的《关山月》中的词进行地名的团组化命名。该海底地名的专名取词自该诗作中的："汉下白登道，胡窥青海湾"。"青海湾"即今青海省青海湖。 A group naming of undersea features after the phrases in the poem *Moon over Fortified Pass* by Li Bai (701–762 A.D.), a famous poet in the Tang Dynasty (618–907 A.D.). The specific term of this undersea feature name "Qinghai", means blue lake, referring to today's Qinghai Lake in Qinghai province, is derived from the poetic lines "The Han army was determined to cross the white mountain, while Tartars coveted for the extensive lake blue".		

4.18 青海南海山群

标准名称 Standard Name	青海南海山群 Qinghainan Haishanqun	类别 Generic Term	海山群 Seamounts
中心点坐标 Center Coordinates	14°27.5'N, 113°16.7'E	规模（千米 × 千米） Dimension（km × km）	50 × 36
最小水深（米） Min Depth (m)	590	最大水深（米） Max Depth (m)	4420
地理实体描述 Feature Description	青海南海山群发育在盆西海岭东北部，平面形态呈不规则多边形。海山群的北坡较陡，南坡较缓，顶部发育多座峰（图 4–9）。 Qinghainan Haishanqun develop in the northeast of Penxi Hailing, with a planform in the shape of an irregular polygon. Their north slopes are relatively steeper while their south slopes are relatively gentle, with multiple peaks developed on top (Fig.4–9).		
命名由来 Origin of Name	该海山群位于青海海山以南，因此得名。 These Seamounts are located to the south of Qinghai Haishan, and "Nan" means south in Chinese, so the word "Qinghainan" was used to name these Seamounts.		

(a)

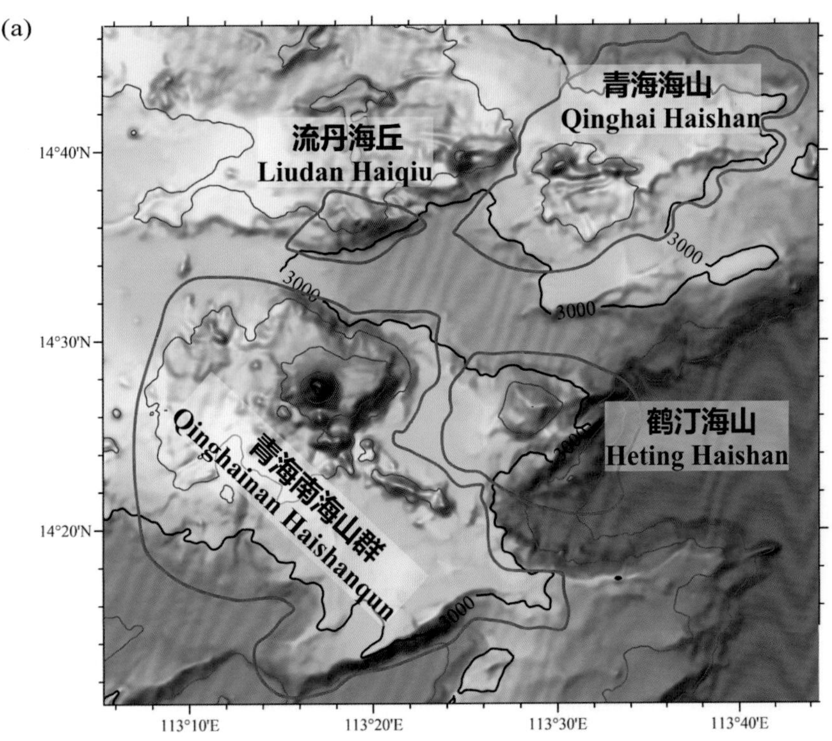

(b)

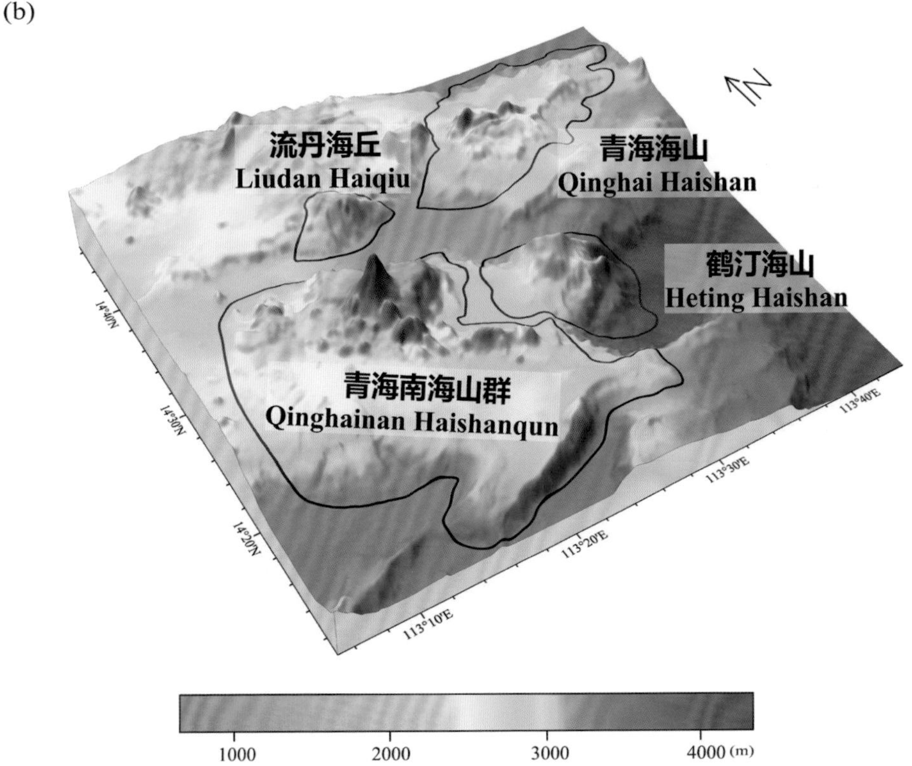

图 4-9 流丹海丘、鹤汀海山、青海海山、青海南海山群

(a) 海底地形图（等深线间隔 500 米）；(b) 三维海底地形图

Fig.4-9 Liudan Haiqiu, Heting Haishan, Qinghai Haishan, Qinghainan Haishanqun

(a) Seafloor topographic map (with contour interval of 500 m); (b) 3-D seafloor topographic map

4.19 青海北海脊

标准名称 Standard Name	青海北海脊 Qinghaibei Haiji	类别 Generic Term	海脊 Ridge
中心点坐标 Center Coordinates	14°51.0'N, 113°33.2'E	规模（千米 × 千米） Dimension（km × km）	54 × 8
最小水深（米） Min Depth (m)	2310	最大水深（米） Max Depth (m)	4480
地理实体描述 Feature Description	青海北海脊发育在盆西海岭东北部，平面形态呈北西西—南东东向的长条形。海脊的北侧较陡，南侧较缓（图 4–10）。 Qinghaibei Haiji develops in the northeast of Penxi Hailing, with a planform in the shape of a long strip in the direction of NWW–SEE. Its north slope is relatively steeper, while its south slope is relatively gentler (Fig.4–10).		
命名由来 Origin of Name	该海脊位于青海海山以北，因此得名。 The Ridge is located to the north of Qinghai Haishan, and “Bei” means north in Chinese, so the word “Qinghaibei” was used to name the Ridge.		

4.20 边色海脊

标准名称 Standard Name	边色海脊 Bianse Haiji	类别 Generic Term	海脊 Ridge
中心点坐标 Center Coordinates	14°52.8’N, 113°14.6’E	规模（千米 × 千米） Dimension（km × km）	45 × 8
最小水深（米） Min Depth (m)	2260	最大水深（米） Max Depth (m)	4380
地理实体描述 Feature Description	边色海脊发育在盆西海岭东北部，中沙南海盆南部，平面形态呈北东—南西向的长条形（图 4–10）。 Bianse Haiji develops in the northeast of Penxi Hailing and on the south of Zhongshanan Haipen, with a planform in the shape of a long strip in the direction of NE–SW (Fig.4–10).		
命名由来 Origin of Name	以唐朝诗人李白的《关山月》中的词进行地名的团组化命名。该海底地名的专名取词自该诗作中的：“戍客望边色，思归多苦颜”。“边色”即边邑，指边城。 A group naming of undersea features after the phrases in the poem *Moon over Fortified Pass* by Li Bai (701–762 A.D.), a famous poet in the Tang Dynasty (618–907 A.D.). The specific term of this undersea feature name “Bianse”, means remote city on border, is derived from the poetic lines “Guards watch the scene of borderland, thinking of home, with wistful eyes”.		

(a)

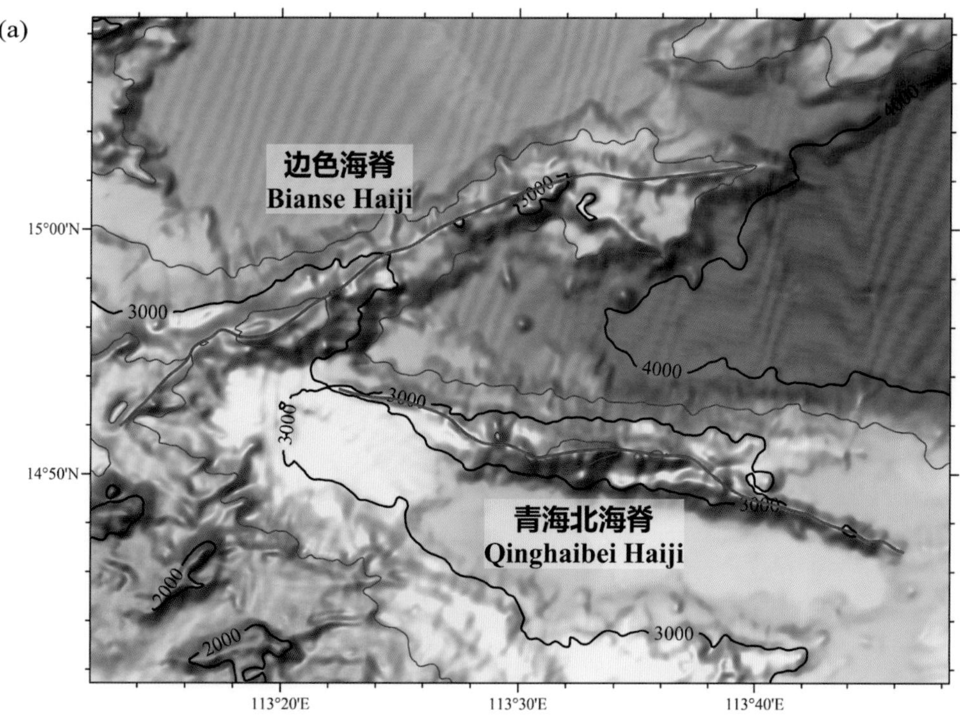

(b)

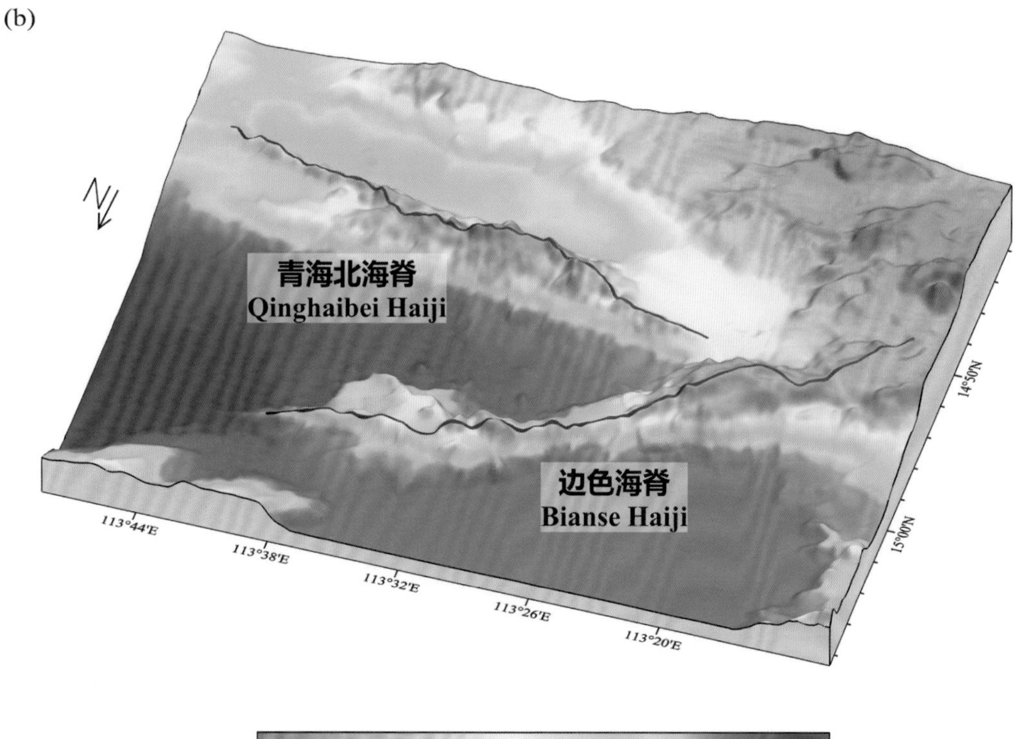

图 4-10　青海北海脊、边色海脊

(a) 海底地形图（等深线间隔 500 米）；(b) 三维海底地形图

Fig.4-10　Qinghaibei Haiji, Bianse Haiji

(a) Seafloor topographic map (with contour interval of 500 m)；(b) 3-D seafloor topographic map

4.21 孤城海山

标准名称 Standard Name	孤城海山 Gucheng Haishan	类别 Generic Term	海山 Seamount
中心点坐标 Center Coordinates	15°05.4'N, 113°09.8'E	规模（千米 × 千米） Dimension（km × km）	31 × 24
最小水深（米） Min Depth (m)	2460	最大水深（米） Max Depth (m)	4030
地理实体描述 Feature Description	孤城海山发育在盆西海岭北部，中沙南海盆内，平面形态呈不规则多边形，顶部发育多座峰（图 4-11）。 Gucheng Haishan develops in the north of Penxi Hailing within Zhongshanan Haipen, with a planform in the shape of an irregular polygons. Several peaks have developed on its top (Fig.4-11).		
命名由来 Origin of Name	以唐朝诗人王之涣的《凉州词》中的词进行地名的团组化命名。该海底地名的专名取词自该诗作中的：“黄河远上白云间，一片孤城万仞山”。“孤城”指孤零零的戍边城堡玉门关。 A group naming of undersea features after the phrases in the poem *Out of the Great Wall*, by Wang Zhihuan (688–742 A.D.), a famous poet in the Tang Dynasty (618–907 A.D.). The specific term of this undersea feature name “Gucheng”, means a lonely town, referring to the Gate of Jade standing alone on the borderland, is derived from the poetic lines “The yellow river uprises as high as cloud, the lonely town is lost among towering mountains”.		

4.22 飞阁海山

标准名称 Standard Name	飞阁海山 Feige Haishan	类别 Generic Term	海山 Seamount
中心点坐标 Center Coordinates	15°00.7'N, 113°00.5'E	规模（千米 × 千米） Dimension（km × km）	14 × 9
最小水深（米） Min Depth (m)	2490	最大水深（米） Max Depth (m)	3730
地理实体描述 Feature Description	飞阁海山发育在盆西海岭北部，平面形态呈不规则多边形（图 4-11）。 Feige Haishan develops in the north of Penxi Hailing, with a planform in the shape of an irregular polygon (Fig.4-11).		
命名由来 Origin of Name	以唐朝诗人王勃的《滕王阁序》中的词进行地名的团组化命名。该海底地名的专名取词自该诗作中的：“飞阁流丹，下临无地”。“飞阁”意为凌空的阁楼。 A group naming of undersea features after the phrases in the poem *A Tribute to King Teng's Tower* by Wang Bo (650–676 A.D.), a famous poet in the Tang Dynasty (618–907 A.D.). The specific term of this undersea feature name “Feige”, means lofty tower, is derived from the poetic lines “The lofty tower is glowing in red, soaring so high that the ground is beyond the sight”.		

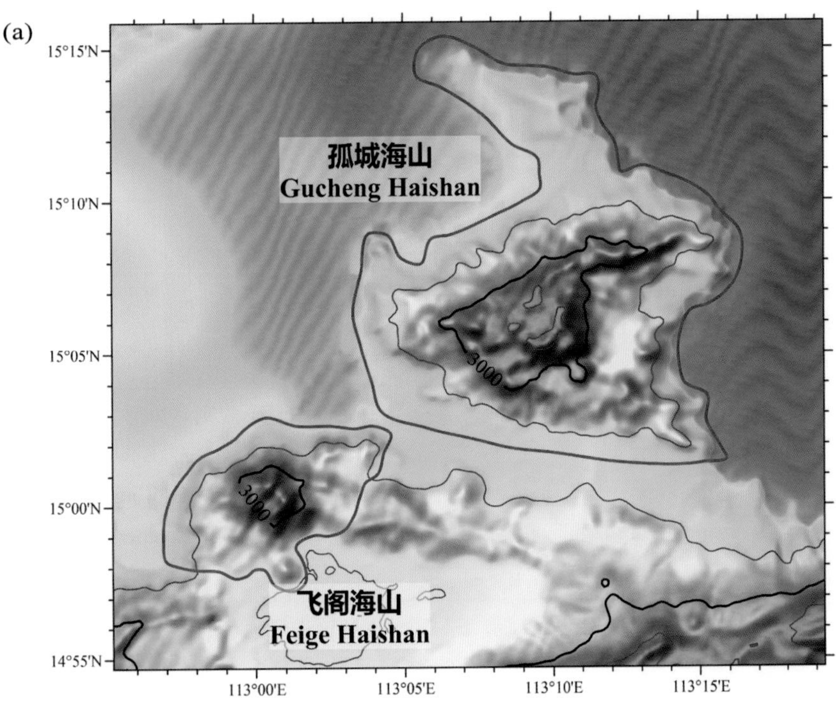

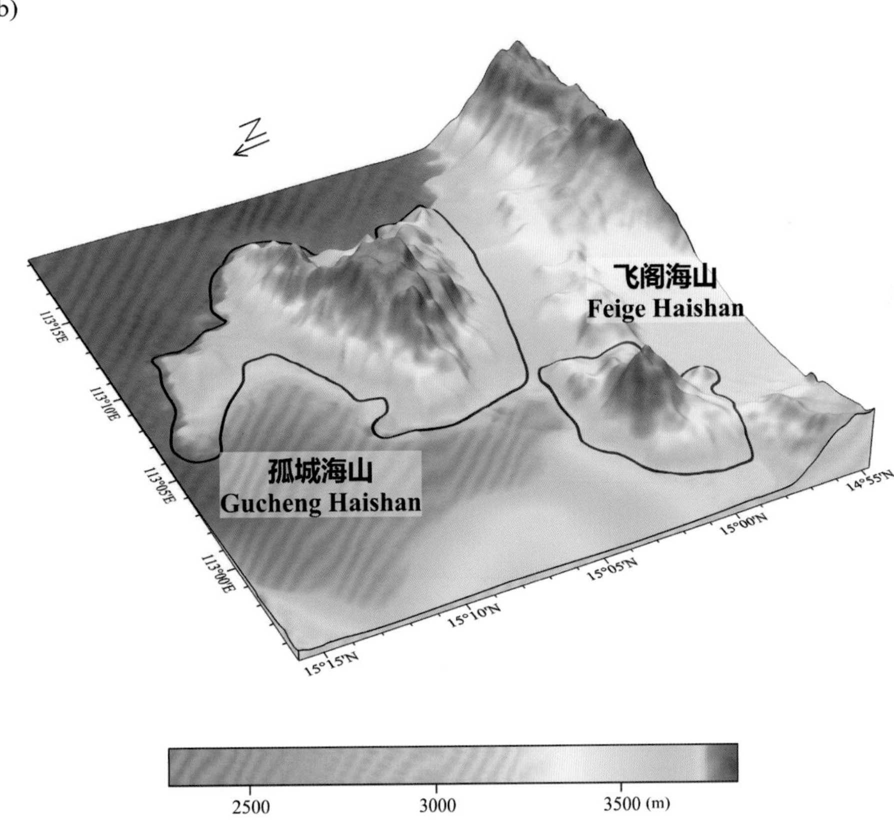

图 4-11　孤城海山、飞阁海山

(a) 海底地形图（等深线间隔 500 米）；(b) 三维海底地形图

Fig.4-11　Gucheng Haishan, Feige Haishan

(a) Seafloor topographic map (with contour interval of 500 m)；(b) 3-D seafloor topographic map

4.23 重霄海山

标准名称 Standard Name	重霄海山 Chongxiao Haishan	类别 Generic Term	海山 Seamount
中心点坐标 Center Coordinates	14°49.0'N, 112°58.8'E	规模（千米 × 千米） Dimension（km × km）	14 × 11
最小水深（米） Min Depth (m)	1360	最大水深（米） Max Depth (m)	3260
地理实体描述 Feature Description	重霄海山发育在盆西海岭北部，平面形态呈不规则多边形，其北坡较陡，南坡较缓（图 4–12）。 Chongxiao Haishan develops in the north of Penxi Hailing, with planform in the shape of an irregular polygon. Its north slope is relatively steeper, while its south slope is relatively gentler (Fig.4–12).		
命名由来 Origin of Name	以唐朝诗人王勃的《滕王阁序》中的词进行地名的团组化命名。该海底地名的专名取词自该诗作中的："层峦耸翠，上出重霄"。"重霄"指极高的天空。 A group naming of undersea features after the phrases in the poem *A Tribute to King Teng's Tower* by Wang Bo (650–676 A.D.), a famous poet in the Tang Dynasty (618–907 A.D.). The specific term of this undersea feature name "Chongxiao", means highest heaven, is derived from the poetic lines "The high peaks rising in emerald verdure, reaching the highest heaven".		

4.24 耸翠海丘

标准名称 Standard Name	耸翠海丘 Songcui Haiqiu	类别 Generic Term	海丘 Hill
中心点坐标 Center Coordinates	14°42.9'N, 112°59.1'E	规模（千米 × 千米） Dimension（km × km）	18 × 11
最小水深（米） Min Depth (m)	1980	最大水深（米） Max Depth (m)	2620
地理实体描述 Feature Description	耸翠海丘发育在盆西海岭北部，重霄海山和云海海山之间，平面形态呈不规则多边形（图 4–12）。 Songcui Haiqiu develops in the north of Penxi Hailing and is between Chongxiao Haishan and Yunhai Haishan, with a planform in the shape of an irregular polygon (Fig.4–12).		
命名由来 Origin of Name	以唐朝诗人王勃的《滕王阁序》中的词进行地名的团组化命名。该海底地名的专名取词自该诗作中的："层峦耸翠，上出重霄"。"耸翠"意为山峦高耸苍翠。 A group naming of undersea features after the phrases in the poem *A Tribute to King Teng's Tower* by Wang Bo (650–676 A.D.), a famous poet in the Tang Dynasty (618–907 A.D.). The specific term of this undersea feature name "Songcui", means peaks rising in emerald verdure, is derived from the poetic lines "The high peaks rising in emerald verdure, reaching the highest heaven".		

4.25 云海海山

标准名称 Standard Name	云海海山 Yunhai Haishan	类别 Generic Term	海山 Seamount
中心点坐标 Center Coordinates	14°33.1'N, 112°57.0'E	规模（千米 × 千米） Dimension（km × km）	22 × 21
最小水深（米） Min Depth (m)	1340	最大水深（米） Max Depth (m)	2980
地理实体描述 Feature Description	云海海山发育在盆西海岭北部，平面形态呈不规则多边形，顶部发育多座峰（图 4–12）。 Yunhai Haishan is developing in the north of Penxi Hailing, with a planform in the shape of an irregular polygon. Multiple peaks developed on its top (Fig.4–12).		
命名由来 Origin of Name	以唐朝诗人李白的《关山月》中的词进行地名的团组化命名。该海底地名的专名取词自该诗作中的："明月出天山，苍茫云海间"。"云海"意为漫无边际的云。 A group naming of undersea features after the phrases in the poem *Moon over Fortified Pass* by Li Bai (701–762 A.D.), a famous poet in the Tang Dynasty (618–907 A.D.). The specific term of this undersea feature name "Yunhai", means boundless sea of cloud, is derived from the poetic lines "From Heaven's Peak the moon rises bright, over a boundless sea of cloud".		

(a)

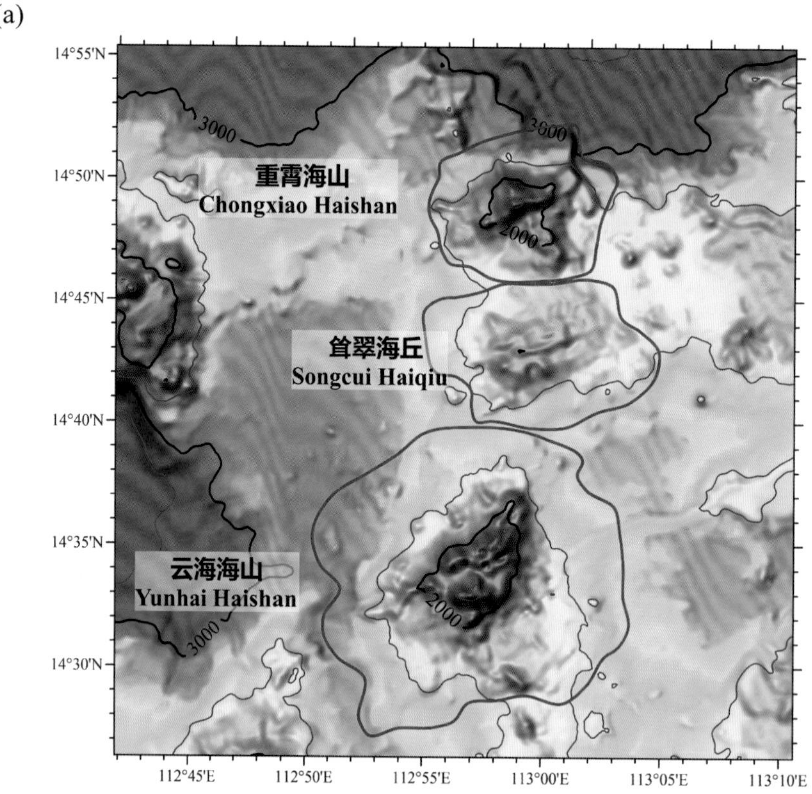

(b)

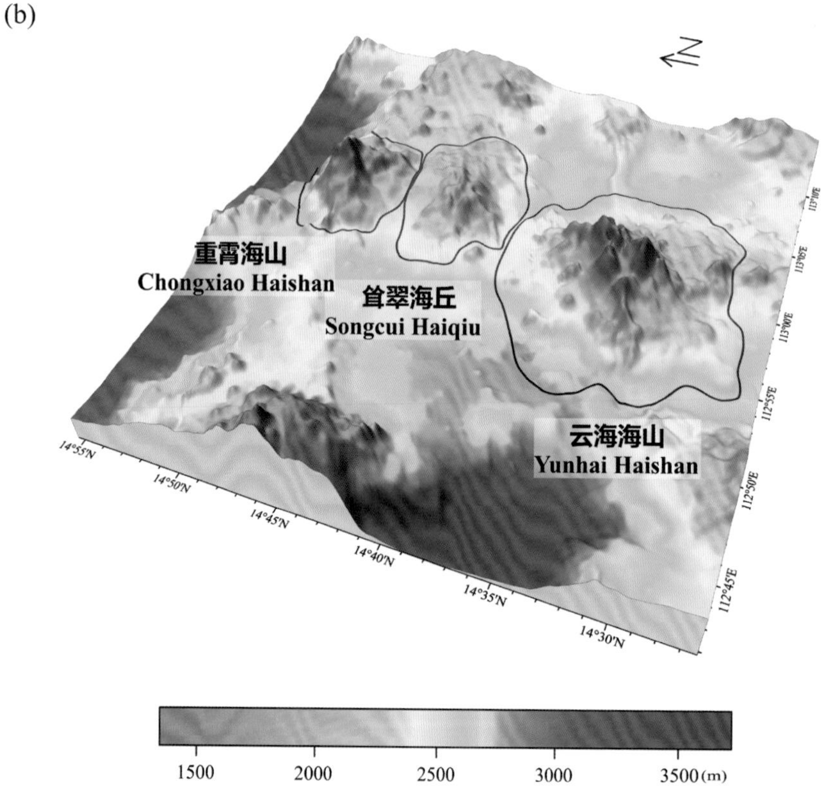

图 4-12　重霄海山、耸翠海丘、云海海山

(a) 海底地形图（等深线间隔 500 米）；(b) 三维海底地形图

Fig.4-12　Chongxiao Haishan, Songcui Haiqiu, Yunhai Haishan

(a) Seafloor topographic map (with contour interval of 500 m)；(b) 3-D seafloor topographic map

5

南海海盆北部海区海底地理实体

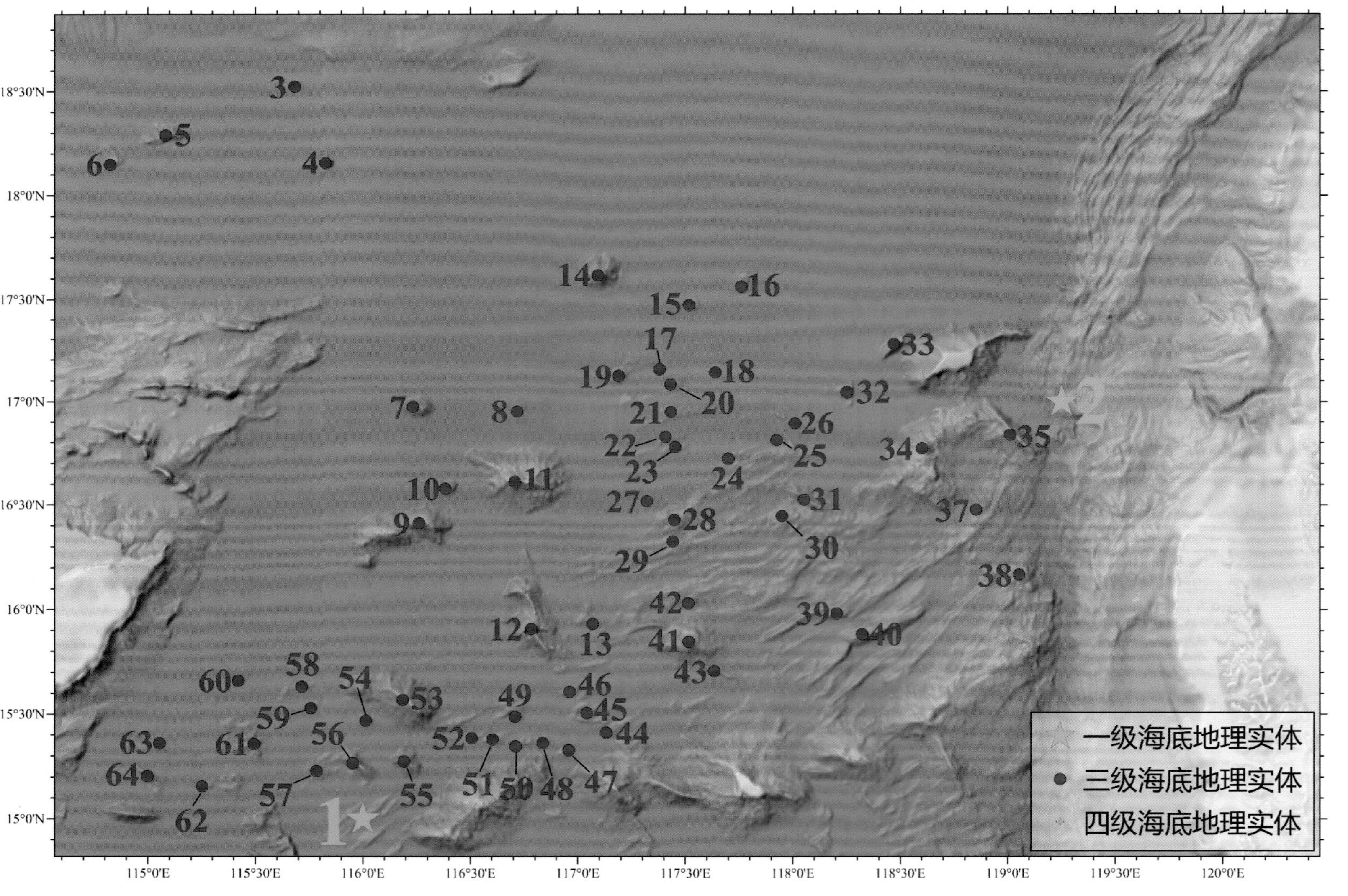

图 5-1 南海海盆北部海区海底地理实体中心点位置示意图，序号含义见表 5-1

Fig.5-1 Location of center coordinates of undersea features in the north of Nanhai Haipen, with the meanings of the serial numbers shown in Tab. 5-1

表 5-1　南海海盆北部海区海底地理实体列表

Tab.5-1　List of undersea features in the north of Nanhai Haipen

序号 No.	标准名称 Standard Name	汉语拼音 Chinese Phonetic Alphabet	类别 Generic Term	中心点坐标 Center Coordinates		实体等级 Order
				纬度 Latitude	经度 Longitude	
1	南海东部岛坡 Nanhaidongbu Daopo	Nánhǎidōngbù Dǎopō	岛坡 Slope	17°00.0'N	119°15.0'E	1
2	南海海盆 Nanhai Haipen	Nánhǎi Hǎipén	海盆 Basin	15°00.0'N	116°00.0'E	1
3	同文海丘 Tongwen Haiqiu	Tóngwén Hǎiqiū	海丘 Hill	18°31.5'N	115°40.9'E	3
4	王祯海山 Wangzhen Haishan	Wángzhēn Hǎishān	海山 Seamount	18°09.5'N	115°49.5'E	3
5	双峰海山 Shuangfeng Haishan	Shuāngfēng Hǎishān	海山 Seamount	18°17.4'N	115°04.9'E	3
6	双峰西海丘群 Shuangfengxi Haiqiuqun	Shuāngfēngxī Hǎiqiūqún	海丘群 Hills	18°08.9'N	114°49.5'E	3
7	一行海山 Yixing Haishan	Yīxíng Hǎishān	海山 Seamount	16°58.6'N	116°13.9'E	3
8	红装海丘 Hongzhuang Haiqiu	Hóngzhuāng Hǎiqiū	海丘 Hill	16°57.2'N	116°43.0'E	3
9	石星海山群 Shixing Haishanqun	Shíxīng Hǎishānqún	海山群 Seamounts	16°24.8'N	116°15.6'E	3
10	石星北海丘 Shixingbei Haiqiu	Shíxīngběi Hǎiqiū	海丘 Hill	16°34.8'N	116°23.0'E	3
11	宪北海山 Xianbei Haishan	Xiànběi Hǎishān	海山 Seamount	16°36.8'N	116°42.4'E	3
12	宪南海山 Xiannan Haishan	Xiànnán Hǎishān	海山 Seamount	15°54.5'N	116°46.9'E	3
13	柱石海丘 Zhushi Haiqiu	Zhùshí Hǎiqiū	海丘 Hill	15°55.9'N	117°04.1'E	3
14	玳瑁海山 Daimao Haishan	Dàimào Hǎishān	海山 Seamount	17°37.0'N	117°05.6'E	3
15	潜龙海丘 Qianlong Haiqiu	Qiánlóng Hǎiqiū	海丘 Hill	17°28.4'N	117°31.0'E	3
16	蛟龙海丘 Jiaolong Haiqiu	Jiāolóng Hǎiqiū	海丘 Hill	17°33.8'N	117°45.6'E	3

续表

序号 No.	标准名称 Standard Name	汉语拼音 Chinese Phonetic Alphabet	类别 Generic Term	中心点坐标 Center Coordinates		实体等级 Order
				纬度 Latitude	经度 Longitude	
17	廷理海丘 Tingli Haiqiu	Tínglǐ Hǎiqiū	海丘 Hill	17°09.3'N	117°22.7'E	3
18	司隶海丘 Sili Haiqiu	Sīlì Hǎiqiū	海丘 Hill	17°08.4'N	117°38.3'E	3
19	飒爽海丘 Sashuang Haiqiu	Sàshuǎng Hǎiqiū	海丘 Hill	17°07.4'N	117°11.3'E	3
20	英姿海丘 Yingzi Haiqiu	Yīngzī Hǎiqiū	海丘 Hill	17°05.0'N	117°25.8'E	3
21	苏颂海丘 Susong Haiqiu	Sūsòng Hǎiqiū	海丘 Hill	16°57.1'N	117°25.9'E	3
22	立法海脊 Lifa Haiji	Lìfǎ Hǎijǐ	海脊 Ridge	16°49.7'N	117°24.4'E	3
23	奇志海丘 Qizhi Haiqiu	Qízhì Hǎiqiū	海丘 Hill	16°46.8'N	117°27.1'E	3
24	叶桂海丘 Yegui Haiqiu	Yèguì Hǎiqiū	海丘 Hill	16°43.4'N	117°41.9'E	3
25	御史海丘 Yushi Haiqiu	Yùshǐ Hǎiqiū	海丘 Hill	16°48.8'N	117°55.4'E	3
26	初照海丘链 Chuzhao Haiqiulian	Chūzhào Hǎiqiūlián	海丘链 Hill Chain	16°53.7'N	118°00.6'E	3
27	司寇海丘 Sikou Haiqiu	Sīkòu Hǎiqiū	海丘 Hill	16°31.2'N	117°19.2'E	3
28	廷尉海丘 Tingwei Haiqiu	Tíngwèi Hǎiqiū	海丘 Hill	16°25.7'N	117°26.9'E	3
29	朱丹溪海丘 Zhudanxi Haiqiu	Zhūdānxī Hǎiqiū	海丘 Hill	16°19.6'N	117°26.5'E	3
30	审刑海丘 Shenxing Haiqiu	Shěnxíng Hǎiqiū	海丘 Hill	16°26.8'N	117°56.9'E	3
31	华佗海山 Huatuo Haishan	Huàtuó Hǎishān	海山 Seamount	16°31.6'N	118°03.1'E	3
32	祖冲之海丘 Zuchongzhi Haiqiu	Zǔchōngzhī Hǎiqiū	海丘 Hill	17°02.8'N	118°15.0'E	3
33	曙光海山 Shuguang Haishan	Shǔguāng Hǎishān	海山 Seamount	17°16.7'N	118°28.1'E	3

续表

序号 No.	标准名称 Standard Name	汉语拼音 Chinese Phonetic Alphabet	类别 Generic Term	中心点坐标 Center Coordinates		实体等级 Order
				纬度 Latitude	经度 Longitude	
34	钱乙海山 Qianyi Haishan	Qiányǐ Hǎishān	海山 Seamount	16°46.4'N	118°36.0'E	3
35	吴有性海山 Wuyouxing Haishan	Wúyǒuxìng Hǎishān	海山 Seamount	16°50.3'N	119°00.6'E	3
36	法明海底峰 Faming Haidifeng	Fǎmíng Hǎidǐfēng	海底峰 Peak	16°55.4'N	118°47.3'E	4
37	孙思邈海山 Sunsimiao Haishan	Sūnsīmiǎo Hǎishān	海山 Seamount	16°28.7'N	118°51.1'E	3
38	新天海山 Xintian Haishan	Xīntiān Hǎishān	海山 Seamount	16°10.2'N	119°03.1'E	3
39	葛洪海丘 Gehong Haiqiu	Gěhóng Hǎiqiū	海丘 Hill	15°59.0'N	118°12.2'E	3
40	李时珍海山 Lishizhen Haishan	Lǐshízhēn Hǎishān	海山 Seamount	15°53.0'N	118°19.3'E	3
41	张仲景海山 Zhangzhongjing Haishan	Zhāngzhòngjǐng Hǎishān	海山 Seamount	15°50.8'N	117°30.9'E	3
42	扁鹊海丘 Bianque Haiqiu	Biǎnquè Hǎiqiū	海丘 Hill	16°01.9'N	117°30.8'E	3
43	基石海丘 Jishi Haiqiu	Jīshí Hǎiqiū	海丘 Hill	15°42.4'N	117°38.0'E	3
44	火石海丘 Huoshi Haiqiu	Huǒshí Hǎiqiū	海丘 Hill	15°24.7'N	117°07.9'E	3
45	飞石海丘 Feishi Haiqiu	Fēishí Hǎiqiū	海丘 Hill	15°30.3'N	117°02.5'E	3
46	鸟蛤海丘 Niaoge Haiqiu	Niǎogé Hǎiqiū	海丘 Hill	15°36.4'N	116°57.7'E	3
47	篮蛤海丘 Lange Haiqiu	Lángé Hǎiqiū	海丘 Hill	15°19.7'N	116°57.4'E	3
48	舟蚶海丘 Zhouhan Haiqiu	Zhōuhān Hǎiqiū	海丘 Hill	15°21.7'N	116°50.1'E	3
49	帽蚶海丘 Maohan Haiqiu	Màohān Hǎiqiū	海丘 Hill	15°29.3'N	116°42.4'E	3
50	青蚶海丘 Qinghan Haiqiu	Qīnghān Hǎiqiū	海丘 Hill	15°20.8'N	116°42.6'E	3

续表

序号 No.	标准名称 Standard Name	汉语拼音 Chinese Phonetic Alphabet	类别 Generic Term	中心点坐标 Center Coordinates 纬度 Latitude	 经度 Longitude	实体等级 Order
51	椿蚶海丘 Chunhan Haiqiu	Chūnhān Hǎiqiū	海丘 Hill	15°22.7'N	116°36.2'E	3
52	贻贝海脊 Yibei Haiji	Yíbèi Hǎijǐ	海脊 Ridge	15°23.1'N	116°30.4'E	3
53	涨中海山 Zhangzhong Haishan	Zhǎngzhōng Hǎishān	海山 Seamount	15°34.1'N	116°11.1'E	3
54	竹溪海丘 Zhuxi Haiqiu	Zhúxī Hǎiqiū	海丘 Hill	15°28.1'N	116°00.8'E	3
55	兰溪海山 Lanxi Haishan	Lánxī Hǎishān	海山 Seamount	15°16.5'N	116°11.4'E	3
56	张先海山 Zhangxian Haishan	Zhāngxiān Hǎishān	海山 Seamount	15°16.0'N	115°57.2'E	3
57	青溪海丘 Qingxi Haiqiu	Qīngxī Hǎiqiū	海丘 Hill	15°13.7'N	115°47.0'E	3
58	鼓角海丘 Gujiao Haiqiu	Gǔjiǎo Hǎiqiū	海丘 Hill	15°37.9'N	115°42.9'E	3
59	壁垒海丘 Bilei Haiqiu	Bìlěi Hǎiqiū	海丘 Hill	15°31.7'N	115°45.4'E	3
60	旌旗海丘 Jingqi Haiqiu	Jīngqí Hǎiqiū	海丘 Hill	15°39.6'N	115°25.2'E	3
61	西涧海脊 Xijian Haiji	Xījiàn Hǎijǐ	海脊 Ridge	15°21.4'N	115°29.6'E	3
62	荆溪海丘 Jingxi Haiqiu	Jīngxī Hǎiqiū	海丘 Hill	15°09.3'N	115°15.1'E	3
63	泥蚶海丘 Nihan Haiqiu	Níhān Hǎiqiū	海丘 Hill	15°21.6'N	115°03.2'E	3
64	双溪海脊 Shuangxi Haiji	Shuāngxī Hǎijǐ	海脊 Ridge	15°12.2'N	114°59.8'E	3

5.1 南海东部岛坡

标准名称 Standard Name	南海东部岛坡 Nanhaidongbu Daopo	类别 Generic Term	岛坡 Slope
中心点坐标 Center Coordinates	17°00.0′N, 119°15.0′E	规模（千米 × 千米） Dimension（km × km）	1000 × 130
最小水深（米） Min Depth (m)	240	最大水深（米） Max Depth (m)	4000
地理实体描述 Feature Description	南海东部岛坡北起台湾岛南部，南至民都洛岛外缘，西以马尼拉海沟为界，东部紧邻菲律宾吕宋岛，为南北方向延伸的长条状地形区，其南北长 1000 千米，东西宽 50 ～ 200 千米。该区地形的典型特征是海脊 / 海岭与海槽发育，自北至南发育恒春海脊、吕宋海脊、北吕宋海槽、南吕宋海槽等。海脊由多个小型海脊组合而成，外形呈现麻花状，海底坡度变化剧烈（图 5−2）。 Nanhaidongbu Daopo starts from the south of Taiwan Dao on the north and ends at the edge of Mindoro Island on the south. The Slope borders on Manila Trench on the west and is adjacent to Luzon Island in Philippines on the east. It is a long strip in the direction of S−N with a length of 1000 km from south to north, and a width of 50~200 km from east to west. The seafloor topography of the area is typically characterized by ridges and troughs, including Hengchun Haiji, Luzon Ridge, North Luzon Trough, and South Luzon Trough developed from north to south. The ridge area consists of multiple small-scale ridges and is in the shape of a twist dough, with dramatic change of slope gradient on the seafloor (Fig.5−2).		
命名由来 Origin of Name	该岛坡位于南海东部，因此得名。 The Slope is located in the eastern part of the the South China Sea, and “Dongbu” means eastern part in Chinese, so the word “Nanhaidongbu” was used to name the Slope.		

(a)

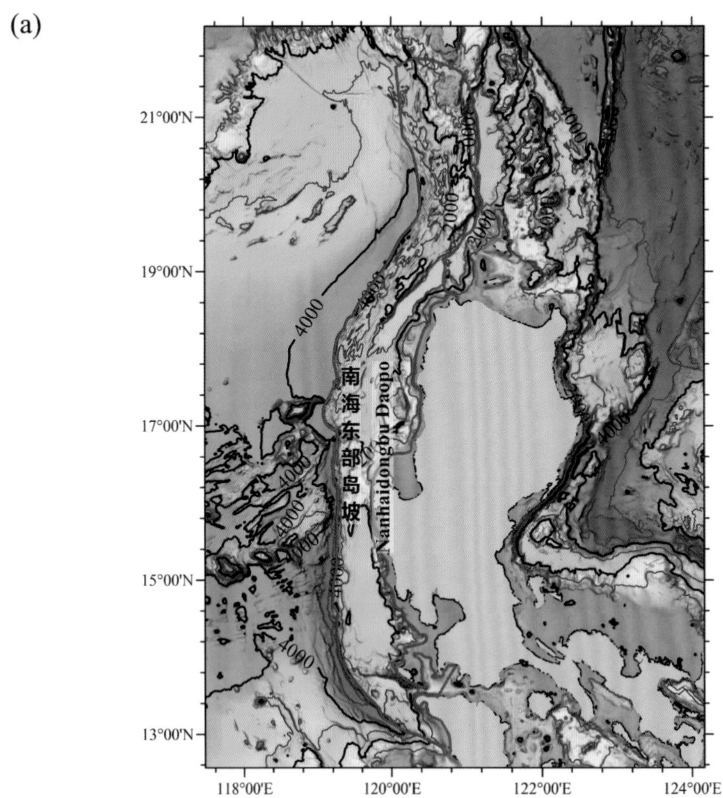

(b)

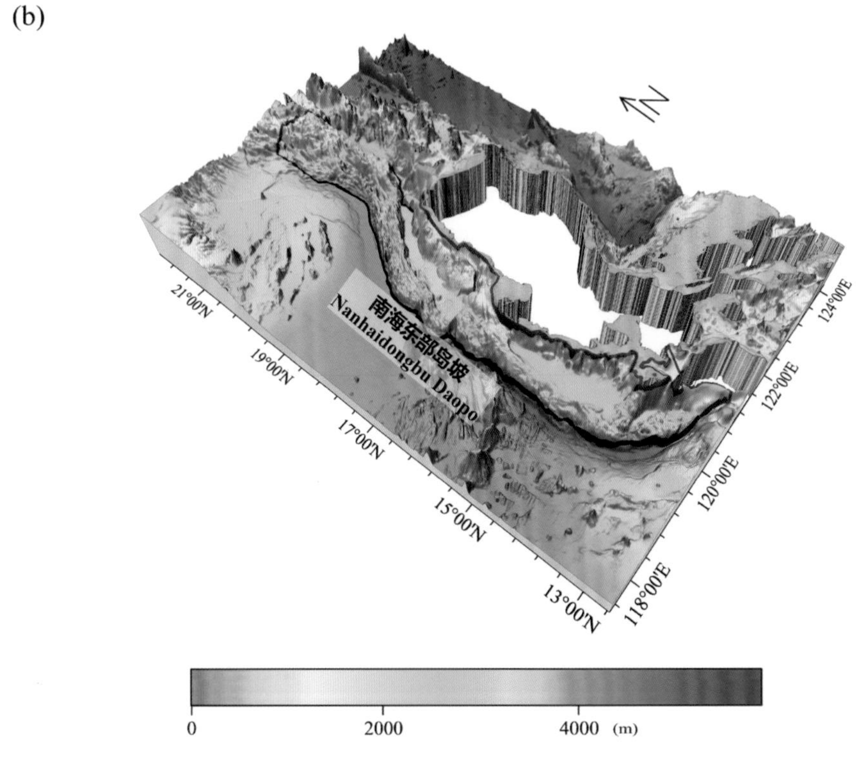

0 2000 4000 (m)

图 5-2 南海东部岛坡

(a) 海底地形图（等深线间隔 500 米）；(b) 三维海底地形图

Fig.5-2 Nanhaidongbu Daopo

(a) Seafloor topographic map (with contour interval of 500 m); (b) 3-D seafloor topographic map

5.2 南海海盆

标准名称 Standard Name	南海海盆 Nanhai Haipen	类别 Generic Term	海盆 Basin
中心点坐标 Center Coordinates	15°00.0′N, 116°00.0′E	规模（千米 × 千米） Dimension（km × km）	1450 × 730
最小水深（米） Min Depth (m)	3400	最大水深（米） Max Depth (m)	4400
地理实体描述 Feature Description	南海海盆位于南海中部，四周被地形复杂多变的陆坡（岛坡）包围，地形低陷而平缓，水深 3400 ~ 4400 米，总体呈北东—南西向菱形展布，长约 2380 千米，面积 44.6×10^4 平方千米。以“珍贝—黄岩海山链”为界，分为南海海盆北部和南海海盆南部两个部分，海底分别自南—北两侧向中央微微倾斜。海盆内发育大量的链状海山、海山和海丘等（图 5-3）。 Nanhai Haipen is located in the middle of the South China Sea and is surrounded by continental slopes with complicate and variable topographies. It has a sunk, flat and gentle terrain and with depth of 3400~4400 m. Generally, the Basin is in the shape of a diamond in the direction of NE−SW, with a length of about 2380 km and an area of 44.6×10^4 km^2. The Basin is divided into southern part and northern part by “Zhenbei−Huangyan Haishanlian”. The seafloor slightly tilts towards the center from the south and north side. A large number of hills, hill chains and seamounts have developed within the Basin (Fig.5−3).		
命名由来 Origin of Name	1986 年，国务院、外交部和中国地名委员会批准了前地矿部第二海洋地质调查大队（现广州海洋地质调查局）对南海 22 个海底地理实体进行的命名，“南海中央海盆”是这 22 个海底地理实体名称之一。但在学术定义上，“南海中央海盆”仅指范围较小的海域，“南海海盆”可准确表达地貌特征，故更名。 In 1986, the State Council, the Ministry of Foreign Affairs and Chinese Toponymy Committee approved 22 undersea feature names in Nanhai named by former Second Marine Geological Survey Brigade of the Ministry of Geology and Mineral Resources (present Guangzhou Marine Geological Survey of China Geological Survey). Nanhaizhongyang Haipen is one of the 22 undersea feature names. However, in the academic definition, Nanhaizhongyang Haipen only refers to a smaller range of sea area, and Nanhai Haipen can accurately express the geomorphological characteristics, so the name was changed.		

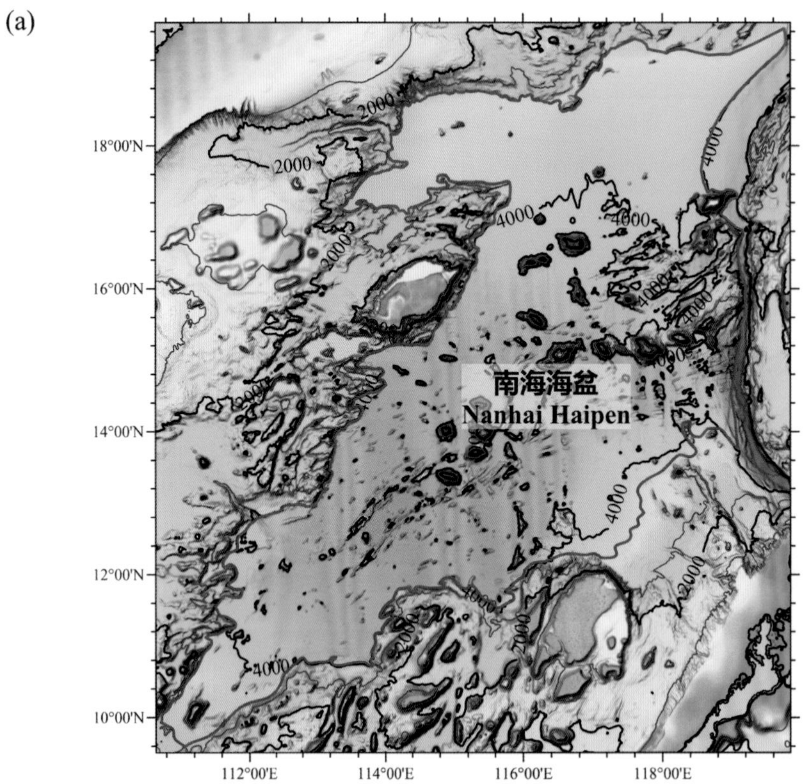

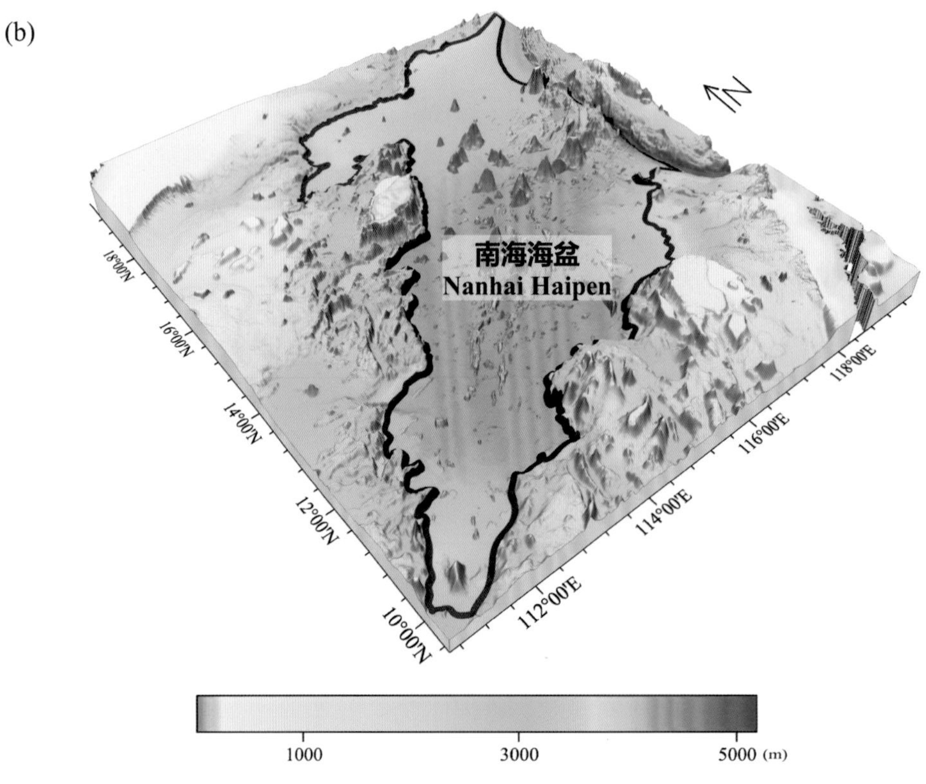

图 5-3 南海海盆

(a) 海底地形图（等深线间隔 1000 米）；(b) 三维海底地形图

Fig.5-3 Nanhai Haipen

(a) Seafloor topographic map (with contour interval of 1000 m)；(b) 3-D seafloor topographic map

5.3 同文海丘

标准名称 Standard Name	同文海丘 Tongwen Haiqiu	类别 Generic Term	海丘 Hill
中心点坐标 Center Coordinates	18°31.5′N, 115°40.9′E	规模（千米 × 千米） Dimension（km × km）	6 × 5
最小水深（米） Min Depth (m)	3500	最大水深（米） Max Depth (m)	3750
地理实体描述 Feature Description	同文海丘位于南海海盆西北部，王祯海山以北，平面形态呈圆形（图 5–4）。 Tongwen Haiqiu is located in the northwest of Nanhai Haipen and is to the north of Wangzhen Haishan, with a circular planform (Fig.5–4).		
命名由来 Origin of Name	以中国古代文学作品中的短语进行地名的团组化命名。该海底地名的专名取词自《礼记》：“书同文”。“同文”即用同一种文字。 A group naming of undersea features after the phrases in ancient Chinese literatures. The specific term of this undersea feature name “Tongwen”, literally meaning the unified languages, is derived from the sentence “Write with unified languages” in *The Book of Rites*.		

5.4 王祯海山

标准名称 Standard Name	王祯海山 Wangzhen Haishan	类别 Generic Term	海山 Seamount
中心点坐标 Center Coordinates	18°09.5′N, 115°49.5′E	规模（千米 × 千米） Dimension（km × km）	10 × 8
最小水深（米） Min Depth (m)	2612	最大水深（米） Max Depth (m)	3819
地理实体描述 Feature Description	王祯海山位于南海海盆西北部，同文海丘以南，平面形态呈圆形，其北侧延伸出小山脊（图 5–4）。 Wangzhen Haishan is located in the northwest of Nanhai Haipen and is to the south of Tongwen Haiqiu, with a circular planform. A small ridge is extending from its north side (Fig.5–4).		
命名由来 Origin of Name	以我国古代科学家、医学家和文人的名字进行地名的团组化命名。该海山以元朝著名科学家王祯命名，纪念他对中国古代农业发展做出的巨大贡献。 A group naming of undersea features after the names of ancient Chinese scientists, medical scientists and literati. The Seamount is named after Wang Zhen (1271–1368 A.D.), a famous scientist in the Yuan Dynasty (1271–1368 A.D.), to commemorate his outstanding contributions to the agricultural development in ancient China.		

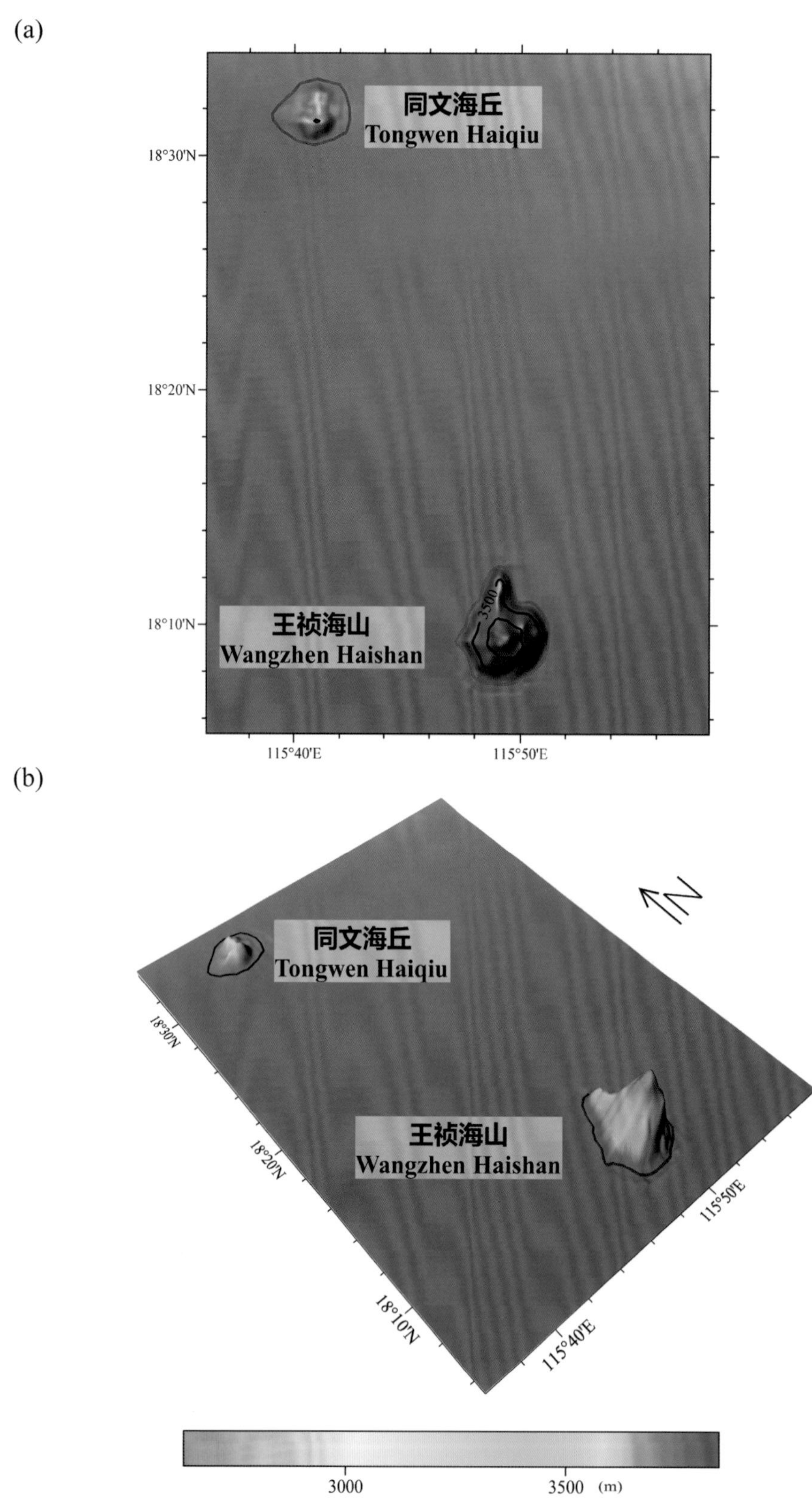

图 5-4　同文海丘、王祯海山

(a) 海底地形图（等深线间隔 500 米）；(b) 三维海底地形图

Fig.5-4　Tongwen Haiqiu, Wangzhen Haishan

(a) Seafloor topographic map (with contour interval of 500 m)；(b) 3-D seafloor topographic map

5.5 双峰海山

标准名称 Standard Name	双峰海山 Shuangfeng Haishan	类别 Generic Term	海山 Seamount
中心点坐标 Center Coordinates	18°17.4′N, 115°04.9′E	规模（千米 × 千米） Dimension（km × km）	22 × 12
最小水深（米） Min Depth (m)	2380	最大水深（米） Max Depth (m)	3730
地理实体描述 Feature Description	双峰海山位于南海海盆西北部，平面形态呈长条形。海山的顶部发育两座峰，其中西侧峰较高（水深约 2400 米），东侧峰较矮（水深约 3000 米）（图 5-5）。 Shuangfeng Haishan is located in the northwest of Nanhai Haipen with a planform in the shape of a long strip. Two peaks have developed on its top, of which, the west peak is higher with depth of about 2400 m and the east peak is lower with depth of about 3000 m (Fig.5-5).		
命名由来 Origin of Name	1986 年，国务院、外交部和中国地名委员会批准了前地矿部第二海洋地质调查大队（现广州海洋地质调查局）对南海 22 个海底地理实体进行的命名，“双峰海山”是这 22 个海底地理实体名称之一。 In 1986, the State Council, the Ministry of Foreign Affairs and Chinese Toponymy Committee approved 22 undersea feature names in Nanhai named by former Second Marine Geological Survey Brigade of the Ministry of Geology and Mineral Resources (present Guangzhou Marine Geological Survey of China Geological Survey). Shuangfeng Haishan is one of the 22 undersea feature names. Named from its shape, as this feature has “two peaks” which translates as Shuangfeng in Chinese.		

5.6 双峰西海丘群

标准名称 Standard Name	双峰西海丘群 Shuangfengxi Haiqiuqun	类别 Generic Term	海丘群 Hills
中心点坐标 Center Coordinates	18°08.9′N, 114°49.5′E	规模（千米 × 千米） Dimension（km × km）	11 × 4
最小水深（米） Min Depth (m)	3030	最大水深（米） Max Depth (m)	3660
地理实体描述 Feature Description	双峰西海丘群位于南海海盆西北部，平面形态呈不规则多边形（图 5-5）。 Shuangfengxi Haiqiuqun is located in the northwest of Nanhai Haipen, with a planform in the shape of an irregular polygon (Fig.5-5).		
命名由来 Origin of Name	该海丘位于双峰海山以西，因此得名。 These Hills is located to the west of Shuangfeng Haishan, and “Xi” means west in Chinese, so the word “Shuangfengxi” was used to name these Hills.		

(a)

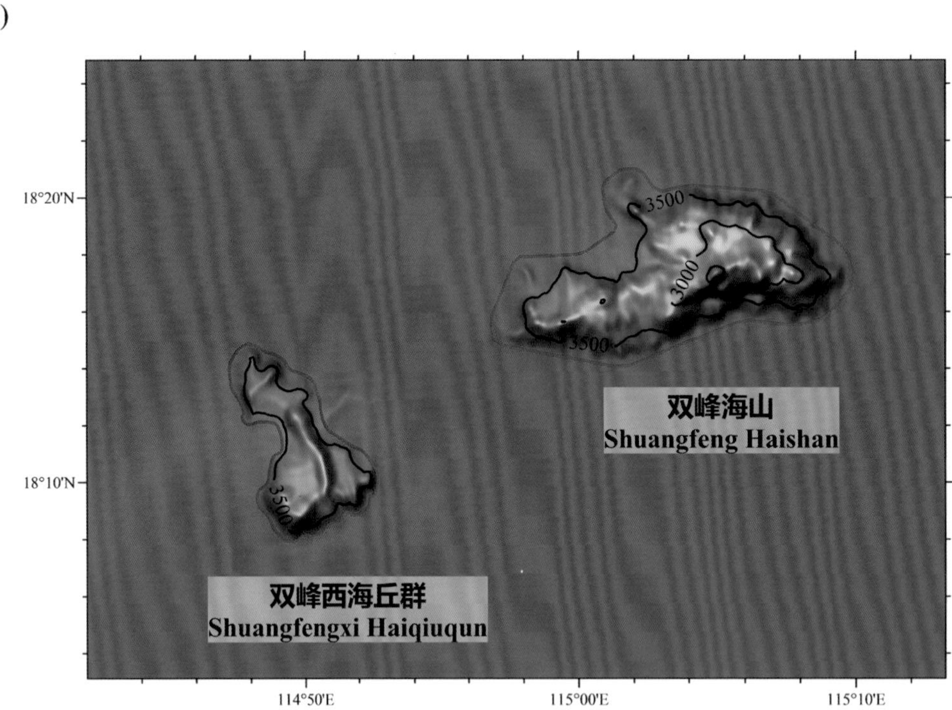

(b)

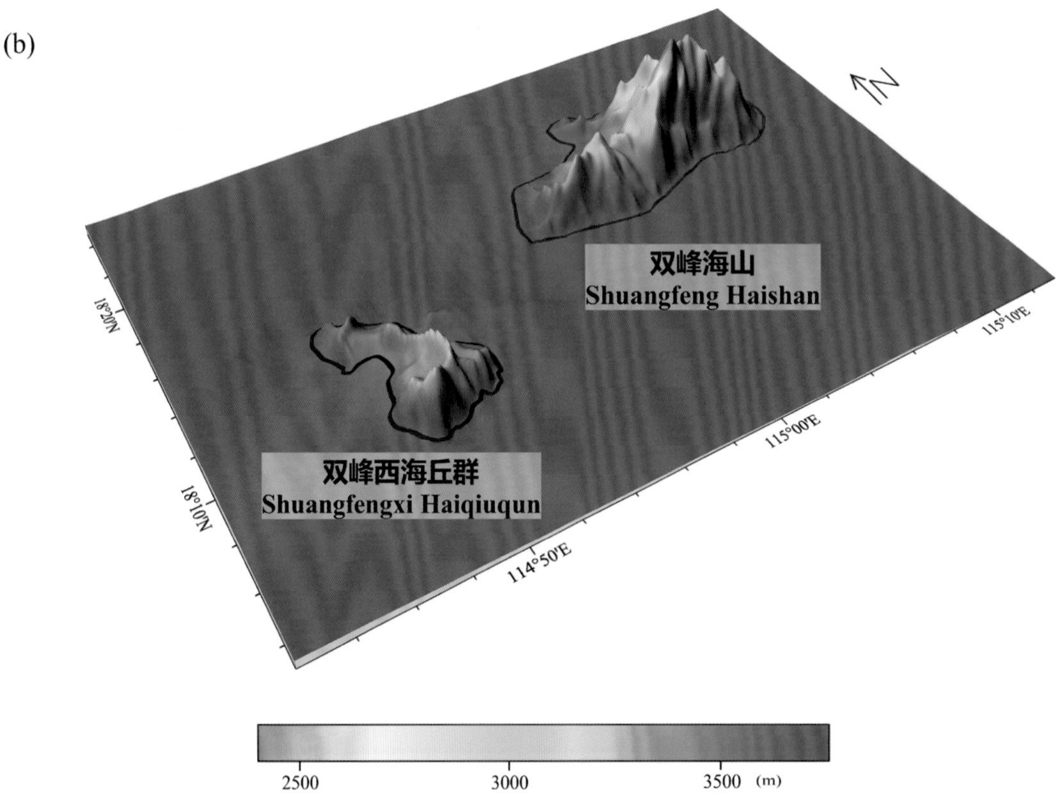

图 5-5　双峰海山、双峰西海丘群

(a) 海底地形图（等深线间隔 500 米）；(b) 三维海底地形图

Fig.5-5　Shuangfeng Haishan, Shuangfengxi Haiqiuqun

(a) Seafloor topographic map (with contour interval of 500 m)；(b) 3-D seafloor topographic map

5.7 一行海山

标准名称 Standard Name	一行海山 Yixing Haishan	类别 Generic Term	海山 Seamount
中心点坐标 Center Coordinates	16°58.6′N, 116°13.9′E	规模（千米 × 千米） Dimension（km × km）	15 × 11
最小水深（米） Min Depth (m)	2307	最大水深（米） Max Depth (m)	4045
地理实体描述 Feature Description	一行海山位于南海海盆北部，石星海山群以北，平面形态近似菱形，其山脊线呈东—西走向（图 5-6）。 Yixing Haishan is located in the north of Nanhai Haipen and is to the north of Shixing Haishanqun, with a planform in the shape of a rough diamond and with a ridge line in the direction of E−W (Fig.5−6).		
命名由来 Origin of Name	以我国古代科学家、医学家和文人的名字进行地名的团组化命名。该海山以唐朝著名佛学家和天文学家一行命名，纪念他在天文学上的重要贡献。 A group naming of undersea features the names of ancient Chinese scientists, medical scientists and literati. The Seamount is named after Yi Xing (683−727 A.D.), a famous Buddhist and astronomer in the Tang Dynasty (618−907 A.D.), to commemorate his important contribution to the astronomy. He translated a variety of Indian Buddhist scriptures and became the leader of Buddhism Tantric.		

5.8 红装海丘

标准名称 Standard Name	红装海丘 Hongzhuang Haiqiu	类别 Generic Term	海丘 Hill
中心点坐标 Center Coordinates	16°57.2′N, 116°43.0′E	规模（千米 × 千米） Dimension（km × km）	6 × 5
最小水深（米） Min Depth (m)	3650	最大水深（米） Max Depth (m)	4000
地理实体描述 Feature Description	红装海丘位于南海海盆北部，宪北海山以北，平面形态近似圆形（图 5-6）。 Hongzhuang Haiqiu is located in the north of Nanhai Haipen and is to the north of Xianbei Haishan with a nearly circular planform (Fig.5−6).		
命名由来 Origin of Name	以毛泽东诗词作品中的词进行地名的团组化命名。该海底地名的专名取词自毛泽东的《七绝・为女民兵题照》："不爱红装爱武装"。"红装"即女性的红色装饰。 A group naming of undersea features after the words in Mao Zedong's poems. The specific term of this undersea feature name "Hongzhuang", means gay feminine attire, is derived from the poetic lines "They love bearing arms, not the gay feminine attire" in the poem *Inscription on Photography of Militia Women- A Verse of Four Seven-character Line* by Mao Zedong (1893−1976).		

5.9 石星海山群

标准名称 Standard Name	石星海山群 Shixing Haishanqun	类别 Generic Term	海山群 Seamounts
中心点坐标 Center Coordinates	16°24.8′N, 116°15.6′E	规模（千米 × 千米） Dimension（km × km）	52 × 20
最小水深（米） Min Depth (m)	1371	最大水深（米） Max Depth (m)	4195
地理实体描述 Feature Description	石星海山群位于南海海盆北部，平面形态近似长条形。海山的顶部发育 3 座峰，其中北侧两座峰较高（水深分别为 1370 米和 1530 米），南侧的山峰较矮（水深约 2600 米）（图 5-6）。 Shixing Haishanqun is located in the north of Nanhai Haipen and its planform is similar to a long strip. Three peaks have developed on its top, of which, the two peaks on the north side are higher (with depth of 1370 m and 1530 m respectively), and the peaks on the south side are relatively lower (with depth of about 2600 m) (Fig.5-6).		
命名由来 Origin of Name	该海山群的名字来自于附近的中沙群岛，在中国明朝著名航海家郑和所使用的海图上被标记为石星。1986 年，国务院、外交部和中国地名委员会批准了前地矿部第二海洋地质调查大队（现广州海洋地质调查局）对南海 22 个海底地理实体进行的命名，“石星海山”是这 22 个海底地理实体名称之一。2017 年，SCUFN 第 30 次会议上认定该地理实体为海山群，故更名为“石星海山群”。 These Seamounts are named from the nearby Zhongsha Qundao (Zhongsha Islands), which were marked as Shixing on the nautical chart of Zhenghe (1371-1433 A.D.), the famous Chinese navigator during the Ming Dynasty (1368-1644 A.D.). In 1986, the State Council, the Ministry of Foreign Affairs and Chinese Toponymy Committee approved 22 undersea feature names in Nanhai named by former Second Marine Geological Survey Brigade of the Ministry of Geology and Mineral Resources (present Guangzhou Marine Geological Survey of China Geological Survey). Shixing Haishan is one of the 22 undersea feature names. In 2017, at the 30th SCUFN meeting , the generic term of the feature was identified as a seamounts (Haishanqun in Chinese), so it was renamed as “Shixing Haishanqun”.		

5.10 石星北海丘

标准名称 Standard Name	石星北海丘 Shixingbei Haiqiu	类别 Generic Term	海丘 Hill
中心点坐标 Center Coordinates	16°34.8′N, 116°23.0′E	规模（千米 × 千米） Dimension（km × km）	11 × 7
最小水深（米） Min Depth (m)	3400	最大水深（米） Max Depth (m)	4100
地理实体描述 Feature Description	石星北海丘位于南海海盆北部，平面形态近似椭圆形，其顶部发育一火山口（图 5–6）。 Shixingbei Haiqiu is located in the north of Nanhai Haipen with a nearly oval planform, and has a volcanic crater developed on the top (Fig.5–6).		
命名由来 Origin of Name	该海丘位于石星海山群以北，因此得名。 The Hill is located to the north of Shixing Haishanqun, and “Bei” means north in Chinese, so the word “Shixingbei” was used to name the Hill.		

5.11 宪北海山

标准名称 Standard Name	宪北海山 Xianbei Haishan	类别 Generic Term	海山 Seamount
中心点坐标 Center Coordinates	16°36.8′N, 116°42.4′E	规模（千米 × 千米） Dimension（km × km）	60 × 46
最小水深（米） Min Depth (m)	216	最大水深（米） Max Depth (m)	4118
地理实体描述 Feature Description	宪北海山位于南海海盆北部，平面形态近似扇贝形，其山脊线呈北东—南西走向（图 5–6）。 Xianbei Haishan is located in the north of Nanhai Haipen and its planform is similar to a scallop in shape, with ridge line of the Seamount in the direction of NE–SW (Fig.5–6).		
命名由来 Origin of Name	1986 年，国务院、外交部和中国地名委员会批准了前地矿部第二海洋地质调查大队（现广州海洋地质调查局）对南海 22 个海底地理实体进行的命名，“宪北海山”是这 22 个海底地理实体名称之一。 In 1986, the State Council, the Ministry of Foreign Affairs and Chinese Toponymy Committee approved 22 undersea feature names in Nanhai named by former Second Marine Geological Survey Brigade of the Ministry of Geology and Mineral Resources (present Guangzhou Marine Geological Survey of China Geological Survey). Xianbei Haishan is one of the 22 undersea feature names. Named from its geographical location, North of Xianfa Shoal (Xianfa Ansha, in Chinese). “Bei” means North in Chinese. Xianfa Ansha was thus named by the China Committee on Geographical Names in 1986.		

(a)

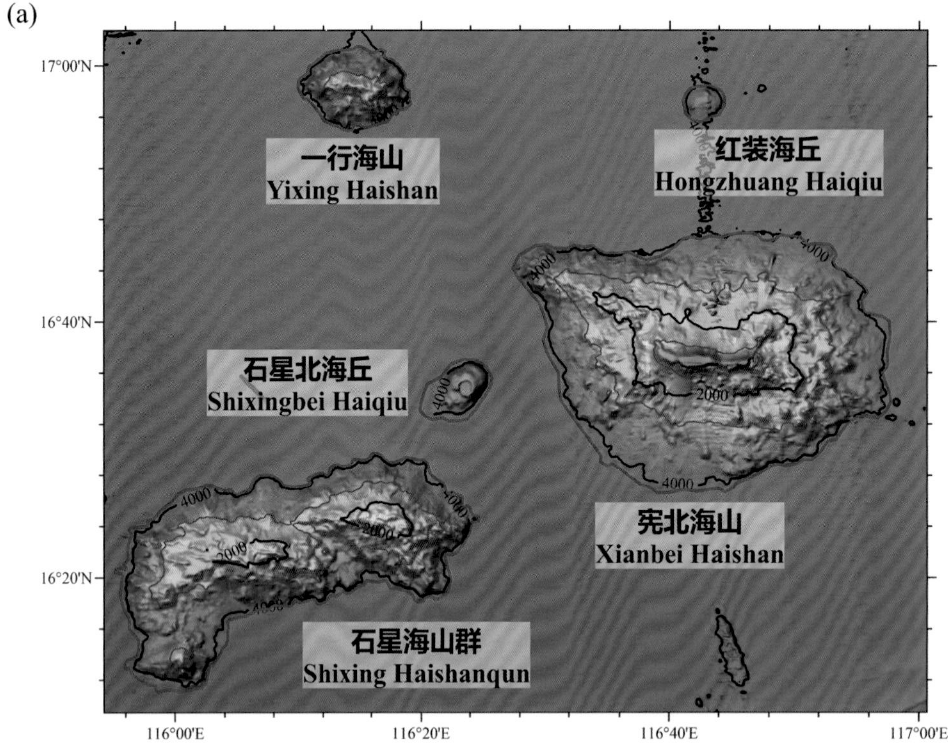

(b)

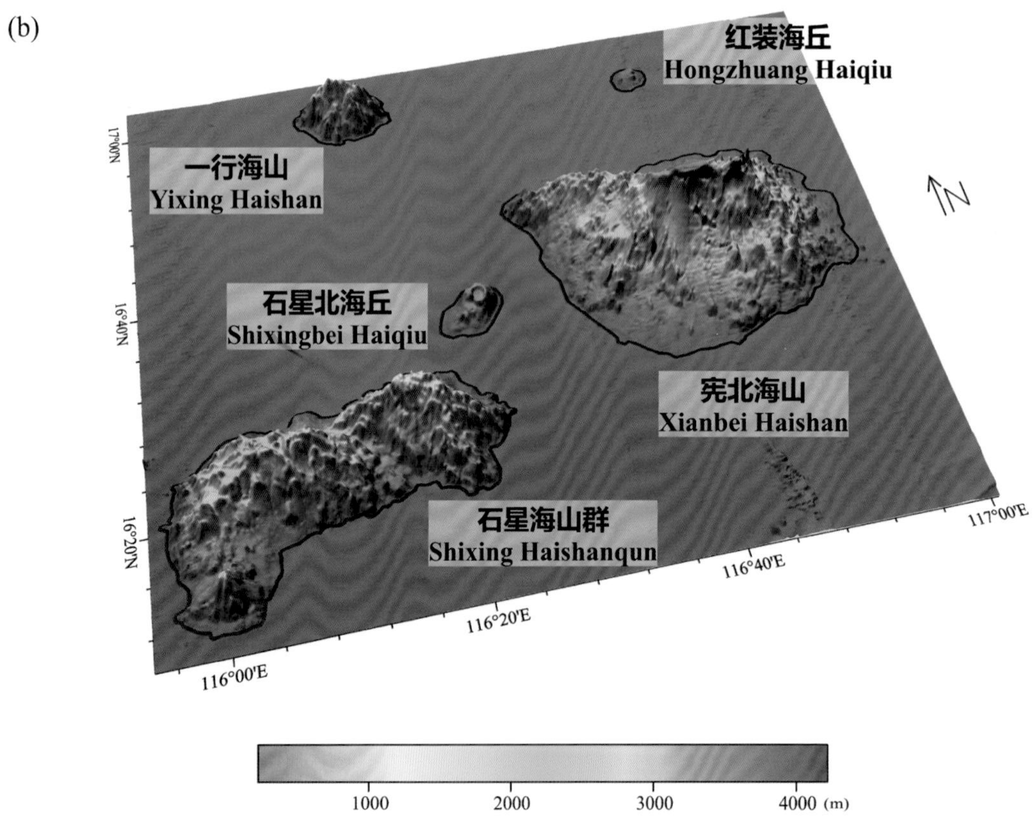

图 5-6　一行海山、红装海丘、石星海山群、石星北海丘、宪北海山

(a) 海底地形图（等深线间隔 1000 米）；(b) 三维海底地形图

Fig.5-6　Yixing Haishan, Hongzhuang Haiqiu, Shixing Haishanqun, Shixingbei Haiqiu, Xianbei Haishan

(a) Seafloor topographic map (with contour interval of 1000 m)；(b) 3-D seafloor topographic map

5.12 宪南海山

标准名称 Standard Name	宪南海山 Xiannan Haishan	类别 Generic Term	海山 Seamount
中心点坐标 Center Coordinates	15°54.5′N, 116°46.9′E	规模（千米 × 千米） Dimension（km × km）	55 × 28
最小水深（米） Min Depth (m)	1544	最大水深（米） Max Depth (m)	4200
地理实体描述 Feature Description	宪南海山位于南海海盆北部，平面形态近似长条形，其山脊线呈近南—北走向。海山顶部发育两座峰，其中南侧峰较高（水深约 1530 米），北侧峰较矮（水深约 1980 米）（图 5-7）。 Xiannan Haishan is located in the north of Nanhai Haipen and the planform is similar to a long strip, with ridge line of the Seamount in the direction of S−N. There are two peaks developed on the top of the Seamount, of which, the southern peak is relatively higher (with depth of about 1530 m) and the northern peak is relative lower (with water depth about 1980 m) (Fig.5−7).		
命名由来 Origin of Name	1986 年，国务院、外交部和中国地名委员会批准了前地矿部第二海洋地质调查大队（现广州海洋地质调查局）对南海 22 个海底地理实体进行的命名，"宪南海山"是这 22 个海底地理实体名称之一。 In 1986, the State Council, the Ministry of Foreign Affairs and Chinese Toponymy Committee approved 22 undersea feature names in Nanhai named by former Second Marine Geological Survey Brigade of the Ministry of Geology and Mineral Resources (present Guangzhou Marine Geological Survey of China Geological Survey). Xiannan Haishan is one of the 22 undersea feature names.		

5.13 柱石海丘

标准名称 Standard Name	柱石海丘 Zhushi Haiqiu	类别 Generic Term	海丘 Hill
中心点坐标 Center Coordinates	15°55.9′N, 117°04.1′E	规模（千米 × 千米） Dimension（km × km）	15 × 10
最小水深（米） Min Depth (m)	3700	最大水深（米） Max Depth (m)	4100
地理实体描述 Feature Description	柱石海丘位于南海海盆北部，平面形态近似三叶草形（图 5-7）。 Zhushi Haiqiu is located in the north of Nanhai Haipen and the planform is similar to the shape of a cloverleaf (Fig.5−7).		
命名由来 Origin of Name	以中国古代文学作品中的短语进行地名的团组化命名。该海底地名的专名取词自唐代诗人元稹的《有鸟》诗之九："大厦虽存柱石倾，暗啮栋梁成蠹木"。"柱石"即顶梁的柱子和垫柱的础石。 A group naming of undersea features after the phrases in ancient Chinese literatures. The specific term of this undersea feature name "Zhushi", means cornerstone for pillar and column, is derived from the poetic lines "The cornerstones of columns are tumbledown though the mansion still remains, the pillars have been gnawed and rotten without notice". in *Some Bird* by Yuan Zhen (779−831 A.D.), a poet in the Tang Dynasty (618−907 A.D.).		

(a)

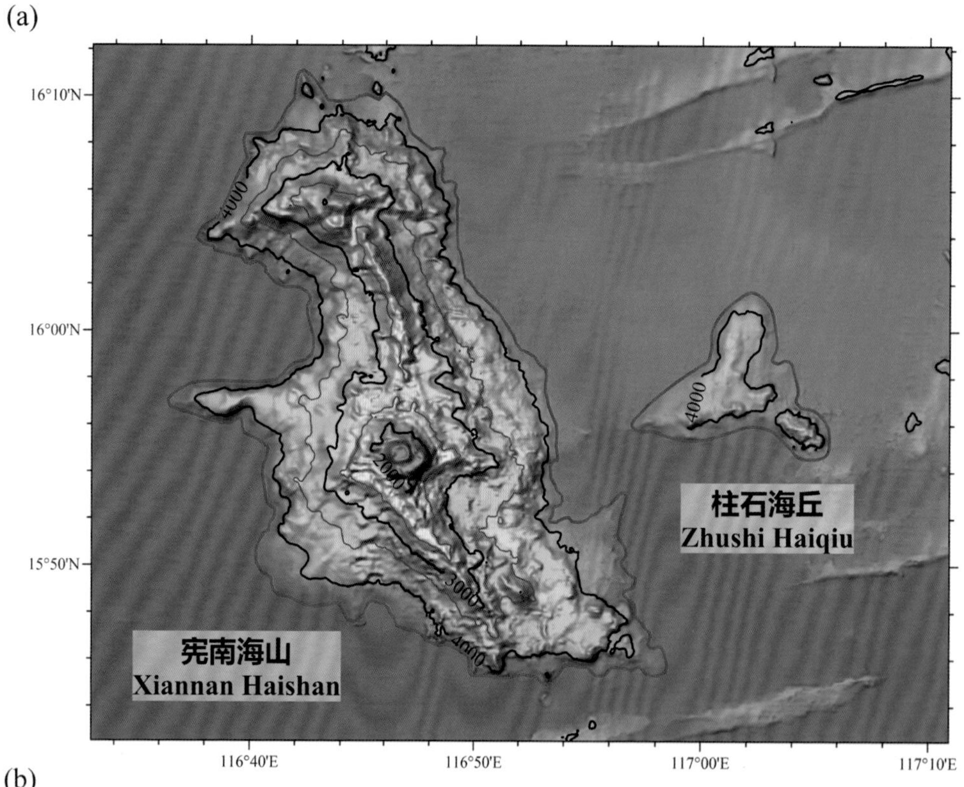

(b)

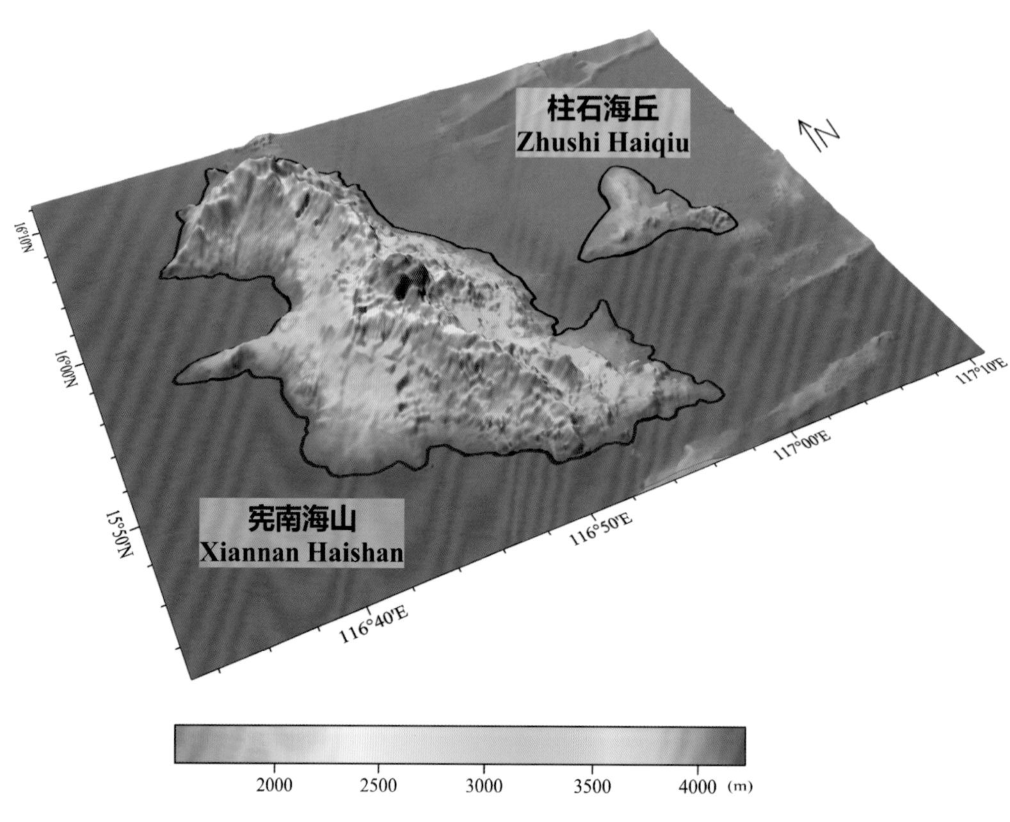

图 5-7 宪南海山、柱石海丘

(a) 海底地形图（等深线间隔 500 米）；(b) 三维海底地形图

Fig.5-7 Xiannan Haishan, Zhushi Haiqiu

(a) Seafloor topographic map (with contour interval of 500 m)；(b) 3-D seafloor topographic map

5.14 玳瑁海山

标准名称 Standard Name	玳瑁海山 Daimao Haishan	类别 Generic Term	海山 Seamount
中心点坐标 Center Coordinates	17°37.0′N, 117°05.6′E	规模（千米 × 千米） Dimension（km × km）	23 × 21
最小水深（米） Min Depth (m)	1648	最大水深（米） Max Depth (m)	3996
地理实体描述 Feature Description	玳瑁海山位于南海海盆北部，平面形态近似圆形（图 5-8）。 Daimao Haishan is located in the north of Nanhai Haipen with a nearly circular planform (Fig.5-8).		
命名由来 Origin of Name	1986 年，国务院、外交部和中国地名委员会批准了前地矿部第二海洋地质调查大队（现广州海洋地质调查局）对南海 22 个海底地理实体进行的命名，“玳瑁海山”是这 22 个海底地理实体名称之一。 In 1986, the State Council, the Ministry of Foreign Affairs and Chinese Toponymy Committee approved 22 undersea feature names in Nanhai named by former Second Marine Geological Survey Brigade of the Ministry of Geology and Mineral Resources (present Guangzhou Marine Geological Survey of China Geological Survey). Daimao Haishan is one of the 22 undersea feature names. “Daimao” is a kind of creatures in the South China Sea.		

5.15 潜龙海丘

标准名称 Standard Name	潜龙海丘 Qianlong Haiqiu	类别 Generic Term	海丘 Hill
中心点坐标 Center Coordinates	17°28.4′N, 117°31.0′E	规模（千米 × 千米） Dimension（km × km）	6 × 6
最小水深（米） Min Depth (m)	3750	最大水深（米） Max Depth (m)	3980
地理实体描述 Feature Description	潜龙海丘位于南海海盆北部，平面形态呈圆形。海丘顶部发育火山口（图 5-8）。 Qianlong Haiqiu is located in the north of Nanhai Haipen with a circular planform. The top of the Hill is developed with a volcanic crater (Fig.5-8).		
命名由来 Origin of Name	以我国自主研发的 AUV“潜龙一号”命名。 The Hill is named after AUV “Qianlong No.1” that was independently researched and developed by China.		

5.16 蛟龙海丘

标准名称 Standard Name	蛟龙海丘 Jiaolong Haiqiu	类别 Generic Term	海丘 Hill
中心点坐标 Center Coordinates	17°33.8′N, 117°45.6′E	规模（千米 × 千米） Dimension（km × km）	10 × 8
最小水深（米） Min Depth (m)	3330	最大水深（米） Max Depth (m)	3950
地理实体描述 Feature Description	蛟龙海丘位于南海海盆北部，平面形态近似椭圆形。海丘顶部发育两座峰，其中南侧山峰较高（水深约 3330 米），为一塌陷火山口，北侧峰较矮（水深约 3700 米）（图 5−8）。 Jiaolong Haiqiu is located in the north of Nanhai Haipen and has a nearly oval planform. Two peaks have developed on its top, of which, the southern peak is relatively higher (with depth of about 3330 m) and is a sunk crater, the northern peak is relatively lower (with depth of about 3700 m) (Fig.5−8).		
命名由来 Origin of Name	2013 年，我国蛟龙号载人深潜器在此下潜考察，故以此命名。 In 2013, China’s Jiaolong deep-sea manned submersible was launched here, so the word “Jiaolong” was used to name the Hill.		

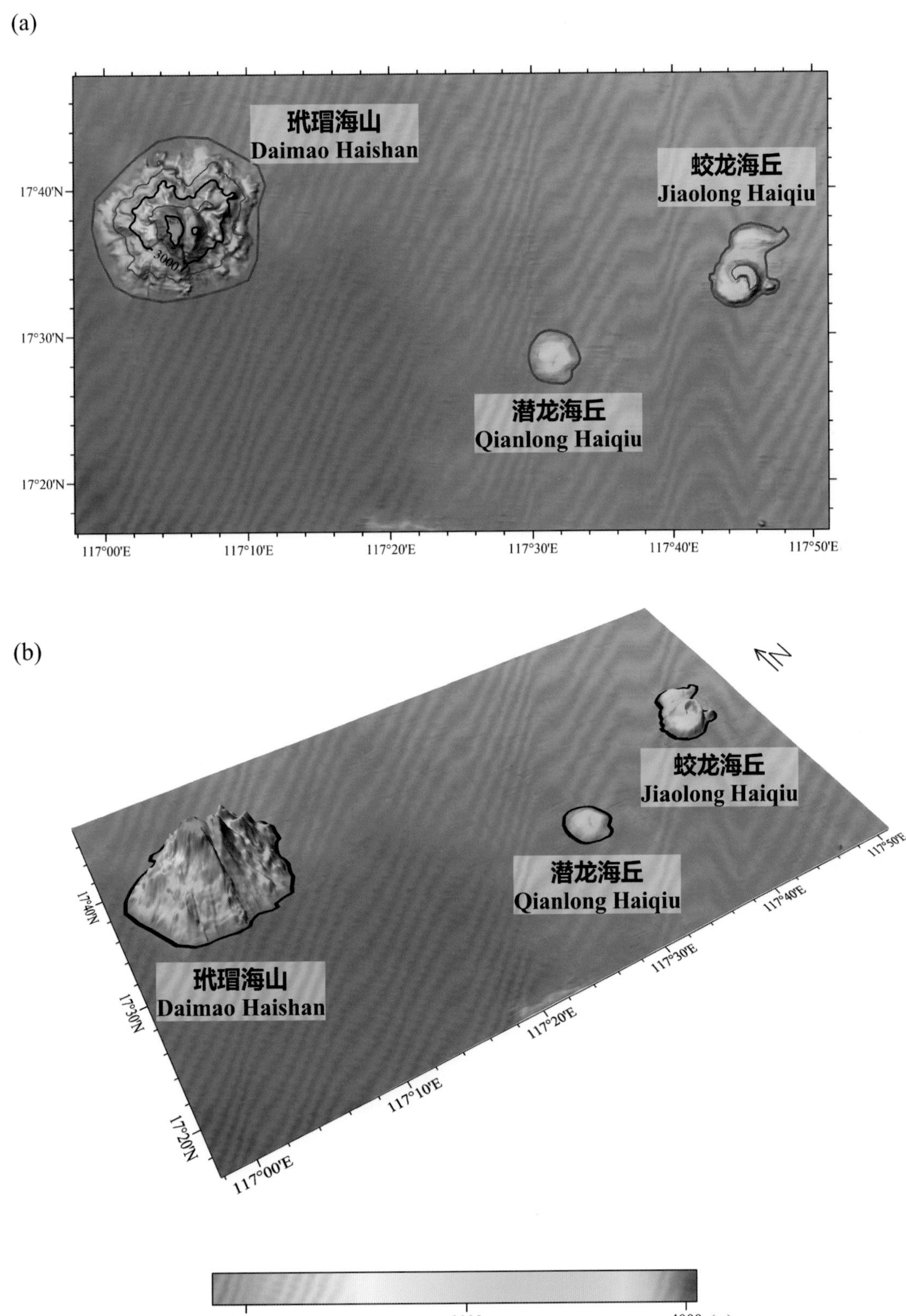

图 5-8　玳瑁海山、潜龙海丘、蛟龙海丘

(a) 海底地形图（等深线间隔 500 米）；(b) 三维海底地形图

Fig.5-8　Daimao Haishan, Qianlong Haiqiu, Jiaolong Haiqiu

(a) Seafloor topographic map (with contour interval of 500 m)；(b) 3-D seafloor topographic map

5.17 廷理海丘

标准名称 Standard Name	廷理海丘 Tingli Haiqiu	类别 Generic Term	海丘 Hill
中心点坐标 Center Coordinates	17°09.3'N, 117°22.7'E	规模（千米 × 千米） Dimension（km × km）	5 × 2
最小水深（米） Min Depth (m)	3780	最大水深（米） Max Depth (m)	4000
地理实体描述 Feature Description	廷理海丘位于南海海盆北部，平面形态近似长椭圆形。海丘顶部发育两座峰，其中东侧峰较高（水深约 3780 米），西侧峰较矮（水深约 3940 米）（图 5–9）。 Tingli Haiqiu is located in the north of Nanhai Haipen and has a nearly oval planform. Two peaks have developed on its top, of which, the eastern peak is relatively higher (with depth of about 3780 m) and the southern peak is relatively lower (with depth of about 3940 m) (Fig.5–9).		
命名由来 Origin of Name	以中国古代司法官职名称进行地名的团组化命名。“廷理”是春秋时期楚国掌管刑狱的官吏名称。 A group naming of undersea features after the titles of ancient Chinese judicial officials. “Tingli” was a title of judicial official in charge of penalty in the Chu Kingdom during the Spring and Autumn Period (770 B.C.–476 B.C.).		

5.18 司隶海丘

标准名称 Standard Name	司隶海丘 Sili Haiqiu	类别 Generic Term	海丘 Hill
中心点坐标 Center Coordinates	17°08.4'N, 117°38.3'E	规模（千米 × 千米） Dimension（km × km）	7 × 2
最小水深（米） Min Depth (m)	3780	最大水深（米） Max Depth (m)	3980
地理实体描述 Feature Description	司隶海丘位于南海海盆北部，平面形态呈长条形，其山脊线呈东—西走向（图 5–9）。 Sili Haiqiu is located in the north of Nanhai Haipen with a planform in the shape of a long strip. The ridge line of the Hill stretches in the direction of E–W (Fig.5–9).		
命名由来 Origin of Name	以中国古代司法官职名称进行地名的团组化命名。“司隶”是汉朝至魏晋时期监督京师和地方的监察官。 A group naming of undersea features after the titles of ancient Chinese judicial officials. “Sili” was a title of judicial official in charge of the supervision of the capital and local government from the Han to the Wei and Jin Dynasty (202 B.C.–420 A.D.).		

5.19 飒爽海丘

标准名称 Standard Name	飒爽海丘 Sashuang Haiqiu	类别 Generic Term	海丘 Hill
中心点坐标 Center Coordinates	17°07.4′N, 117°11.3′E	规模（千米 × 千米） Dimension（km × km）	22 × 16
最小水深（米） Min Depth (m)	3600	最大水深（米） Max Depth (m)	4000
地理实体描述 Feature Description	飒爽海丘位于南海海盆北部，其平面形态像一只游动的章鱼（图 5–9）。 Sashuang Haiqiu is located in the north of Nanhai Haipen and its planform is in the shape of a swimming octopus (Fig.5–9).		
命名由来 Origin of Name	以毛泽东诗词作品中的词进行地名的团组化命名。该海底地名的专名取词自毛泽东的《七绝 · 为女民兵题照》：“飒爽英姿五尺枪，曙光初照演兵场”。“飒爽”形容女子豪迈矫健。 A group naming of undersea features after the words in Mao Zedong's poems. The specific term of this undersea feature name “Sashuang”, means bright and inspired bearing, is derived from the poetic lines “There march women with rifles, bright, brave and inspired, when the morn casts its first light on the range of trial fire” in the poem *Inscription on Photography of Militia Women- A Verse of Four Seven-character Line* by Mao Zedong (1893–1976).		

5.20 英姿海丘

标准名称 Standard Name	英姿海丘 Yingzi Haiqiu	类别 Generic Term	海丘 Hill
中心点坐标 Center Coordinates	17°05.0′N, 117°25.8′E	规模（千米 × 千米） Dimension（km × km）	14 × 11
最小水深（米） Min Depth (m)	3670	最大水深（米） Max Depth (m)	4000
地理实体描述 Feature Description	英姿海丘位于南海海盆北部，平面形态呈北东—南西向的不规则多边形。其顶部发育两座峰，其中西北侧峰较高（水深约 3670 米），东南侧峰较矮（水深约 3890 米）（图 5–9）。 Yingzi Haiqiu is located in the north of Nanhai Haipen, with an overall planform in the shape of an irregular polygon in the direction of NE–SW. Two peaks have developed on its top, of which, the northeast peak is relatively higher (with depth of about 3670 m) while southeast peak is relatively lower (with depth of about 3980 m) (Fig.5–9).		
命名由来 Origin of Name	以毛泽东诗词作品中的词进行地名的团组化命名。该海底地名的专名取词自毛泽东的《七绝 · 为女民兵题照》：“飒爽英姿五尺枪，曙光初照演兵场”。“英姿”指英俊勇武的风姿。 A group naming of undersea features after the words in Mao Zedong’s poems. The specific term of this undersea feature name “Yingzi”, means handsome and brave bearing of a women, is derived from the poetic lines “There march women with rifles, bright, brave and inspired, when the morn casts its first light on the range of trial fire” in the poem *Inscription on Photography of Militia Women- A Verse of Four Seven-character Line* by Mao Zedong (1893–1976).		

5.21 苏颂海丘

标准名称 Standard Name	苏颂海丘 Susong Haiqiu	类别 Generic Term	海丘 Hill
中心点坐标 Center Coordinates	16°57.1′N, 117°25.9′E	规模（千米 × 千米） Dimension（km × km）	70 × 18
最小水深（米） Min Depth (m)	3750	最大水深（米） Max Depth (m)	4050
地理实体描述 Feature Description	苏颂海丘位于南海海盆北部，整体平面形态呈树枝状，山脊线呈近东—西走向。海丘顶部发育两座峰，其中东侧峰较高（水深约 3730 米），西侧峰较矮（水深约 3780 米）（图 5–9）。 Susong Haiqiu is located in the north of Nanhai Haipen with a dendritic planform. The ridge line of the Hill stretches nearly in the direction of E–W. Two peaks have developed on its top, of which, the southern peak is relatively higher (with depth of about 3730 m) while the western peak is relatively lower (with depth of about 3780 m) (Fig.5–9).		
命名由来 Origin of Name	以我国古代科学家、医学家和文人的名字进行地名的团组化命名。该海丘以宋朝著名天文学家苏颂命名，纪念他在天文学上的重要贡献。 A group naming of undersea features after the names of ancient Chinese scientists, medical scientists and literati. The Hill is named after Su Song (1020–1101 A.D.), a famous astronomer in the Song Dynasty (960–1279 A.D.), to commemorate his important contribution to the astronomy.		

(a)

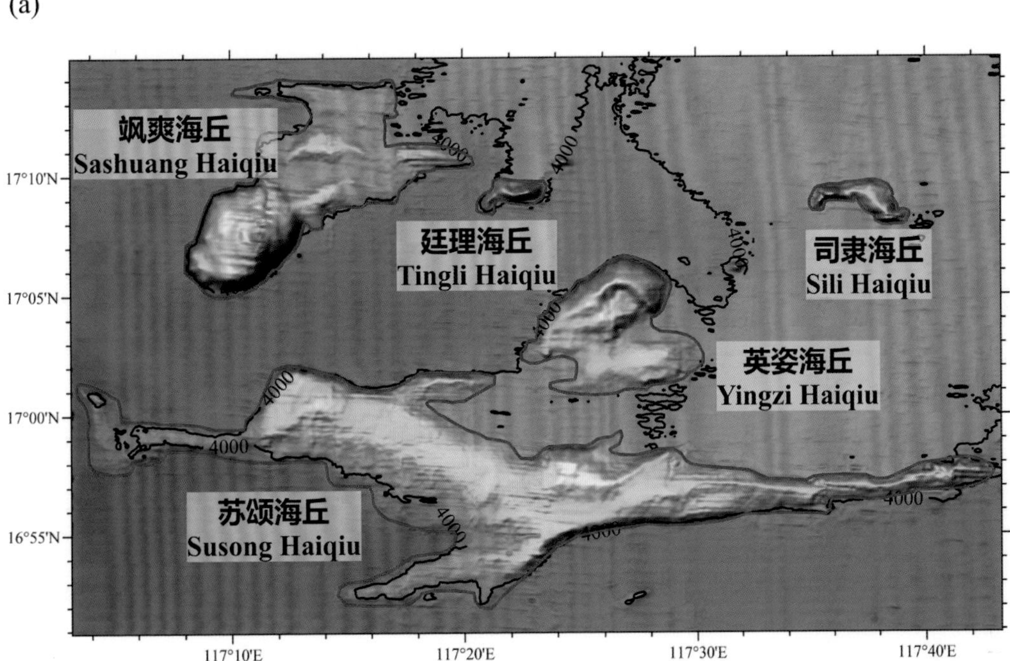

(b)

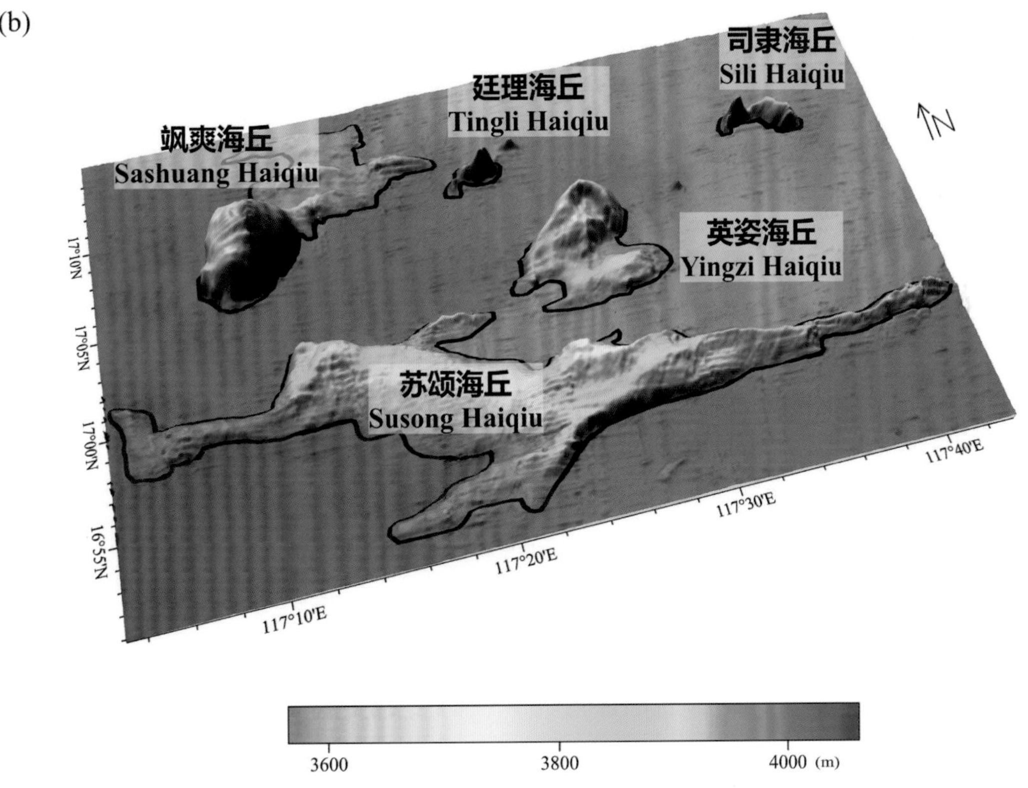

图 5-9　廷理海丘、司隶海丘、飒爽海丘、英姿海丘、苏颂海丘

(a) 海底地形图（等深线间隔 500 米）；(b) 三维海底地形图

Fig.5-9　Tingli Haiqiu, Sili Haiqiu, Sashuang Haiqiu, Yingzi Haiqiu, Susong Haiqiu

(a) Seafloor topographic map (with contour interval of 500 m); (b) 3-D seafloor topographic map

5.22 立法海脊

标准名称 Standard Name	立法海脊 Lifa Haiji	类别 Generic Term	海脊 Ridge
中心点坐标 Center Coordinates	16°49.7′N, 117°24.4′E	规模（千米 × 千米） Dimension（km × km）	44 × 5
最小水深（米） Min Depth (m)	3840	最大水深（米） Max Depth (m)	4088
地理实体描述 Feature Description	立法海脊位于南海海盆北部，紧邻奇志海丘，其脊线西侧为东—西走向，东侧为北东—南西走向（图 5-10）。 Lifa Haiji is located in the north of Nanhai Haipen and is adjacent to Qizhi Haiqiu. It stretches in the direction of E-W on the west side and is in the direction of NE-SW on the east side (Fig.5-10).		
命名由来 Origin of Name	以中国古代文学作品中的短语进行地名的团组化命名。该海底地名的专名取词自战国时期法家著作《商君书·算地》："圣人为国也，观俗立法则治，察国事本则宜"。"立法"即制定法律。 A group naming of undersea features after the phrases in ancient Chinese literatures. The specific term of this undersea feature name "Lifa", means formulating laws, is derived from the quote "Competent people govern a country by studying its history and customs and then formulating laws and regulations to ensure orderly governance and peaceful country" in *Book of Lord Shang- Calculation of Land*, a work of School of Legalism in the Warring States Period (453 B.C.-221 B.C.).		

5.23 奇志海丘

标准名称 Standard Name	奇志海丘 Qizhi Haiqiu	类别 Generic Term	海丘 Hill
中心点坐标 Center Coordinates	16°46.8′N, 117°27.1′E	规模（千米 × 千米） Dimension（km × km）	15 × 4
最小水深（米） Min Depth (m)	3820	最大水深（米） Max Depth (m)	4020
地理实体描述 Feature Description	奇志海丘位于南海海盆北部，紧邻立法海脊，整体平面形态呈东—西向的长条形（图 5-10）。 Qizhi Haiqiu is located in the north of Nanhai Haipen and is adjacent to Lifa Haiji, with an overall planform in the shape of a long strip in the direction of E-W (Fig.5-10).		
命名由来 Origin of Name	以毛泽东诗词作品中的词进行地名的团组化命名。该海底地名的专名取词自毛泽东的《七绝·为女民兵题照》："中华儿女多奇志，不爱红装爱武装"。"奇志"指极不平凡的志向。 A group naming of undersea features after the words in Mao Zedong's poems. The specific term of this undersea feature name "Qizhi", means lofty ideal, is derived from the poetic lines "Most China's sons and daughters have lofty ideal, they love bearing arms, not the gay feminine attire" in the poem *Inscription on Photography of Militia Women- A Verse of Four Seven-character Line* by Mao Zedong (1893-1976).		

(a)

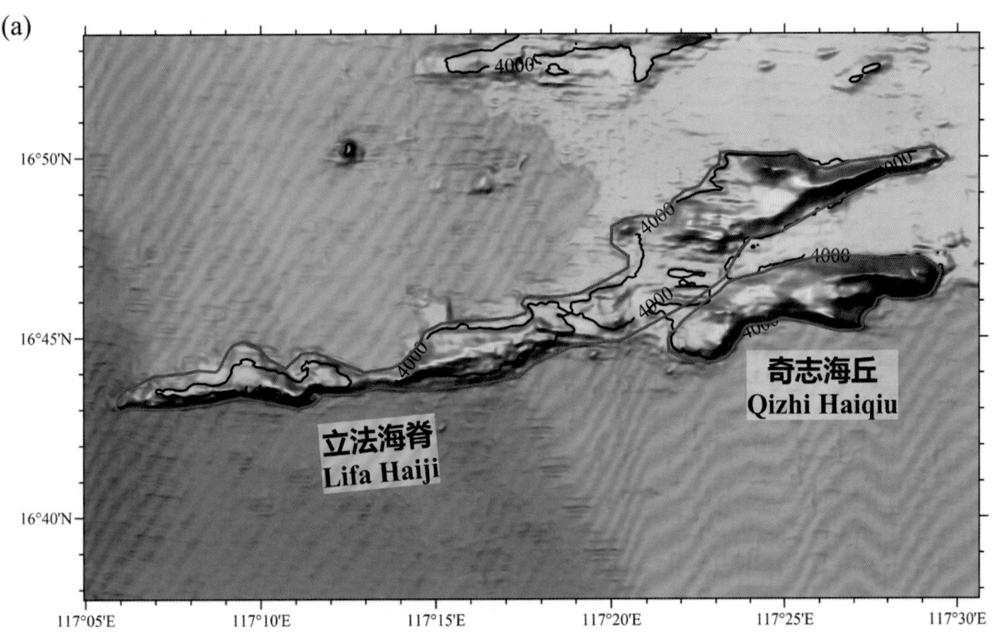

(b)

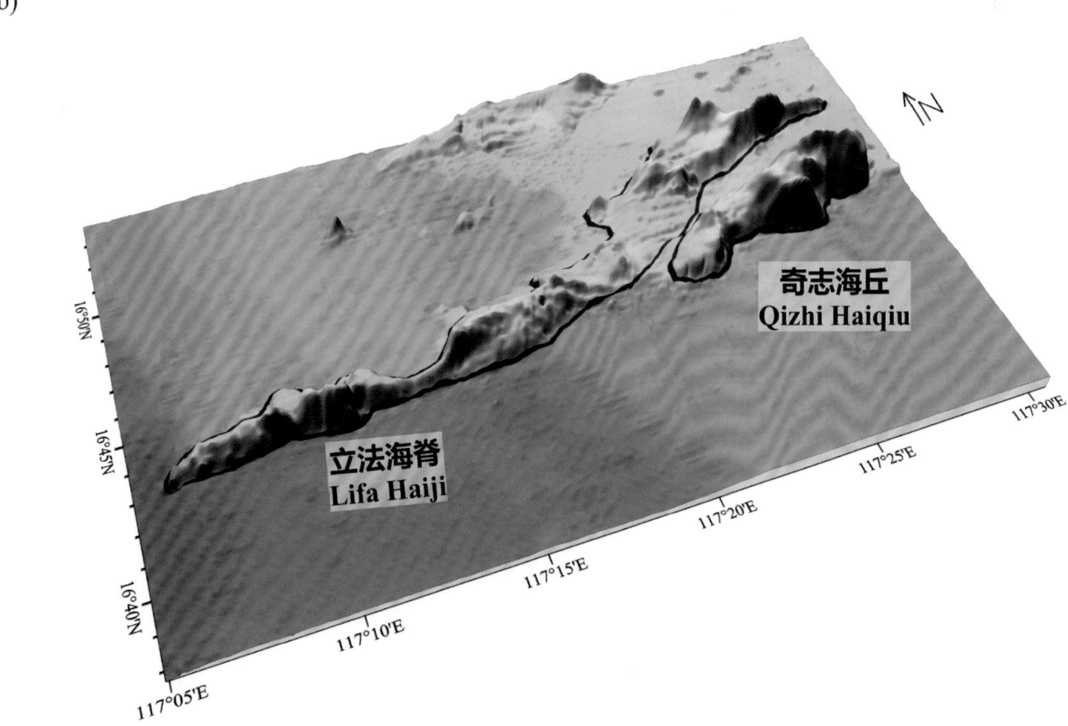

图 5-10 立法海脊、奇志海丘

(a) 海底地形图（等深线间隔 500 米）；(b) 三维海底地形图

Fig.5-10 Lifa Haiji, Qizhi Haiqiu

(a) Seafloor topographic map (with contour interval of 500 m); (b) 3-D seafloor topographic map

5.24 叶桂海丘

标准名称 Standard Name	叶桂海丘 Yegui Haiqiu	类别 Generic Term	海丘 Hill
中心点坐标 Center Coordinates	16°43.4′N, 117°41.9′E	规模（千米 × 千米） Dimension（km × km）	12 × 3
最小水深（米） Min Depth (m)	3500	最大水深（米） Max Depth (m)	4340
地理实体描述 Feature Description	叶桂海丘位于南海海盆北部，平面形态呈北东—南西向的长条形，其主峰位于海丘的东北侧（图 5–11）。 Yegui Haiqiu is located in the north of Nanhai Haipen, with a planform in the shape of a long strip in the direction of NE–SW. Its main peak is located in the northeast side of the Hill (Fig.5–11).		
命名由来 Origin of Name	以我国古代科学家、医学家和文人的名字进行地名的团组化命名。该海丘以清朝著名医学家叶桂命名，纪念他在医学上的重要贡献。 A group naming of undersea features after the names of ancient Chinese scientists, medical scientists and literati. The Hill is named after Ye Gui (1666–1745 A.D.), a famous medical scientist in the Qing Dynasty (1636–1912 A.D.), to commemorate his important contribution to the medical science.		

5.25 御史海丘

标准名称 Standard Name	御史海丘 Yushi Haiqiu	类别 Generic Term	海丘 Hill
中心点坐标 Center Coordinates	16°48.8'N, 117°55.4'E	规模（千米 × 千米） Dimension（km × km）	13 × 5
最小水深（米） Min Depth (m)	3560	最大水深（米） Max Depth (m)	4020
地理实体描述 Feature Description	御史海丘位于南海海盆北部，整体平面形态呈北西—南东向的长条形。海丘的顶部发育多座峰，其中主峰位于海丘的西北侧（图 5–11）。 Yushi Haiqiu is located in the north of Nanhai Haipen, with an overall planform in the shape of a long strip in the direction of NW–SE. Multiple peaks have developed on its top, of which, the main peak is located in the northwest side of the Hill (Fig.5–11).		
命名由来 Origin of Name	以中国古代司法官职名称进行地名的团组化命名。“御史”是中国古代掌管监察的官吏名称。 A group naming of undersea features after the titles of ancient Chinese judicial officials. “Yushi” was a title of judicial official mainly in charge of supervision.		

5.26 初照海丘链

标准名称 Standard Name	初照海丘链 Chuzhao Haiqiulian	类别 Generic Term	海丘链 Hill Chain
中心点坐标 Center Coordinates	16°53.7′N, 118°00.6′E	规模（千米 × 千米） Dimension（km × km）	18 × 4
最小水深（米） Min Depth (m)	3650	最大水深（米） Max Depth (m)	4000
地理实体描述 Feature Description	初照海丘链位于南海海盆北部，由多个呈北西—南东向、直线排列的海丘所组成。海丘链顶部发育多座峰，其中主峰位于东北处（图 5-11）。 Chuzhao Haiqiulian is located in the north of Nanhai Haipen and is composed of multiple linear arranged hills in the direction of NW–SE. Multiple peaks have developed on the top, of which, the main peak is located in the northeast (Fig.5–11).		
命名由来 Origin of Name	以毛泽东诗词作品中的词进行地名的团组化命名。该海底地名的专名取词自毛泽东的《七绝·为女民兵题照》："飒爽英姿五尺枪，曙光初照演兵场"。"初照"即刚刚照到。 A group naming of undersea features after the words in Mao Zedong's poems. The specific term of this undersea feature name "Chuzhao", means the first light of dawn, is derived from the poetic lines "There march women with rifles, bright, brave and inspired, when the morn casts its first light on the range of trial fire" in the poem *Inscription on Photography of Militia Women- A Verse of Four Seven-character Line* by Mao Zedong (1893–1976).		

(a)

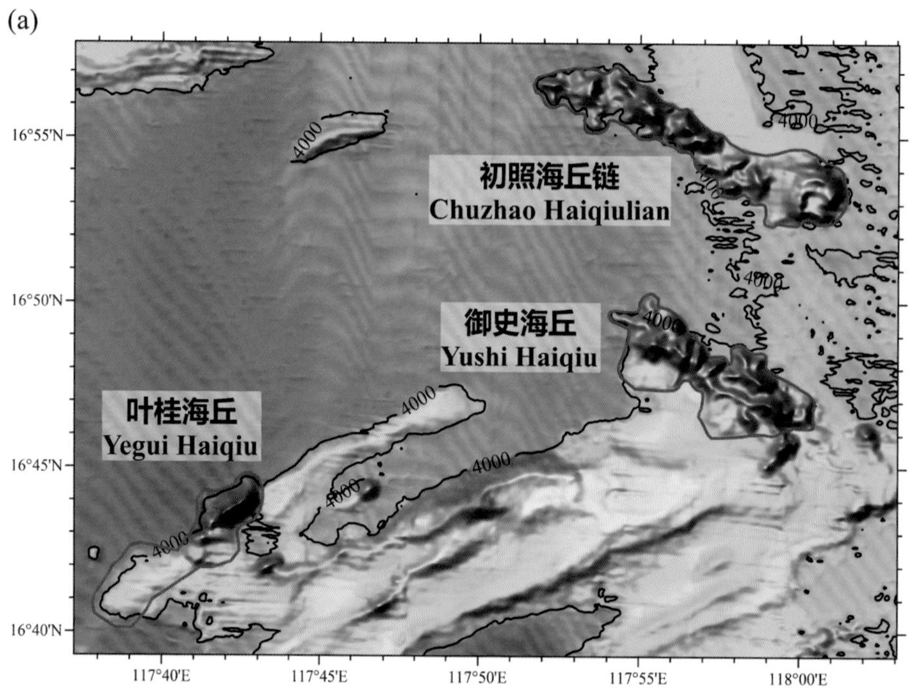

(b)

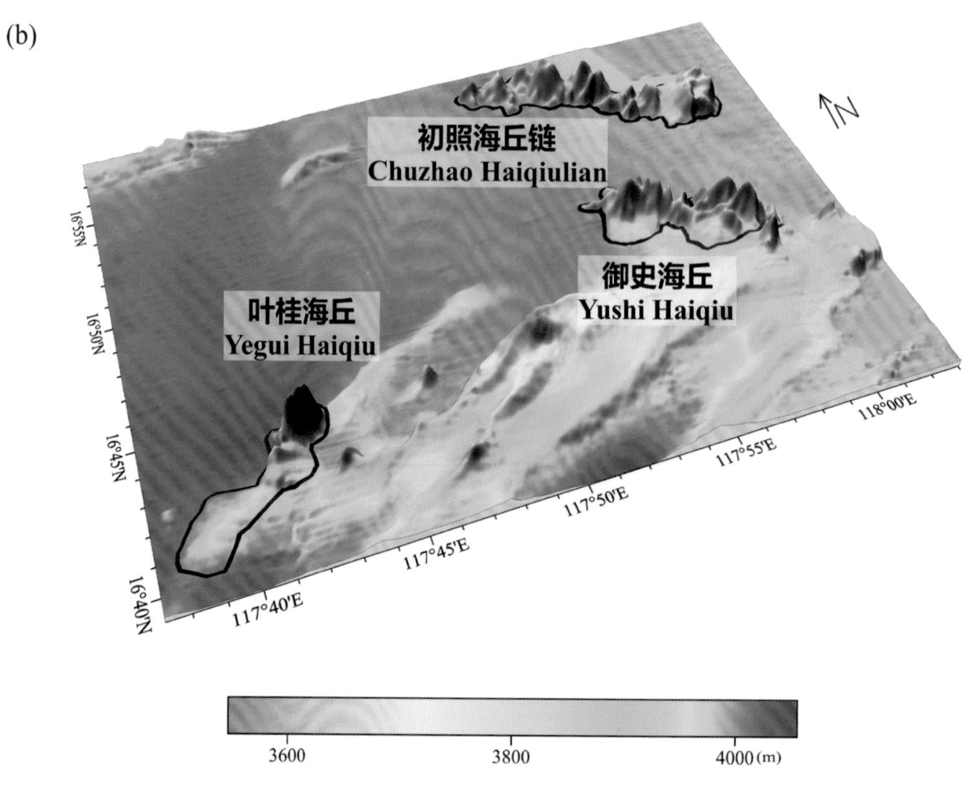

图 5-11　叶桂海丘、御史海丘、初照海丘链

(a) 海底地形图（等深线间隔 500 米）；(b) 三维海底地形图

Fig.5-11　Yegui Haiqiu, Yushi Haiqiu, Chuzhao Haiqiulian

(a) Seafloor topographic map (with contour interval of 500 m)；(b) 3-D seafloor topographic map

5.27 司寇海丘

标准名称 Standard Name	司寇海丘 Sikou Haiqiu	类别 Generic Term	海丘 Hill
中心点坐标 Center Coordinates	16°31.2'N, 117°19.2'E	规模（千米 × 千米） Dimension（km × km）	10 × 8
最小水深（米） Min Depth (m)	3800	最大水深（米） Max Depth (m)	4020
地理实体描述 Feature Description	司寇海丘位于南海海盆北部，平面形态呈北西—南东向的多边形（图 5-12）。 Sikou Haiqiu is located in the north of Nanhai Haipen, with a planform in the shape of an polygon in the direction of NW−SE (Fig.5−12).		
命名由来 Origin of Name	以中国古代司法官职名称进行地名的团组化命名。“司寇”是中国古代掌管刑狱的官吏名称。 A group naming of undersea features after the titles of ancient Chinese judicial officials. “Sikou” is a title of judicial official mainly in charge of penalty and imprisonment.		

5.28 廷尉海丘

标准名称 Standard Name	廷尉海丘 Tingwei Haiqiu	类别 Generic Term	海丘 Hill
中心点坐标 Center Coordinates	16°25.7'N, 117°26.9'E	规模（千米 × 千米） Dimension（km × km）	17 × 12
最小水深（米） Min Depth (m)	3480	最大水深（米） Max Depth (m)	4000
地理实体描述 Feature Description	廷尉海丘位于南海海盆北部，平面形态呈向南侧突出的椭圆形，其顶部发育两座峰（图 5-12）。 Tingwei Haiqiu is located in the north of Nanhai Haipen, with a planform in the shape of an oval protruding towards the south, and has two peaks developed on its top (Fig.5−12).		
命名由来 Origin of Name	以中国古代司法官职名称进行地名的团组化命名。“廷尉”是中国古代自秦汉至北齐时期主管司法的官吏名称。 A group naming of undersea features after the titles of ancient Chinese judicial officials. “Tingwei” is a title of judicial official in charge of judiciary from the Qin and Han Dynasty (221 B.C.−220 A.D.) to the Northern Qi Dynasty (550−577 A.D.).		

5.29 朱丹溪海丘

标准名称 Standard Name	朱丹溪海丘 Zhudanxi Haiqiu	类别 Generic Term	海丘 Hill
中心点坐标 Center Coordinates	16°19.6′N, 117°26.5′E	规模（千米 × 千米） Dimension（km × km）	14 × 9
最小水深（米） Min Depth (m)	3820	最大水深（米） Max Depth (m)	4350
地理实体描述 Feature Description	朱丹溪海丘位于南海海盆北部，平面形态呈不规则多边形，其西南部有延伸发育的小山脊（图 5−12）。 Zhudanxi Haiqiu is located in the north of Nanhai Haipen, with a planform in the shape of an irregular polygon. A small ridges have developed in the southwest of the Hill (Fig.5−12).		
命名由来 Origin of Name	以我国古代科学家、医学家和文人的名字进行地名的团组化命名。该海丘以元朝著名医学家朱丹溪命名，纪念他在医学上的重要贡献。 A group naming of undersea features after the names of ancient Chinese scientists, medical scientists and literati. The Hill is named after Zhu Danxi (1281−1358 A.D.), a famous medical scientist in the Yuan Dynasty (1271−1368 A.D.), to commemorate his outstanding contribution to the medical science.		

(a)

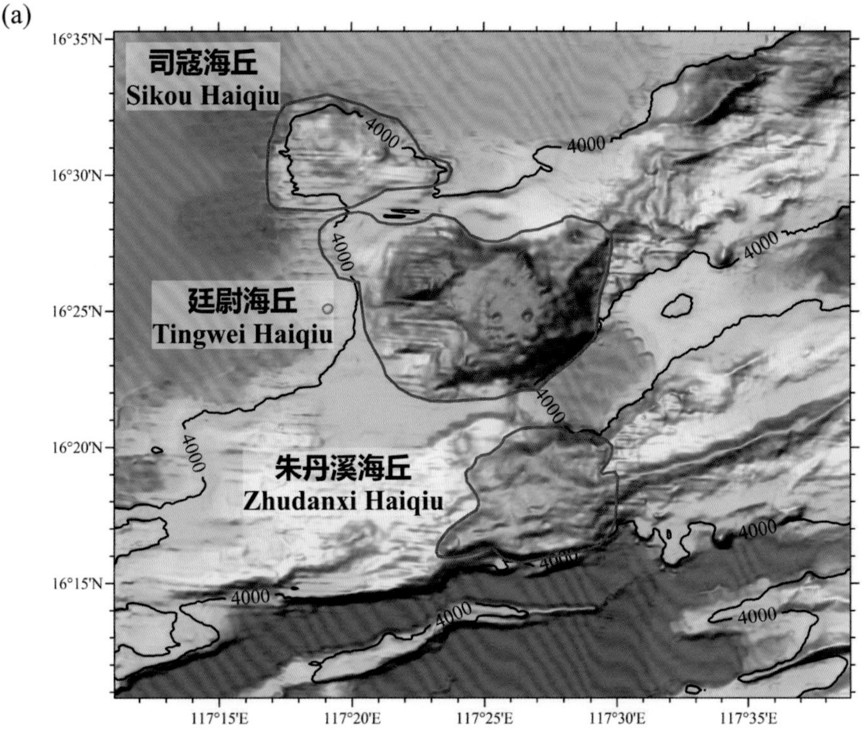

(b)

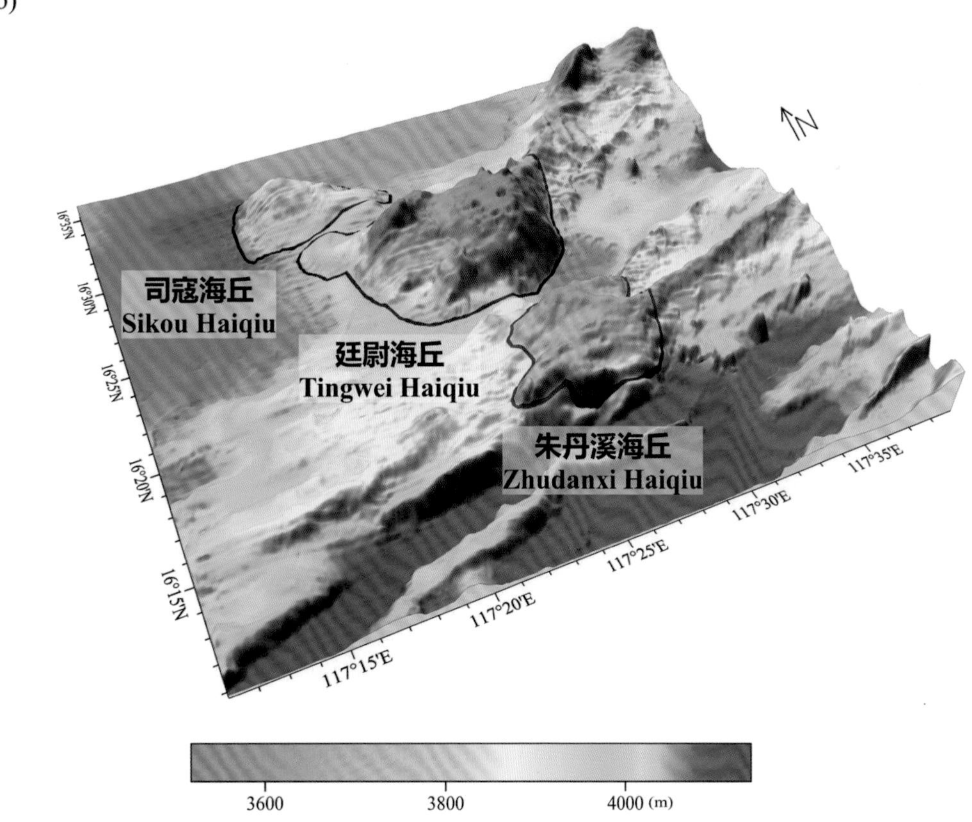

图 5-12　司寇海丘、廷尉海丘、朱丹溪海丘

(a) 海底地形图（等深线间隔 500 米）；(b) 三维海底地形图

Fig.5-12　Sikou Haiqiu, Tingwei Haiqiu, Zhudanxi Haiqiu

(a) Seafloor topographic map (with contour interval of 500 m)；(b) 3-D seafloor topographic map

5.30 审刑海丘

标准名称 Standard Name	审刑海丘 Shenxing Haiqiu	类别 Generic Term	海丘 Hill
中心点坐标 Center Coordinates	16°26.8'N, 117°56.9'E	规模（千米 × 千米） Dimension（km × km）	16 × 7
最小水深（米） Min Depth (m)	2970	最大水深（米） Max Depth (m)	3880
地理实体描述 Feature Description	审刑海丘位于南海海盆北部，平面形态呈不规则多边形，顶部发育多座峰（图 5−13）。 Shenxing Haiqiu is located in the north of Nanhai Haipen, with a planform in the shape of an irregular polygon. Multiple peaks have developed on its top (Fig.5−13).		
命名由来 Origin of Name	以中国古代司法官职名称进行地名的团组化命名。“审刑”是中国古代负责复核案件的司法官吏名称。 A group naming of undersea features after the titles of ancient Chinese judicial officials. “Shenxing” is a title of judicial official mainly in charge of case review in ancient China.		

5.31 华佗海山

标准名称 Standard Name	华佗海山 Huatuo Haishan	类别 Generic Term	海山 Seamount
中心点坐标 Center Coordinates	16°31.6′N, 118°03.1′E	规模（千米 × 千米） Dimension（km × km）	27 × 6
最小水深（米） Min Depth (m)	3020	最大水深（米） Max Depth (m)	4430
地理实体描述 Feature Description	华佗海山位于南海海盆北部，平面形态呈北西—南东向的长条形。海山的顶部发育多座峰，其中主峰位于中部（图 5−13）。 Huatuo Haishan is located in the north of Nanhai Haipen, with a planform in the shape of a long strip in the direction of NW−SE. Multiple peaks have developed on the top, of which, the main peak is in the central part (Fig.5−13).		
命名由来 Origin of Name	以我国古代科学家、医学家和文人的名字进行地名的团组化命名。该海山以东汉末年著名医学家华佗命名，纪念他在医学上的重要贡献。 A group naming of undersea features after the names of ancient Chinese scientists, medical scientists and literati. The Seamount is named after Hua Tuo (145−208 A.D.), a famous medical scientists in the late Eastern Han Dynasty (184−220 A.D.), to commemorate his important contribution to the medical science.		

(a)

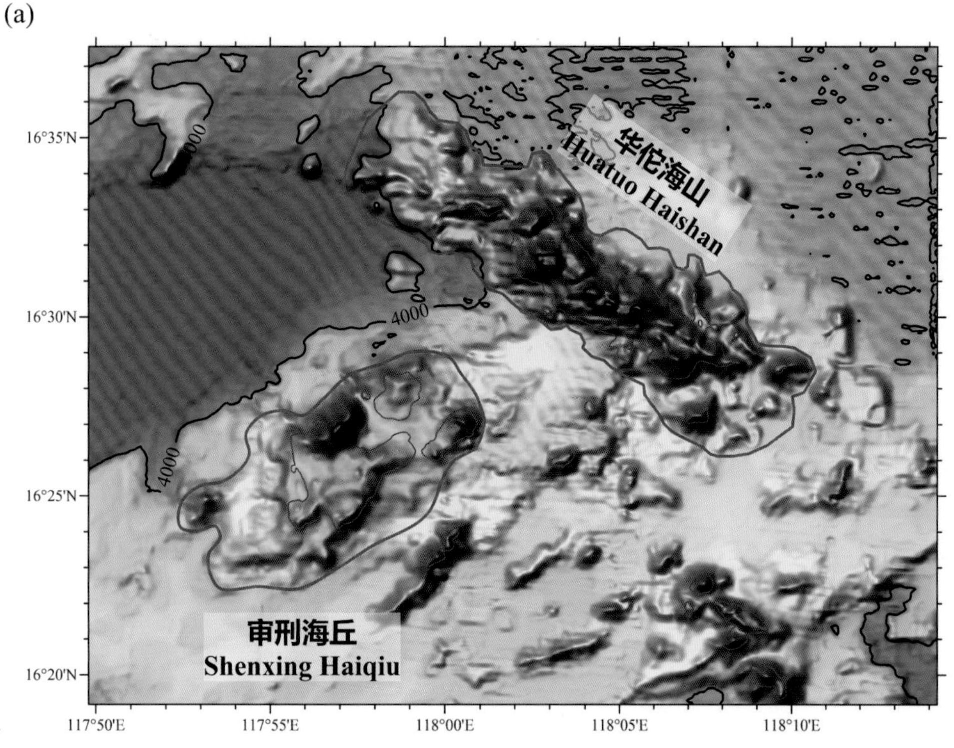

(b)

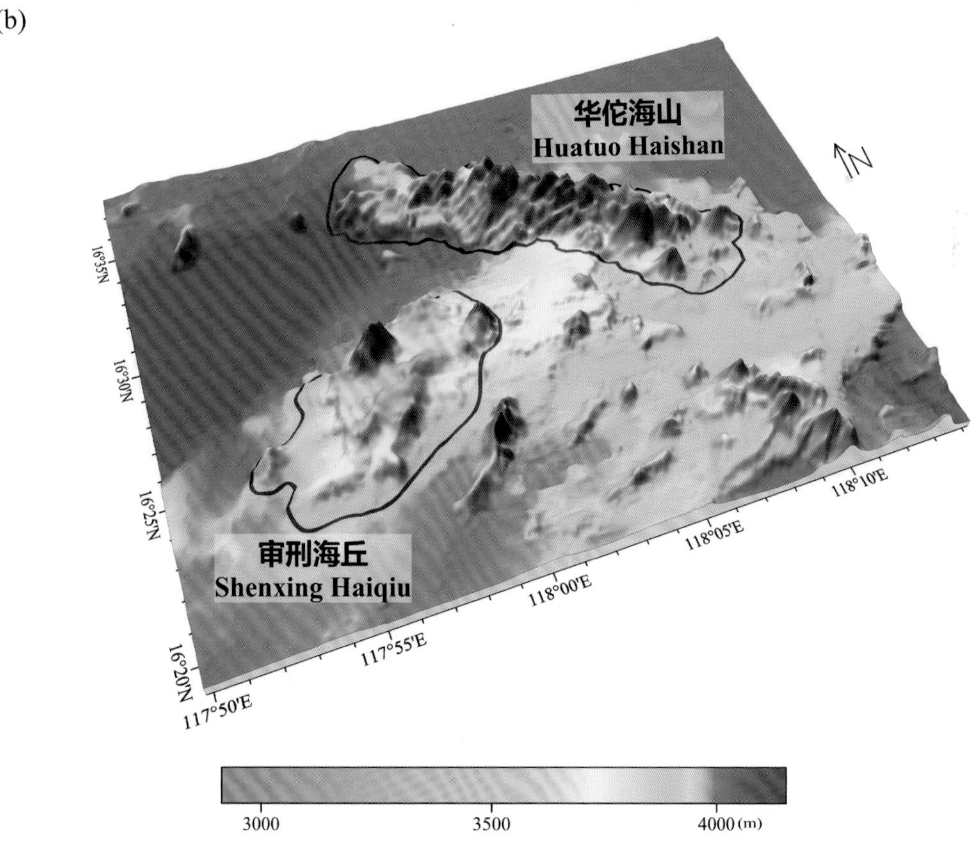

图 5-13　审刑海丘、华佗海山

(a) 海底地形图（等深线间隔 500 米）；(b) 三维海底地形图

Fig.5-13　Shenxing Haiqiu, Huatuo Haishan

(a) Seafloor topographic map (with contour interval of 500 m); (b) 3-D seafloor topographic map

5.32 祖冲之海丘

标准名称 Standard Name	祖冲之海丘 Zuchongzhi Haiqiu	类别 Generic Term	海丘 Hill
中心点坐标 Center Coordinates	17°02.8′N, 118°15.0′E	规模（千米 × 千米） Dimension（km × km）	16 × 11
最小水深（米） Min Depth (m)	3110	最大水深（米） Max Depth (m)	4000
地理实体描述 Feature Description	祖冲之海丘位于南海海盆北部，平面形态呈不规则多边形。海丘的顶部发育两座峰，其中北侧峰较高（水深约 3110 米），南侧峰较矮（水深约 3650 米）（图 5–14）。 Zuchongzhi Haiqiu is located in the north of Nanhai Haipen, with a planform in the shape of an irregular polygon. Two peaks have developed on its top, of which, the northern peak is relatively higher (with depth of about 3110 m) while the southern peak is relatively lower (with depth of about 3650 m) (Fig.5–14).		
命名由来 Origin of Name	以我国古代科学家、医学家和文人的名字进行地名的团组化命名。该海丘以南北朝时期著名数学家祖冲之命名，纪念他在数学和天文学上的重要贡献。 A group naming of undersea features after the names of the names of ancient Chinese scientists, medical scientists and literati. The Hill is named after Zu Chongzhi (429–500 A.D.), a famous mathematician and astronomer in the Southern and Northern Dynasties (420–589 A.D.).		

5.33 曙光海山

标准名称 Standard Name	曙光海山 Shuguang Haishan	类别 Generic Term	海山 Seamount
中心点坐标 Center Coordinates	17°16.7′N, 118°28.1′E	规模（千米 × 千米） Dimension（km × km）	20 × 10
最小水深（米） Min Depth (m)	2420	最大水深（米） Max Depth (m)	4000
地理实体描述 Feature Description	曙光海山位于南海海盆北部，平面形态呈北东—南西走的长条形（图 5–14）。 Shuguang Haishan is locate in the north of Nanhai Haipen, with a planform in the shape of a long strip in the direction of NE–SW (Fig.5–14).		
命名由来 Origin of Name	以毛泽东诗词作品中的词进行地名的团组化命名。该海底地名的专名取词自毛泽东的《七绝・为女民兵题照》：“飒爽英姿五尺枪，曙光初照演兵场”。“曙光”即破晓时的阳光。 A group naming of undersea features after the words in Mao Zedong’s poems. The specific term of this undersea feature name “Shuguang”, means dawn, is derived from the poetic lines “There march women with rifles, bright, brave and inspired, when the morn casts its first light on the range of trial fire” in the poem *Inscription on Photography of Militia Women- A Verse of Four Seven-character Line* by Mao Zedong (1893–1976).		

(a)

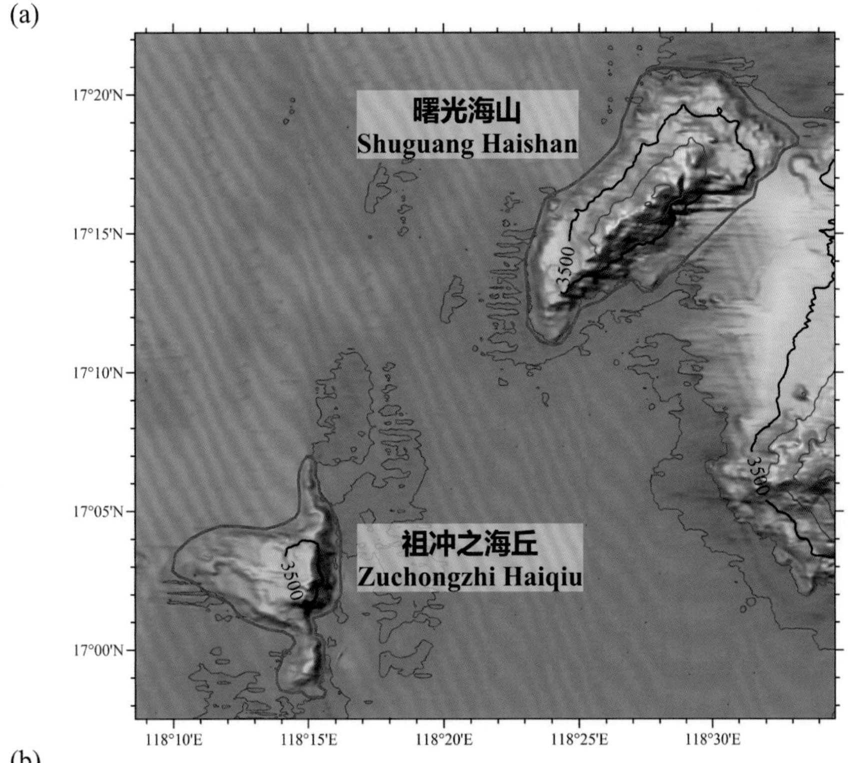

(b)

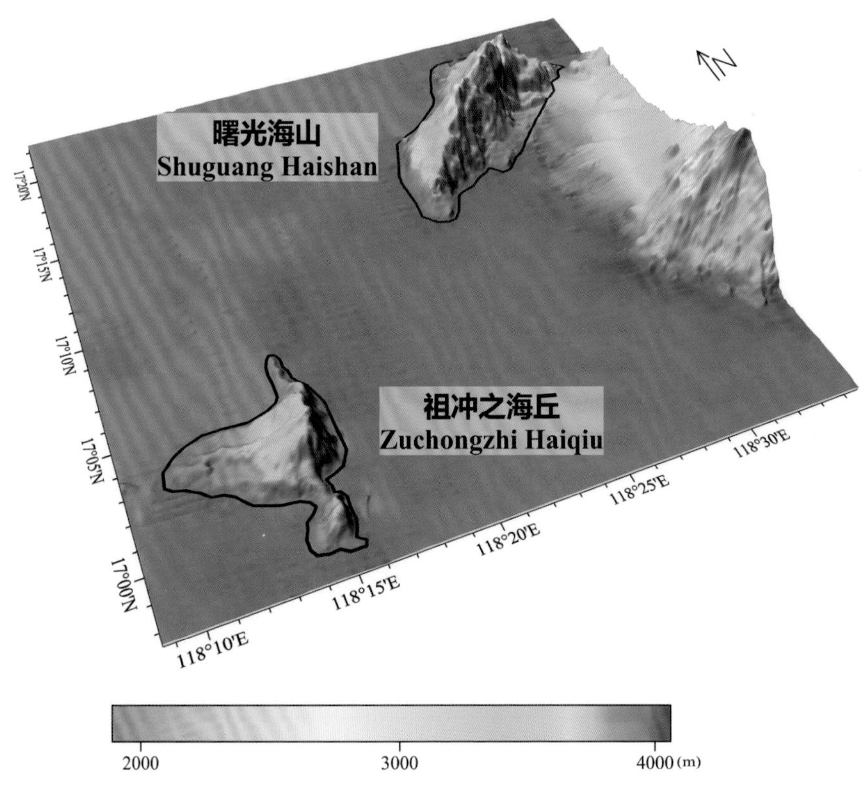

图 5-14 祖冲之海丘、曙光海山

(a) 海底地形图（等深线间隔 500 米）；(b) 三维海底地形图

Fig.5-14 Zuchongzhi Haiqiu, Shuguang Haishan

(a) Seafloor topographic map (with contour interval of 500 m); (b) 3-D seafloor topographic map

5.34 钱乙海山

标准名称 Standard Name	钱乙海山 Qianyi Haishan	类别 Generic Term	海山 Seamount
中心点坐标 Center Coordinates	16°46.4′N, 118°36.0′E	规模（千米 × 千米） Dimension（km × km）	85 × 50
最小水深（米） Min Depth (m)	670	最大水深（米） Max Depth (m)	4270
地理实体描述 Feature Description	钱乙海山位于南海海盆北部，马尼拉海沟西侧，平面形态呈北东—南西向的不规则多边形（图 5−15）。 Qianyi Haishan is located in the north of Nanhai Haipen and is to the west of the Manila Trench, with a planform in the shape of an irregular polygon in the direction of NE−SW (Fig.5−15).		
命名由来 Origin of Name	以我国古代科学家、医学家和文人的名字进行地名的团组化命名。该海山以宋朝著名医学家钱乙命名，纪念他在医学上的重要贡献。 A group naming of undersea features after the names of ancient Chinese scientists, medical scientists and literati. The Seamount is named after Qian Yi (1032−1113 A.D.), a famous medical scientist of the Song Dynasty (960−1279 A.D.), to commemorate his important contribution to the medical science.		

5.35 吴有性海山

标准名称 Standard Name	吴有性海山 Wuyouxing Haishan	类别 Generic Term	海山 Seamount
中心点坐标 Center Coordinates	16°50.3′N, 119°00.6′E	规模（千米 × 千米） Dimension（km × km）	46 × 23
最小水深（米） Min Depth (m)	2090	最大水深（米） Max Depth (m)	4680
地理实体描述 Feature Description	吴有性海山位于南海海盆北部，马尼拉海沟西侧，平面形态呈北西—南东向的长条形，其顶部发育多座峰（图 5−15）。 Wuyouxing Haishan is located in the north of Nanhai Haipen and on the west of the Manila Trench, with a planform in the shape of a long strip in the direction of NW−SE. Multiple peaks have developed on its top (Fig.5−15).		
命名由来 Origin of Name	以我国古代科学家、医学家和文人的名字进行地名的团组化命名。该海山以明末清初著名医学家吴有性命名，纪念他在医学上的重要贡献。 A group naming of undersea features after the names of ancient Chinese scientists, medical scientists and literati. The Seamount is named after Wu Youxing (1582−1652 A.D.), a famous medical scientist in the Late Ming and early Qing Dynasty (1600−1644 A.D.), to commemorate his outstanding contribution to the medical science.		

5.36 法明海底峰

标准名称 Standard Name	法明海底峰 Faming Haidifeng	类别 Generic Term	海底峰 Peak
中心点坐标 Center Coordinates	16°55.4'N, 118°47.3'E	规模（千米 × 千米） Dimension（km × km）	17 × 15
最小水深（米） Min Depth (m)	2310	最大水深（米） Max Depth (m)	3970
地理实体描述 Feature Description	法明海底峰为吴有性海山西北角的一个山峰，平面形态呈圆形（图 5–15）。 Faming Haidifeng is a peak in the northwest of Wuyouxing Haishan, with a circular planform (Fig.5–15).		
命名由来 Origin of Name	以中国古代文学作品中的短语进行地名的团组化命名。该海底地名的专名取词自战国时期法家著作《商君书·画策》："圣王者不贵义而贵法，法必明，令必行，则已矣"。"法明"即法律严明。 A group naming of undersea features after the phrases in ancient Chinese literatures. The specific term of this undersea feature name "Faming", means strict and impartial law, is derived from the quote "A judicial monarch pays closer attention to law than to morality. Laws must be strict and impartial and the orders must be well executed" in *The Book of Lord Shang-Strategies*, a work of the school of legalism in the Warring State Period (453 B.C.–221 B.C.).		

5.37 孙思邈海山

标准名称 Standard Name	孙思邈海山 Sunsimiao Haishan	类别 Generic Term	海山 Seamount
中心点坐标 Center Coordinates	16°28.7′N, 118°51.1′E	规模（千米 × 千米） Dimension（km × km）	65 × 18
最小水深（米） Min Depth (m)	2830	最大水深（米） Max Depth (m)	4890
地理实体描述 Feature Description	孙思邈海山位于南海海盆北部，马尼拉海沟西侧，平面形态呈北西—南东向的长条形（图 5–15）。 Sunsimiao Haishan is located in the north of Nanhai Haipen and on the west side of the Manila Trench, with a planform in the shape of a long strip in the direction of NW–SE (Fig.5–15).		
命名由来 Origin of Name	以我国古代科学家、医学家和文人的名字进行地名的团组化命名。该海山以唐朝著名医学家孙思邈命名，纪念他在医学上的重要贡献。 A group naming of undersea features after the names of ancient Chinese scientists, medical scientists and literati. The Seamount is named after Sun Simiao (541–682 A.D.), a famous medical scientist in the Tang Dynasty (618–907 A.D.), to commemorate his outstanding contribution to the medical science.		

5.38 新天海山

标准名称 Standard Name	新天海山 Xintian Haishan	类别 Generic Term	海山 Seamount
中心点坐标 Center Coordinates	16°10.2'N, 119°03.1'E	规模（千米 × 千米） Dimension（km × km）	18 × 13
最小水深（米） Min Depth (m)	2650	最大水深（米） Max Depth (m)	3940
地理实体描述 Feature Description	新天海山位于南海海盆北部，马尼拉海沟西侧，平面形态呈三角形（图 5−15）。 Xintian Haishan is located in the north of Nanhai Haipen and on the west side of the Manila Trench, with a triangle-shaped planform (Fig.5−15).		
命名由来 Origin of Name	以毛泽东诗词作品中的词进行地名的团组化命名。该海底地名的专名取词自毛泽东的《七律・到韶山》："为有牺牲多壮志，敢教日月换新天"。"新天"即新世界。 A group naming of undersea features after the words in Mao Zedong's poems. The specific term of this undersea feature name "Xintian", means a new world, is derived from the poetic lines "With support ambitious comrades who are ready to sacrifice, we are capable of building a new world with the new sun and moon" in the poem *Arrival at Shaoshan- A Verse of Eight Seven-character Lines* by Mao Zedong (1893−1976).		

(a)

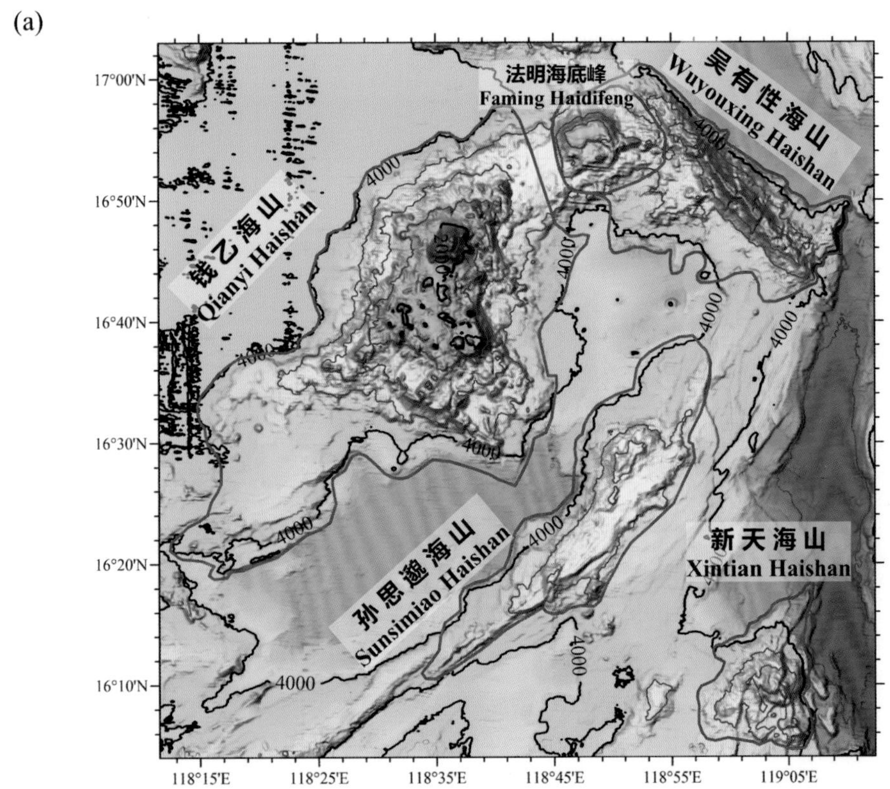

(b)

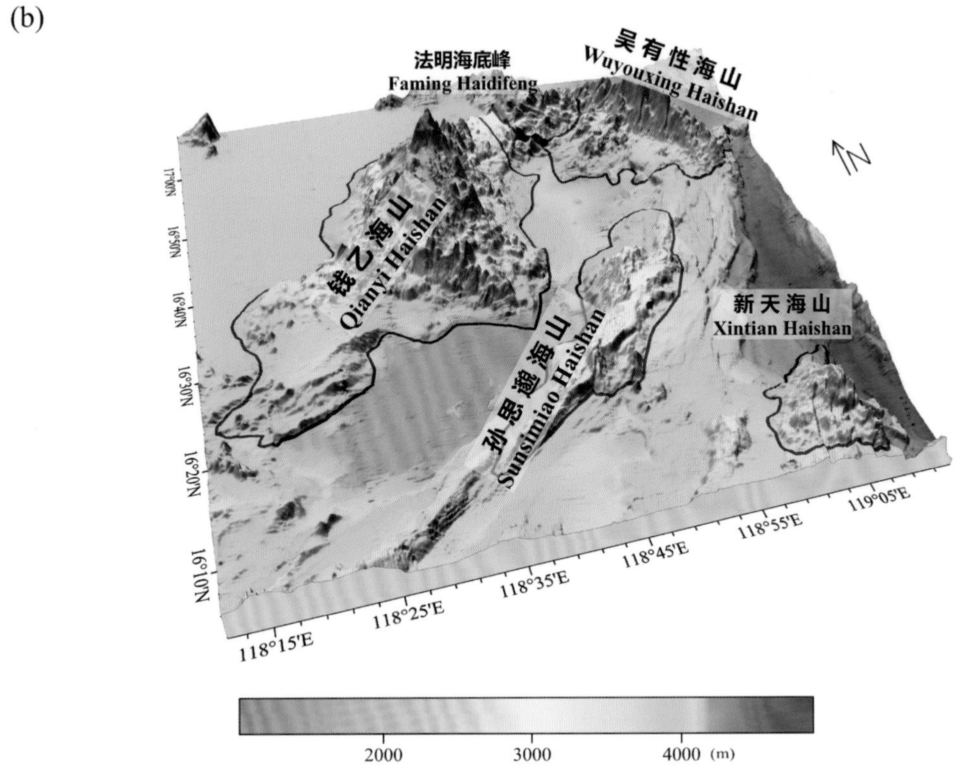

图 5-15 钱乙海山、吴有性海山、法明海底峰、孙思邈海山、新天海山

(a) 海底地形图（等深线间隔 500 米）；(b) 三维海底地形图

Fig.5-15 Qianyi Haishan, Wuyouxing Haishan, Faming Haidifeng, Sunsimiao Haishan, Xintian Haishan

(a) Seafloor topographic map (with contour interval of 500 m); (b) 3-D seafloor topographic map

5.39 葛洪海丘

标准名称 Standard Name	葛洪海丘 Gehong Haiqiu	类别 Generic Term	海丘 Hill
中心点坐标 Center Coordinates	15°59.0′N, 118°12.2′E	规模（千米 × 千米） Dimension（km × km）	95 × 15
最小水深（米） Min Depth (m)	3550	最大水深（米） Max Depth (m)	4440
地理实体描述 Feature Description	葛洪海丘位于南海海盆北部，马尼拉海沟西侧，平面形态呈北东—南西向的长条形，与李时珍海山近似平行排布（图 5−16）。 Gehong Haiqiu is located in the north of Nanhai Haipen and is to the west of Manila Trench, with a planform in the shape of a long strip in the direction of NE−SW. It is nearly parallel with Lishizhen Haishan (Fig.5−16).		
命名由来 Origin of Name	以我国古代科学家、医学家和文人的名字进行地名的团组化命名。该海丘以东晋时期著名医学家葛洪命名，纪念他在医学上的重要贡献。 A group naming of undersea features after the names of ancient Chinese scientists, medical scientists and literati. The Hill is named after Ge Hong (283−363 A.D.), a famous medical scientist in the Eastern Jin Dynasty (317−420 A.D.), to commemorate his outstanding contribution to the medical science.		

5.40 李时珍海山

标准名称 Standard Name	李时珍海山 Lishizhen Haishan	类别 Generic Term	海山 Seamount
中心点坐标 Center Coordinates	15°53.0′N, 118°19.3′E	规模（千米 × 千米） Dimension（km × km）	84 × 14
最小水深（米） Min Depth (m)	3100	最大水深（米） Max Depth (m)	4740
地理实体描述 Feature Description	李时珍海山位于南海海盆北部，马尼拉海沟西侧，平面形态呈北东—南西向的长条形，与葛洪海丘近似平行排布（图 5−16）。 Lishizhen Haishan is located in the north of Nanhai Haipen and is on the west of the Manila Trench, with a planform in the shape of a long strip in the direction of NE−SW. It is nearly parallel with Gehong Haiqiu (Fig.5−16).		
命名由来 Origin of Name	以我国古代科学家、医学家和文人的名字进行地名的团组化命名。该海山以明朝著名医学家李时珍命名，纪念他在医学上的重要贡献。 A group naming of undersea features after the names of ancient Chinese scientists, medical scientists and literati. The Seamount is named after Li Shizhen (1518−1593 A.D.), a famous medical scientist in the Ming Dynasty (1368−1644 A.D.), to commemorate his outstanding contribution to the medical science.		

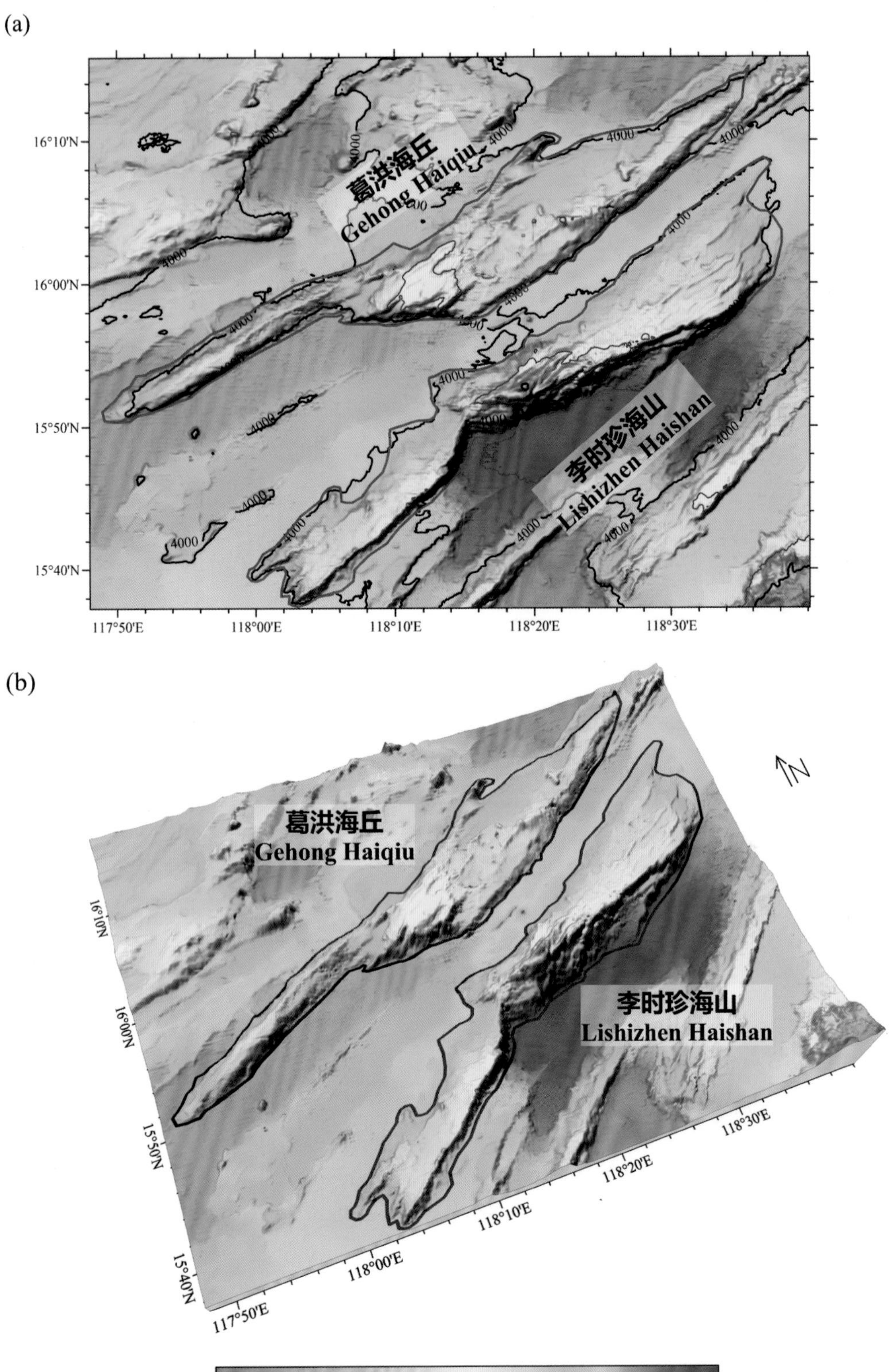

图 5-16 葛洪海丘、李时珍海山

(a) 海底地形图（等深线间隔 500 米）；(b) 三维海底地形图

Fig.5-16 Gehong Haiqiu, Lishizhen Haishan

(a) Seafloor topographic map (with contour interval of 500 m)；(b) 3-D seafloor topographic map

5.41 张仲景海山

标准名称 Standard Name	张仲景海山 Zhangzhongjing Haishan	类别 Generic Term	海山 Seamount
中心点坐标 Center Coordinates	15°50.8′N, 117°30.9′E	规模（千米 × 千米） Dimension（km × km）	45 × 26
最小水深（米） Min Depth (m)	1670	最大水深（米） Max Depth (m)	4370
地理实体描述 Feature Description	张仲景海山位于黄岩岛北部，平面形态近似椭圆，其北坡较陡，南坡较缓（图 5–17）。 Zhangzhongjing Haishan is located to the north of Huangyan Dao with a nearly oval planform. Its north slope is relatively steeper, while its south slope is relatively gentler (Fig.5–17).		
命名由来 Origin of Name	以我国古代科学家、医学家和文人的名字进行地名的团组化命名。该海山以东汉末年著名医学家张仲景命名，纪念他在医学上的重要贡献。 A group naming of undersea features after the names of ancient Chinese scientists, medical scientists and literati. The Seamount is named after Zhang Zhongjing (150–215 A.D.), a famous medical scientist in late Eastern Han Dynasty (184–220 A.D.), to commemorate his outstanding contribution to the medical science.		

5.42 扁鹊海丘

标准名称 Standard Name	扁鹊海丘 Bianque Haiqiu	类别 Generic Term	海丘 Hill
中心点坐标 Center Coordinates	16°01.9′N, 117°30.8′E	规模（千米 × 千米） Dimension（km × km）	8 × 8
最小水深（米） Min Depth (m)	3630	最大水深（米） Max Depth (m)	4430
地理实体描述 Feature Description	扁鹊海丘位于张仲景海山北部，平面形态呈圆形（图 5–17）。 Bianque Haiqiu is located to the north of Zhangzhongjing Haishan with a circular planform (Fig.5–17).		
命名由来 Origin of Name	以我国古代科学家、医学家和文人的名字进行地名的团组化命名。该海丘以春秋时期著名医学家扁鹊命名，纪念他在医学上的重要贡献。 A group naming of undersea features after the names of ancient Chinese scientists, medical scientists and literati. The Hill is named after Bian Que, a famous medical scientist in the Spring and Autumn Period (770 B.C.–476 B.C.), to commemorate his outstanding contribution to the medical science.		

5.43 基石海丘

标准名称 Standard Name	基石海丘 Jishi Haiqiu	类别 Generic Term	海丘 Hill
中心点坐标 Center Coordinates	15°42.4'N, 117°38.0'E	规模（千米 × 千米） Dimension（km × km）	27 × 4
最小水深（米） Min Depth (m)	3580	最大水深（米） Max Depth (m)	4220
地理实体描述 Feature Description	基石海丘位于张仲景海山东南侧，平面形态呈北西—南东向的长条形（图 5–17）。 Jishi Haiqiu is located on the southeast of Zhangzhongjing Haishan, with a planform in the shape of a long strip in the direction of NW–SE (Fig.5–17).		
命名由来 Origin of Name	以中国古代文学作品中的短语进行地名的团组化命名。该海底地名的专名取词自北宋诗人郭祥正的《寄题留二君仪田园基石亭二首其一》："春来成几章，想有基石铭"。"基石"即用作建筑物基础的石头。 A group naming of undersea features after the phrases in ancient Chinese literatures. The specific term of this undersea feature name "Jishi", means the cornerstone used as the foundation of a construction, is derived from the poetic lines "There are some sentences about spring, and the related inscriptions on the stone are unforgettable." in the poem *One of Two Poems for Ji Ti Liu Er Jun Yi Tian Yuan Ji Shi Ting* by Guo Xiangzheng (1035–1113 A.D.), a poet in the Northern Song Dynasty (960–1127 A.D.).		

(a)

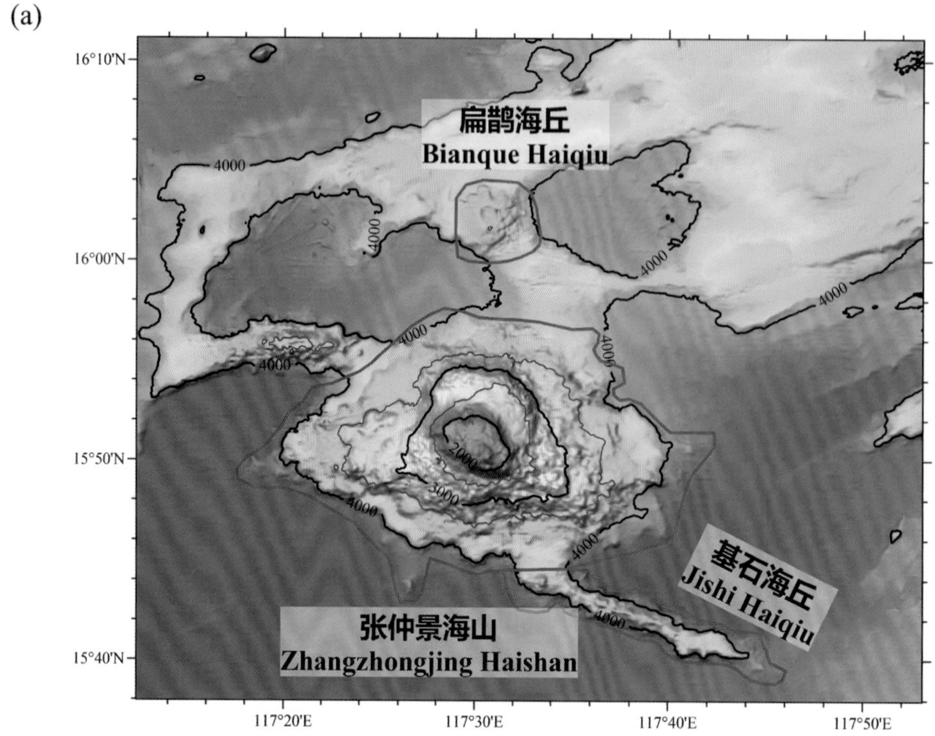

(b)

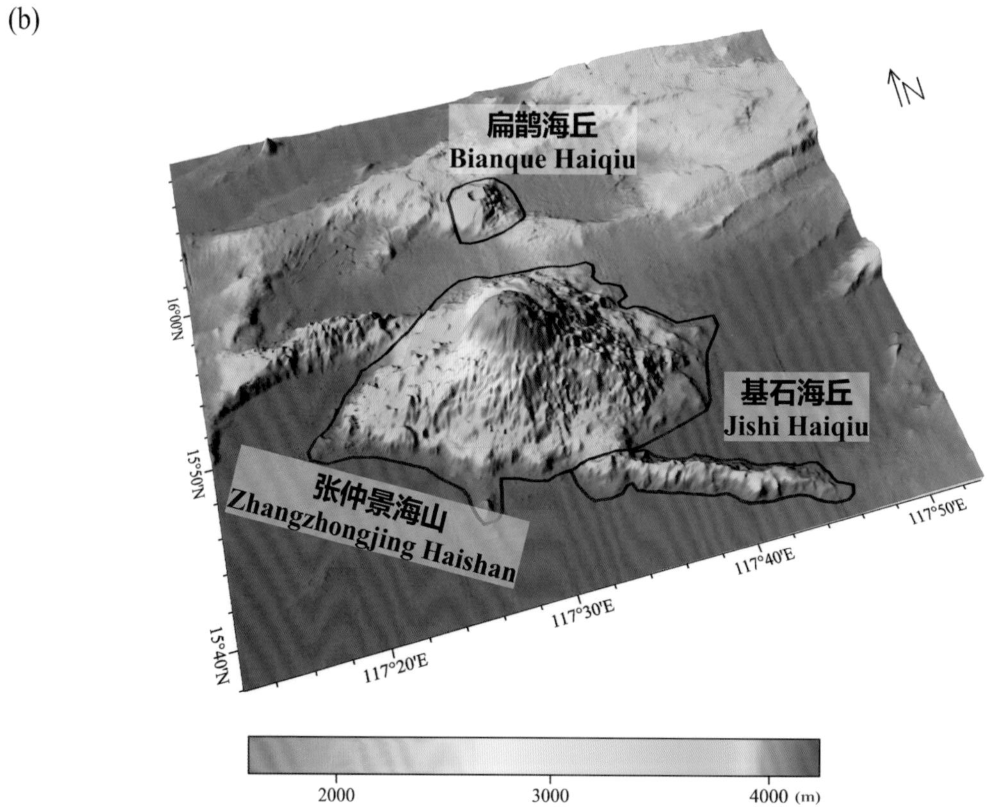

图 5-17　张仲景海山、扁鹊海丘、基石海丘

(a) 海底地形图（等深线间隔 500 米）；(b) 三维海底地形图

Fig.5-17　Zhangzhongjing Haishan, Bianque Haiqiu, Jishi Haiqiu

(a) Seafloor topographic map (with contour interval of 500 m)；(b) 3-D seafloor topographic map

5.44 火石海丘

标准名称 Standard Name	火石海丘 Huoshi Haiqiu	类别 Generic Term	海丘 Hill
中心点坐标 Center Coordinates	15°24.7'N, 117°07.9'E	规模（千米 × 千米） Dimension（km × km）	15 × 6
最小水深（米） Min Depth (m)	3600	最大水深（米） Max Depth (m)	4220
地理实体描述 Feature Description	火石海丘位于黄岩西海山北部，平面形态呈北东—南西向的长条形，其西南侧较高，东北侧较矮（图 5–18）。 Huoshi Haiqiu is located to the north of Huangyanxi Haishan, with a planform in the shape of a long strip in the direction of NE–SW. Its southwest side is relatively higher, while its northeast side is relatively lower (Fig.5–18).		
命名由来 Origin of Name	以中国古代文学作品中的短语进行地名的团组化命名。该海底地名的专名取词自北宋诗人苏舜元的《钓鳖石》："汲泉沙胍动，敲火石痕斜"。"火石"即燧石，敲击时能迸发火星，用来取火。 A group naming of undersea features after the phrases in ancient Chinese literatures. The specific term of this undersea feature name "Huoshi", means flint, which gives off sparks when being struck to make fire, is derived from the poetic lines "The spring water is gurgling, and the flint is struck to produce sparks." in the poem *Turtle-catching Stone* by Su Shunyuan (1006–1054 A.D.), a poet in the Song Dynasty (960–1279 A.D.).		

5.45 飞石海丘

标准名称 Standard Name	飞石海丘 Feishi Haiqiu	类别 Generic Term	海丘 Hill
中心点坐标 Center Coordinates	15°30.3'N, 117°02.5'E	规模（千米 × 千米） Dimension（km × km）	21 × 15
最小水深（米） Min Depth (m)	3712	最大水深（米） Max Depth (m)	4265
地理实体描述 Feature Description	飞石海丘位于黄岩西海山北部，平面形态呈不规则多边形，其内部发育一洼地（图 5–18）。 Feishi Haiqiu is located to the north of Huangyanxi Haishan with a planform in the shape of an irregular polygon. A depression has been developed inside the hill (Fig.5–18).		
命名由来 Origin of Name	以中国古代文学作品中的短语进行地名的团组化命名。该海底地名的专名取词自南宋诗人范成大的《桑岭》："俯惊危栈穿，仰诧飞石落"。"飞石"即古代通过投掷石头来杀伤敌人的武器。 A group naming of undersea features after the phrases in ancient Chinese literatures. The specific term of this undersea feature name "Feishi", was kind of weapon used to kill the enemy by flinging stone in the ancient time, is derived from the poetic lines "Looking down in shock and passing through dangerous stacks, looking up in amazement and flying stones falling." in *Sang Ling* by Fan Chengda (1126–1193 A.D.), a poet in the Southern Song Dynasty (1127–1279 A.D.).		

5.46 鸟蛤海丘

标准名称 Standard Name	鸟蛤海丘 Niaoge Haiqiu	类别 Generic Term	海丘 Hill
中心点坐标 Center Coordinates	15°36.4'N, 116°57.7'E	规模（千米 × 千米） Dimension（km × km）	19 × 8
最小水深（米） Min Depth (m)	3840	最大水深（米） Max Depth (m)	4210
地理实体描述 Feature Description	鸟蛤海丘位于黄岩西海山北部，平面形态似鸟的两翼，其顶部发育多座峰（图 5−18）。 Niaoge Haiqiu is located to the north of Huangyanxi Haiqiu and its planform is in the shape of two wings of a bird, with multiple peaks developed on its top (Fig.5−18).		
命名由来 Origin of Name	以海洋软体动物名进行地名的团组化命名。“鸟蛤”为大型、深水埋栖的双壳贝类，是一种生活在海洋里的软体动物。 A group naming of undersea features after the names of marine mollusks. “Niaoge” refers to cockle, is a large-scale bivalve shellfish dwells in deep water and is a species of marine mollusk.		

(a)

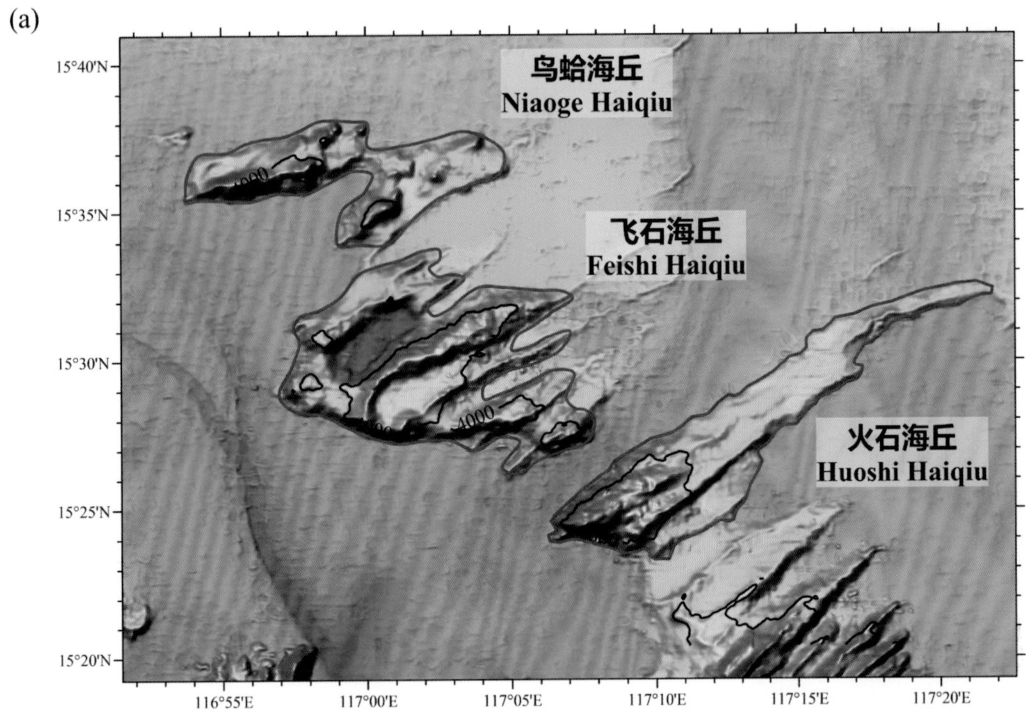

(b)

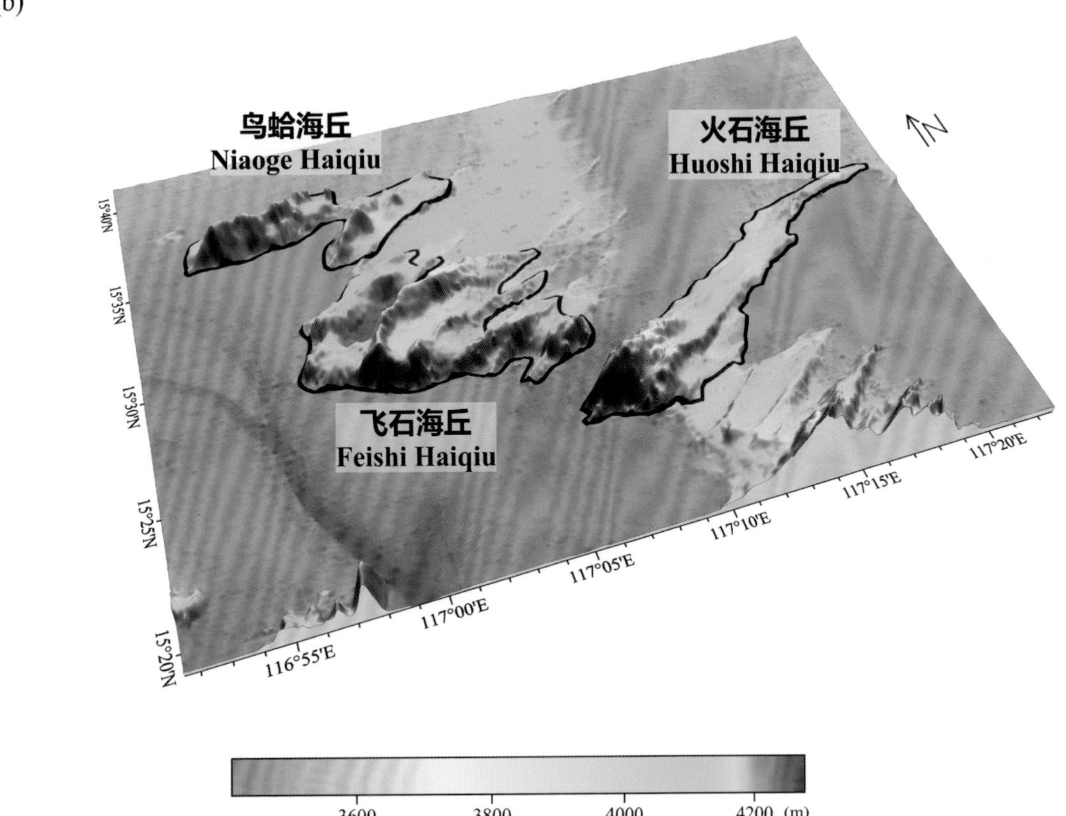

图 5-18 火石海丘、飞石海丘、鸟蛤海丘

(a) 海底地形图（等深线间隔 500 米）；(b) 三维海底地形图

Fig.5-18 Huoshi Haiqiu, Feishi Haiqiu, Niaoge Haiqiu

(a) Seafloor topographic map (with contour interval of 500 m)；(b) 3-D seafloor topographic map

5.47 篮蛤海丘

标准名称 Standard Name	篮蛤海丘 Lange Haiqiu	类别 Generic Term	海丘 Hill
中心点坐标 Center Coordinates	15°19.7'N, 116°57.4'E	规模（千米 × 千米） Dimension（km × km）	9 × 5
最小水深（米） Min Depth (m)	3730	最大水深（米） Max Depth (m)	4230
地理实体描述 Feature Description	篮蛤海丘位于珍贝海山东北部，平面形态呈不规则多边形，其顶部发育多座峰（图 5-19）。 Lange Haiqiu is located to the northeast of Zhenbei Haishan, with a planform in the shape of an irregular polygon. Multiple peaks developed on its top (Fig.5-19).		
命名由来 Origin of Name	以海洋软体动物名进行地名的团组化命名。"篮蛤"是一种生活在海洋里的软体动物。 A group naming of undersea features after the names of marine mollusks. "Lange" refers to corbula, is a species of marine mollusk.		

5.48 舟蚶海丘

标准名称 Standard Name	舟蚶海丘 Zhouhan Haiqiu	类别 Generic Term	海丘 Hill
中心点坐标 Center Coordinates	15°21.7'N, 116°50.1'E	规模（千米 × 千米） Dimension（km × km）	8 × 4
最小水深（米） Min Depth (m)	3870	最大水深（米） Max Depth (m)	4230
地理实体描述 Feature Description	舟蚶海丘位于珍贝海山东北部，平面形态呈半月形（图 5-19）。 Zhouhan Haiqiu is located to the northeast of Zhenbei Haishan with a semilune-shaped planform (Fig.5-19).		
命名由来 Origin of Name	以海洋软体动物名进行地名的团组化命名。"舟蚶"学名为鹰翼魁蛤，是一种生活在海洋里的软体动物。 A group naming of undersea features after the names of marine mollusks. "Zhouhan" refers to scapharca, is a species of mollusk living in ocean.		

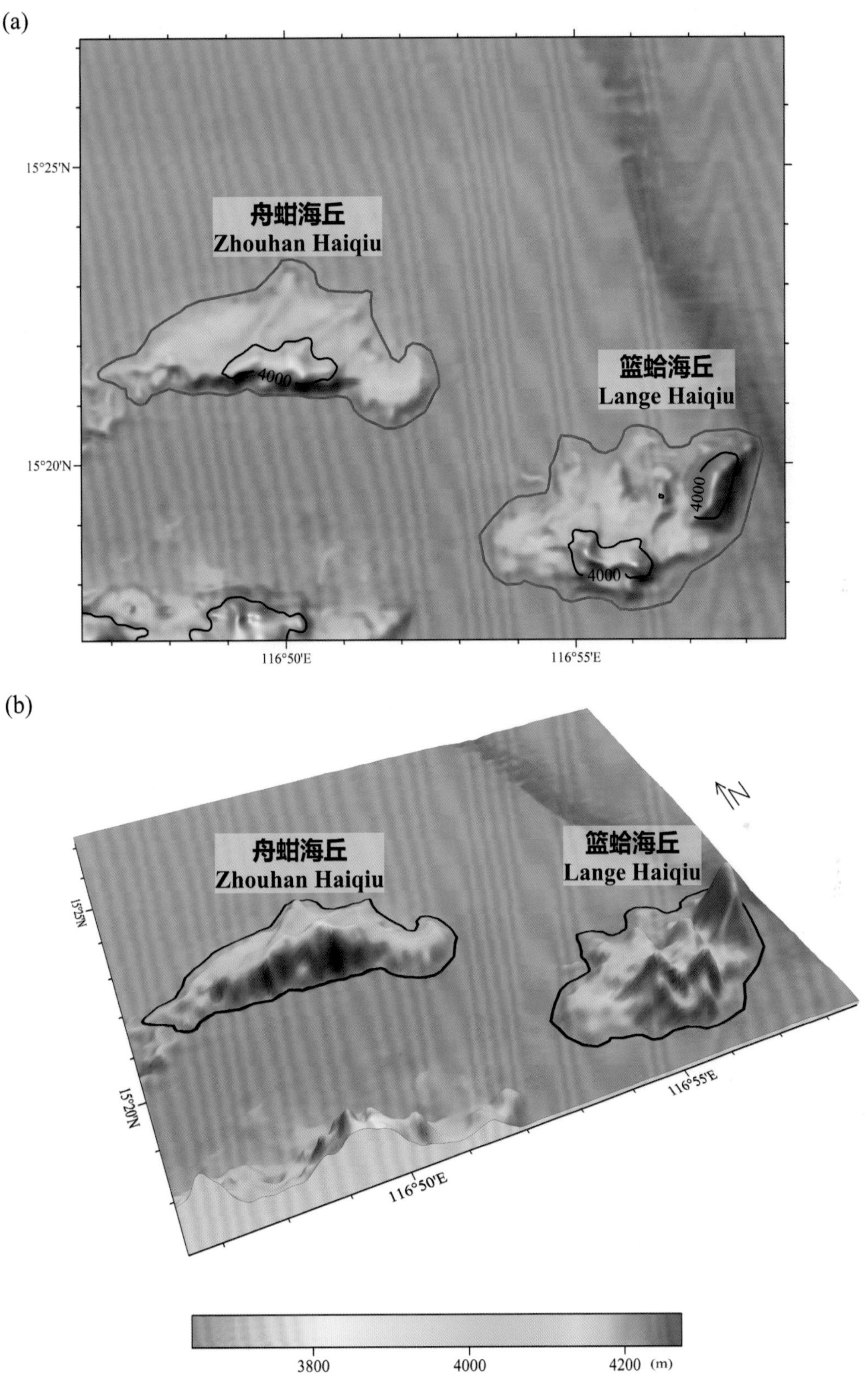

图 5-19　篮蛤海丘、舟蚶海丘

(a) 海底地形图（等深线间隔 500 米）；(b) 三维海底地形图

Fig.5-19　Lange Haiqiu, Zhouhan Haiqiu

(a) Seafloor topographic map (with contour interval of 500 m)；(b) 3-D seafloor topographic map

5.49 帽蚶海丘

标准名称 Standard Name	帽蚶海丘 Maohan Haiqiu	类别 Generic Term	海丘 Hill
中心点坐标 Center Coordinates	15°29.3'N, 116°42.4'E	规模（千米 × 千米） Dimension（km × km）	24 × 14
最小水深（米） Min Depth (m)	3750	最大水深（米） Max Depth (m)	4250
地理实体描述 Feature Description	帽蚶海丘位于珍贝海山北部，平面形态似三叉戟状，由 3 条北东—南西向的脊状海丘组成，其顶部发育多座峰（图 5-20）。 Maohan Haiqiu is located to the north of Zhenbei Haishan, with a planform in the shape of a trident.The Hill is composed of three NE−SW oriented ridge-shaped hills, and has multiple peaks developed on its top (Fig.5−20).		
命名由来 Origin of Name	以海洋软体动物名进行地名的团组化命名。“帽蚶”是一种生活在海洋里的蚶目帽蚶科软体动物。 A group naming of undersea features after the names of marine mollusk. “Maohan” refers to cucullaea labiate, is a species of marine mollusk.		

5.50 青蚶海丘

标准名称 Standard Name	青蚶海丘 Qinghan Haiqiu	类别 Generic Term	海丘 Hill
中心点坐标 Center Coordinates	15°20.8'N, 116°42.6'E	规模（千米 × 千米） Dimension（km × km）	15 × 5
最小水深（米） Min Depth (m)	4060	最大水深（米） Max Depth (m)	4260
地理实体描述 Feature Description	青蚶海丘位于珍贝海山北部，平面形态呈北东东—南西西向的长条形（图 5-20）。 Qinghan Haiqiu is located to the north of Zhenbei Haishan, with a planform in the shape of a long strip in the direction of NEE−SWW (Fig.5−20).		
命名由来 Origin of Name	以海洋软体动物名进行地名的团组化命名。“青蚶”学名为青胡魁蛤，是一种生活在海洋里的软体动物。 A group naming of undersea features after the names of marine mollusk. “Qinghan” refers to barbatia, is a species of marine mollusk.		

5.51 椿蚶海丘

标准名称 Standard Name	椿蚶海丘 Chunhan Haiqiu	类别 Generic Term	海丘 Hill
中心点坐标 Center Coordinates	15°22.7'N, 116°36.2'E	规模（千米 × 千米） Dimension（km × km）	19 × 5
最小水深（米） Min Depth (m)	4030	最大水深（米） Max Depth (m)	4280
地理实体描述 Feature Description	椿蚶海丘位于珍贝海山北部，平面形态呈北东东—南西西向的长条形（图 5–20）。 Chunhan Haiqiu is located to the north of Zhenbei Haishan, with a planform in the shape of a long strip in the direction of NEE–SWW (Fig.5–20).		
命名由来 Origin of Name	以海洋软体动物名进行地名的团组化命名。“椿蚶”学名为船魁蛤，是一种生活在海洋里的软体动物。 A group naming of undersea features after the names of marine mollusks. “Chunhan” refers to arca avellana, is a species of marine mollusk.		

5.52 贻贝海脊

标准名称 Standard Name	贻贝海脊 Yibei Haiji	类别 Generic Term	海脊 Ridge
中心点坐标 Center Coordinates	15°23.1'N, 116°30.4'E	规模（千米 × 千米） Dimension（km × km）	10 × 3
最小水深（米） Min Depth (m)	4020	最大水深（米） Max Depth (m)	4280
地理实体描述 Feature Description	贻贝海脊位于珍贝海山北部，平面形态呈北东东—南西西向的长条形（图 5–20）。 Yibei Haiji is located to the north of Zhenbei Haishan, with a planform in the shape of a long stripe in the direction of NEE–SWW (Fig.5–20).		
命名由来 Origin of Name	以海洋软体动物名进行地名的团组化命名。“贻贝”别名为船海虹，是一种生活在海洋里的软体动物。 A group naming of undersea features after the names of marine mollusk. “Yibei” refers to mussel, is a species of marine mollusk.		

(a)

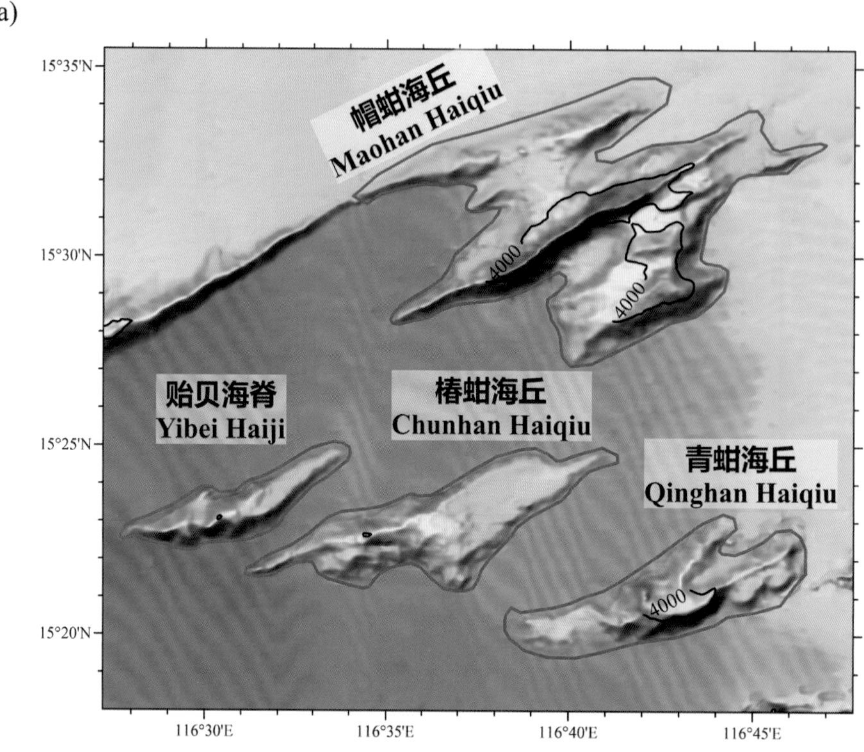

(b)

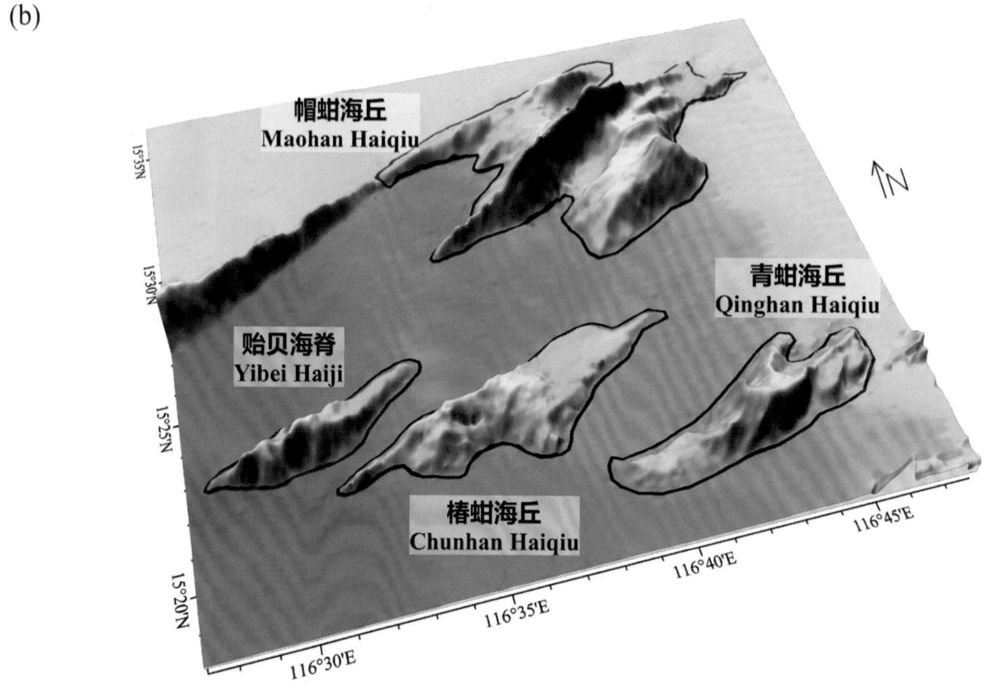

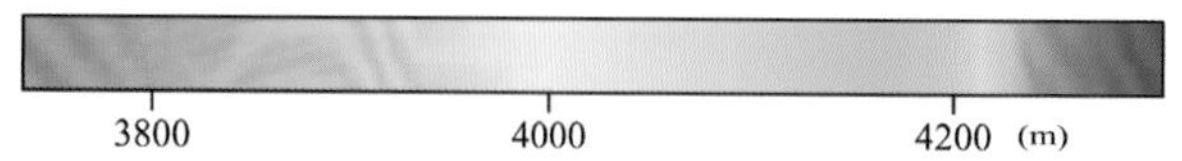

图 5-20 帽蚶海丘、青蚶海丘、椿蚶海丘、贻贝海脊

(a) 海底地形图（等深线间隔 500 米）；(b) 三维海底地形图

Fig.5-20 Maohan Haiqiu, Qinghan Haiqiu, Chunhan Haiqiu, Yibei Haiji

(a) Seafloor topographic map (with contour interval of 500 m); (b) 3-D seafloor topographic map

5.53 涨中海山

标准名称 Standard Name	涨中海山 Zhangzhong Haishan	类别 Generic Term	海山 Seamount
中心点坐标 Center Coordinates	15°34.1′N, 116°11.1′E	规模（千米 × 千米） Dimension（km × km）	55 × 27
最小水深（米） Min Depth (m)	642	最大水深（米） Max Depth (m)	4305
地理实体描述 Feature Description	涨中海山位于南海海盆北部，平面形态似一只在水中遨游的龟。海山顶部发育两座峰，靠东侧的主峰水深约 650 米，靠西侧的次峰水深约 2480 米（图 5-21）。 Zhangzhong Haishan is located in the north of Nanhai Haipen, with a planform in the shape of a swimming fish. Two peaks have developed on its top, with the depth of the main peak on the east side of 650 m and the depth of the second peak on the west side of 2480 m (Fig.5-21).		
命名由来 Origin of Name	1986 年，国务院、外交部和中国地名委员会批准了前地矿部第二海洋地质调查大队（现广州海洋地质调查局）对南海 22 个海底地理实体进行的命名，“涨中海山”是这 22 个海底地理实体名称之一。南海古名之一为涨海，此海山位于南海中部，因此得名。 In 1986, the State Council, the Ministry of Foreign Affairs and Chinese Toponymy Committee approved 22 undersea feature names in Nanhai named by former Second Marine Geological Survey Brigade of the Ministry of Geology and Mineral Resources (present Guangzhou Marine Geological Survey of China Geological Survey). Zhangzhong Haishan is one of the 22 undersea feature names. The Seamount is located in the middle of the South China Sea, which was once named “Zhanghai” in ancient Chinese, so the word “Zhangzhong” was used to name the Seamount.		

5.54 竹溪海丘

标准名称 Standard Name	竹溪海丘 Zhuxi Haiqiu	类别 Generic Term	海丘 Hill
中心点坐标 Center Coordinates	15°28.1'N, 116°00.8'E	规模（千米 × 千米） Dimension（km × km）	10 × 5
最小水深（米） Min Depth (m)	4020	最大水深（米） Max Depth (m)	4240
地理实体描述 Feature Description	竹溪海丘位于涨中海山西南部，平面形态呈不规则多边形（图 5-21）。 Zhuxi Haiqiu is located to the southwest of Zhangzhong Haishan, with a planform in the shape of an irregular polygon (Fig.5-21).		
命名由来 Origin of Name	以中国古代文学作品中的短语进行地名的团组化命名。该海底地名的专名取词自北宋诗人苏轼的《虞美人·波声拍枕长淮晓》：“竹溪花浦曾同醉，酒味多于泪”。“竹溪”即竹溪河。 A group naming of undersea features after the phrases in ancient Chinese literatures. The specific term of this undersea feature name “Zhuxi”, means bamboo brook, is derived from the sentence “You and I ever drunk among flowers along bamboo brook, when the flavor of wine was stronger than the taste of tear from parting” in *A Poem to the Tune of Yu Mei Ren* by Su Shi (1037–1101 A.D.), a famous poet in the Northern Song Dynasty (960–1279 A.D.).		

(a)

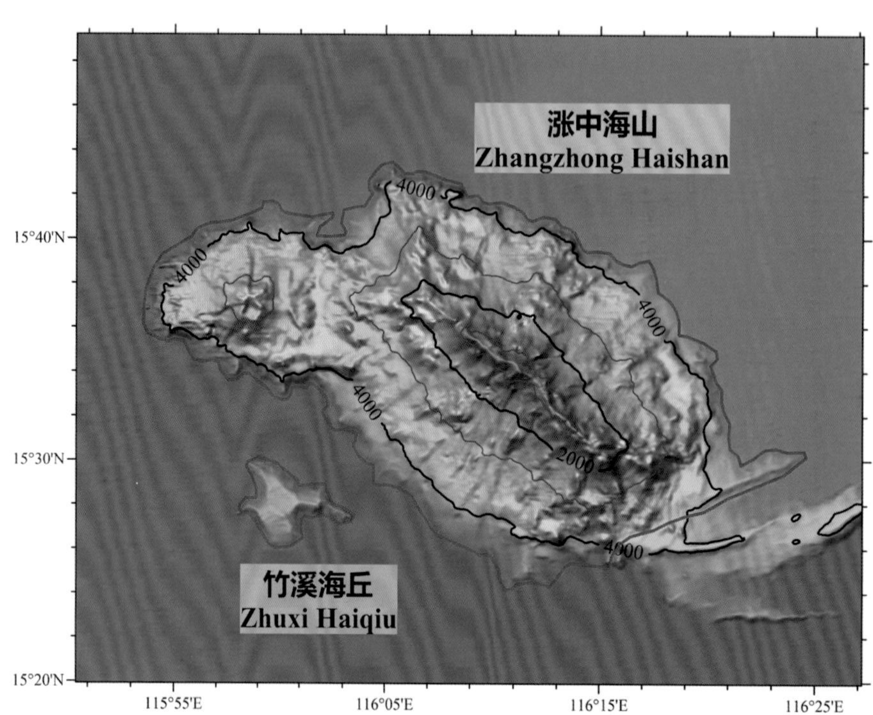

(b)

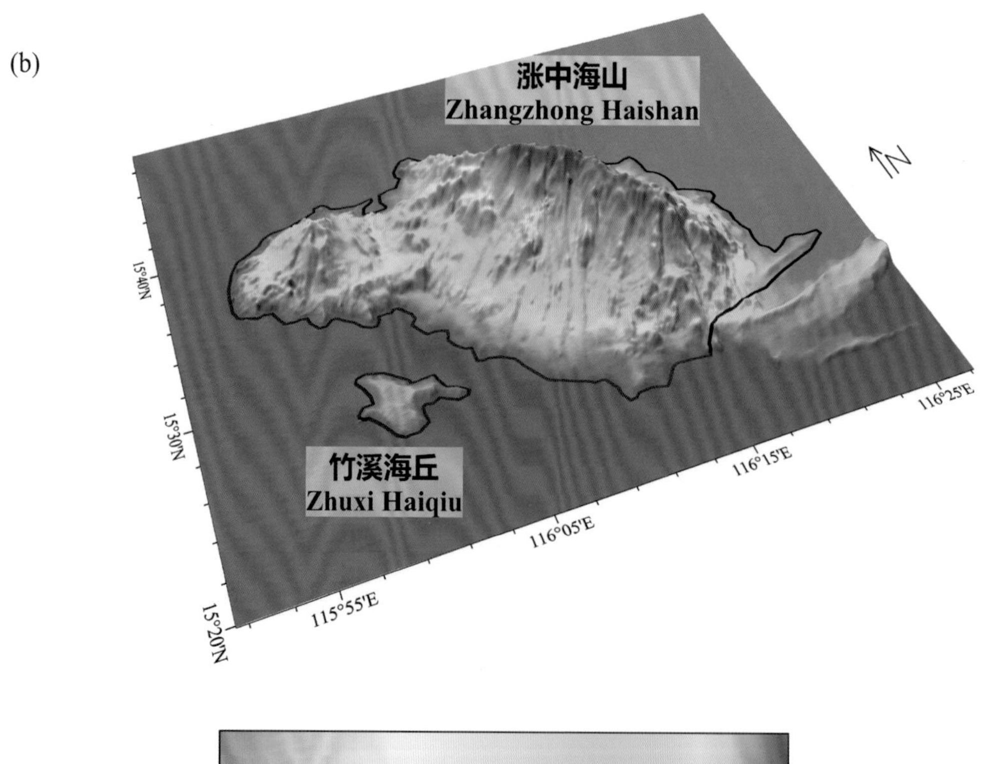

图 5-21　涨中海山、竹溪海丘

(a) 海底地形图（等深线间隔 1000 米）；(b) 三维海底地形图

Fig.5-21　Zhangzhong Haishan, Zhuxi Haiqiu

(a) Seafloor topographic map (with contour interval of 1000 m)；(b) 3-D seafloor topographic map

5.55 兰溪海山

标准名称 Standard Name	兰溪海山 Lanxi Haishan	类别 Generic Term	海山 Seamount
中心点坐标 Center Coordinates	15°16.5'N, 116°11.4'E	规模（千米 × 千米） Dimension（km × km）	20 × 13
最小水深（米） Min Depth (m)	2900	最大水深（米） Max Depth (m)	4300
地理实体描述 Feature Description	兰溪海山位于珍贝海山西北部，平面形态似一只在水中遨游的龟（图 5–22）。 Lanxi Haishan is located to the northwest of Zhenbei Haishan, with a planform in the shape of a swimming turtle (Fig.5–22).		
命名由来 Origin of Name	以中国古代文学作品中的短语进行地名的团组化命名。该海底地名的专名取词自北宋诗人苏轼的《浣溪沙 · 游蕲水清泉寺》：“寺临兰溪，溪水西流”。“兰溪”在今湖北浠水县的一条河流。 A group naming of undersea features after the phrases in ancient Chinese literatures. The specific term of this undersea feature name “Lanxi”, means Orchid River which is located in today’s Xishui county in Hubei province, is derived from the sentence “The temple is along the Orchid River, the water of which runs towards the west” in *A Visit to Clear Spring Temple in Qishui to the Tune of Huan Xi Sha* by Su Shi (1037–1101 A.D.), a famous poet during the Northern Song Dynasty (960–1279 A.D.).		

5.56 张先海山

标准名称 Standard Name	张先海山 Zhangxian Haishan	类别 Generic Term	海山 Seamount
中心点坐标 Center Coordinates	15°16.0′N, 115°57.2′E	规模（千米 × 千米） Dimension（km × km）	22 × 10
最小水深（米） Min Depth (m)	2888	最大水深（米） Max Depth (m)	4316
地理实体描述 Feature Description	张先海山位于珍贝海山西北部，平面形态呈北西—南东走向的长条形（图 5–22）。 Zhangxian Haishan is located to the northwest of Zhenbei Haishan, with a planform in the shape of a long strip in the direction of NW–SE (Fig.5–22).		
命名由来 Origin of Name	以我国古代科学家、医学家和文人的名字团组化命名。该海山以宋朝著名词人张先命名，纪念他在中国诗词史上的重要贡献。在他的作品中，有 180 多首优雅的歌词被保存下来，历久弥新。 A group naming of undersea features after the names of ancient Chinese scientists, medical scientists and literati. The feature is named after Zhang Xian (990–1078 A.D.), a famous lyricist during the Song Dynasty (960–1279 A.D.), to commemorate his outstanding contribution to Chinese poem and lyrics history. Of his works, more than 180 elegant lyrics have been preserved over the ages.		

5.57 青溪海丘

标准名称 Standard Name	青溪海丘 Qingxi Haiqiu	类别 Generic Term	海丘 Hill
中心点坐标 Center Coordinates	15°13.7'N, 115°47.0'E	规模（千米 × 千米） Dimension（km × km）	22 × 15
最小水深（米） Min Depth (m)	3410	最大水深（米） Max Depth (m)	4280
地理实体描述 Feature Description	青溪海丘位于珍贝海山西北部，平面形态呈长条形，其山脊线呈北西—南东走向（图 5−22）。 Qingxi Haiqiu is located to the northwest of Zhenbei Haishan, with a planform in the shape of a long strip. The ridge line of the Hill stretches in the direction of NW−SE. (Fig.5−22).		
命名由来 Origin of Name	以中国古代文学作品中的短语进行地名的团组化命名。该海底地名的专名取词自唐朝诗人王维的《青溪》："言入黄花川，每逐青溪水"。"青溪"指的是今陕西勉县之东的一条河流。 A group naming of undersea features after the phrases in ancient Chinese literatures. The specific term of this undersea feature name "Qingxi", means Blue Brook, which is located in the east of today's Mian county in Shanxi province, is derived from the poetic lines "To view the Yellow-flower Stream, always I follow the course of Blue Brooks' flow" in the poem *Blue Brook* by Wang Wei (701−761 A.D.), a poet in the Tang Dynasty (618−907 A.D.).		

(a)

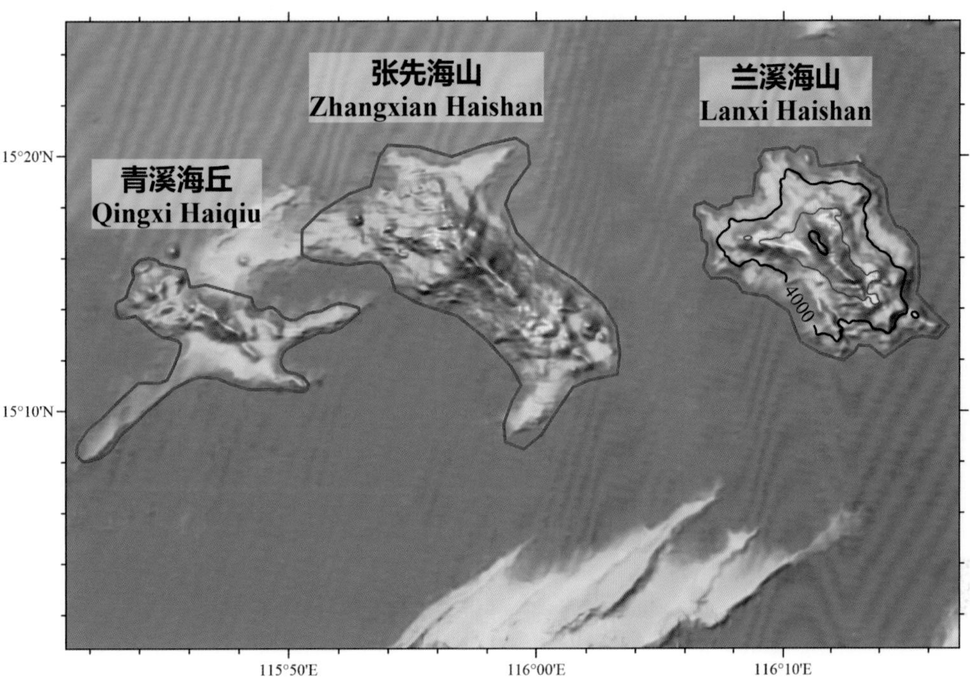

(b)

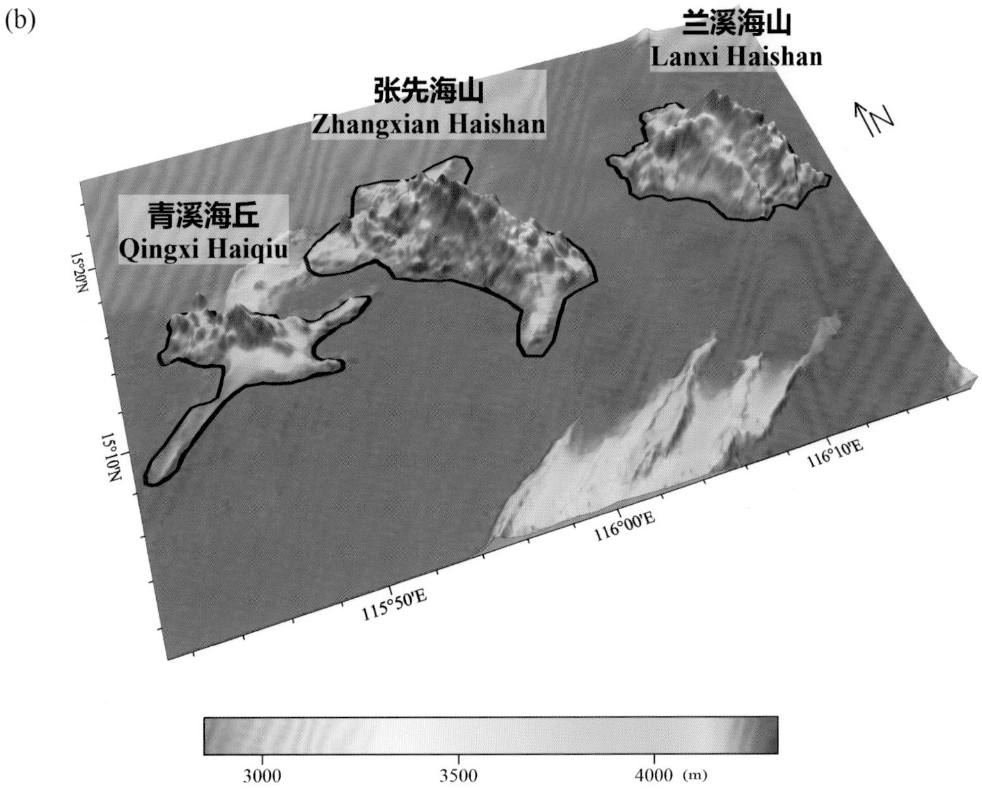

图 5-22　兰溪海山、张先海山、青溪海丘

(a) 海底地形图（等深线间隔 500 米）；(b) 三维海底地形图

Fig.5-22　Lanxi Haishan, Zhangxian Haishan, Qingxi Haiqiu

(a) Seafloor topographic map (with contour interval of 500 m); (b) 3-D seafloor topographic map

5.58 鼓角海丘

标准名称 Standard Name	鼓角海丘 Gujiao Haiqiu	类别 Generic Term	海丘 Hill
中心点坐标 Center Coordinates	15°37.9'N, 115°42.9'E	规模（千米 × 千米） Dimension（km × km）	20 × 3
最小水深（米） Min Depth (m)	3850	最大水深（米） Max Depth (m)	4220
地理实体描述 Feature Description	鼓角海丘位于涨中海山西部，平面形态呈长条形。自西向东，海丘的山脊线由北东—南西向转为东—西向（图 5−23）。 Gujiao Haiqiu is located to the west of Zhangzhong Haishan, with a planform in the shape of a long strip. The ridge line of the Hill turns from the direction of NE−SW to the direction of E−W from west to east (Fig.5−23).		
命名由来 Origin of Name	以毛泽东诗词作品中的词进行地名的团组化命名。该海底地名的专名取词自毛泽东的《西江月·井冈山》："山下旌旗在望，山头鼓角相闻"。"鼓角"即战鼓和号角。 A group naming of undersea features after the words in Mao Zedong's poems. The specific term of this undersea feature name "Gujiao", means war drum and bulge, is derived from the poetic lines "Below the hill fly banners and flags, Above the hilltop sound the drums and bulges" in the poem *Mount Jinggang to the Tune of Xi Jiang Yue* by Mao Zedong (1893−1976).		

5.59 壁垒海丘

标准名称 Standard Name	壁垒海丘 Bilei Haiqiu	类别 Generic Term	海丘 Hill
中心点坐标 Center Coordinates	15°31.7'N, 115°45.4'E	规模（千米 × 千米） Dimension（km × km）	10 × 4
最小水深（米） Min Depth (m)	4060	最大水深（米） Max Depth (m)	4240
地理实体描述 Feature Description	壁垒海丘位于涨中海山西部，平面形态呈近北东—南西向的菱形。海丘的西北坡较缓，东南坡较陡（图 5−23）。 Bilei Haiqiu is located to the west of Zhangzhong Haishan, with a planform in the shape of a diamond nearly in the direction of NE−SW. Its northwest slope is relatively gentler, while its southeast slope is relatively steeper (Fig.5−23).		
命名由来 Origin of Name	以毛泽东诗词作品中的词进行地名的团组化命名。该海底地名的专名取词自毛泽东的《西江月·井冈山》："早已森严壁垒，更加众志成城"。"壁垒"即古时军营的围墙。 A group naming of undersea features after the words in Mao Zedong's poems. The specific term of this undersea feature name "Bilei", means the defence wall of barracks in the ancient time, is derived from the poetic lines "Already the defence wall is iron-clad, yet our united will is all the more fortified" in the poem *Mount Jinggang to the Tune of Xi Jiang Yue* by Mao Zedong (1893−1976).		

5.60 旌旗海丘

标准名称 Standard Name	旌旗海丘 Jingqi Haiqiu	类别 Generic Term	海丘 Hill
中心点坐标 Center Coordinates	15°39.6'N, 115°25.2'E	规模（千米 × 千米） Dimension（km × km）	9 × 3
最小水深（米） Min Depth (m)	3820	最大水深（米） Max Depth (m)	4220
地理实体描述 Feature Description	旌旗海丘位于涨中海山西部，平面形态近似北东东—南西西向的长条形。海丘的西北坡较缓，东南坡较陡（图 5–23）。 Jingqi Haiqiu is located to the west of Zhangzhong Haishan, with a planform in the shape of a long strip nearly in the direction of NEE–SWW. Its northwest slope is relatively gentler, while its southwestern slope is relatively steeper (Fig.5–23).		
命名由来 Origin of Name	以毛泽东诗词作品中的词进行地名的团组化命名。该海底地名的专名取词自毛泽东的《西江月・井冈山》：“山下旌旗在望，山头鼓角相闻”。“旌旗”即旗帜。 A group naming of undersea features after the words in Mao Zedong's poems. The specific term of this undersea feature name “Jingqi”, means banner and flag, is derived from the poetic lines “Below the hill fly banners and flags, Above the hilltop sound the drums and bulges” in the poem *Mount Jinggang to the Tune of Xi Jiang Yue* by Mao Zedong (1893–1976).		

(a)

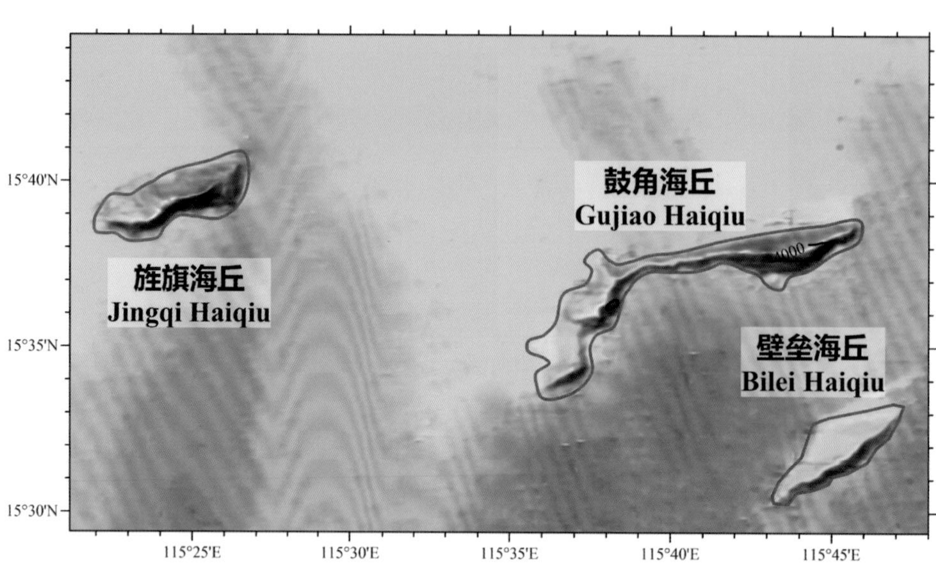

(b)

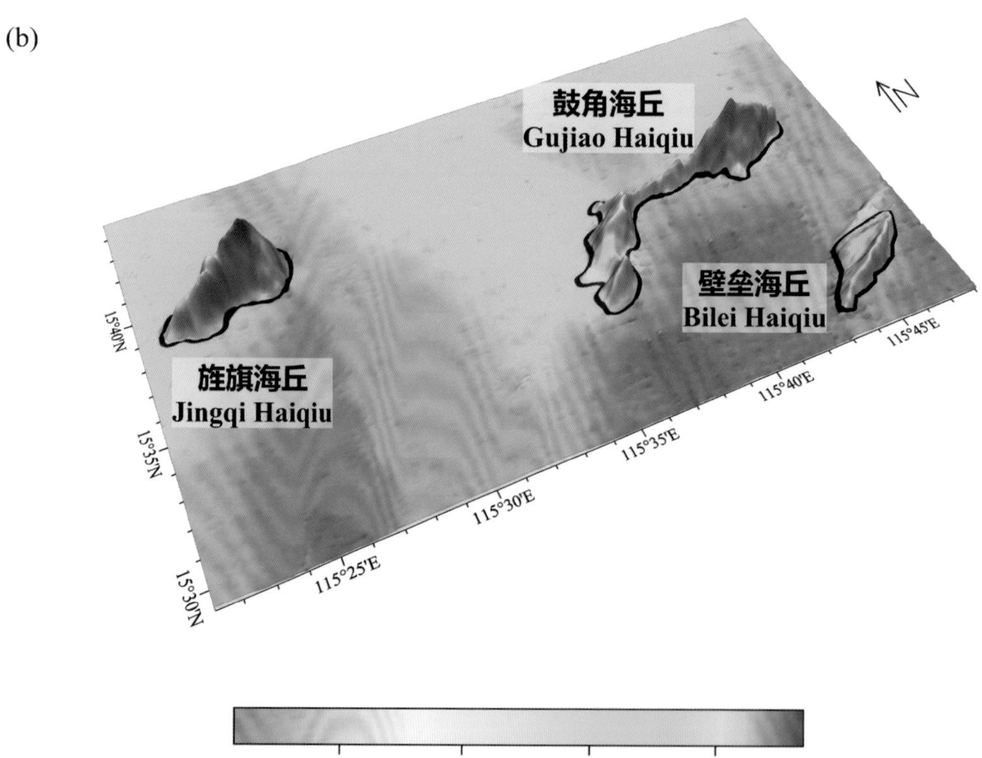

图 5-23　鼓角海丘、壁垒海丘、旌旗海丘

(a) 海底地形图（等深线间隔 500 米）；(b) 三维海底地形图

Fig.5-23　Gujiao Haiqiu, Bilei Haiqiu, Jingqi Haiqiu

(a) Seafloor topographic map (with contour interval of 500 m)；(b) 3-D seafloor topographic map

5.61 西涧海脊

标准名称 Standard Name	西涧海脊 Xijian Haiji	类别 Generic Term	海脊 Ridge
中心点坐标 Center Coordinates	15°21.4'N, 115°29.6'E	规模（千米 × 千米） Dimension（km × km）	20 × 3
最小水深（米） Min Depth (m)	3500	最大水深（米） Max Depth (m)	4280
地理实体描述 Feature Description	西涧海脊位于中沙海台东部，平面形态呈北东—南西向的长条形（图 5-24）。 Xijian Haiji is located to the east of Zhongsha Haitai, with a planform in the shape of a long strip in the direction of NE–SW (Fig.5–24).		
命名由来 Origin of Name	以中国古代文学作品中的短语进行地名的团组化命名。该海底地名的专名取词自唐朝诗人韦应物《滁州西涧》的题名。“西涧”即滁州城西，俗名称上马河。 A group naming of undersea features after the phrases in ancient Chinese literatures. The specific term of this undersea feature name “Xijian”, means West Stream, commonly known as Shangma River in Chuzhou, is derived from the title of the poem *On the West Stream in Chuzhou* by Wei Yingwu (737–792 A.D.), a poet in the Tang Dynasty (618–907 A.D.).		

5.62 荆溪海丘

标准名称 Standard Name	荆溪海丘 Jingxi Haiqiu	类别 Generic Term	海丘 Hill
中心点坐标 Center Coordinates	15°09.3'N, 115°15.1'E	规模（千米 × 千米） Dimension（km × km）	23 × 15
最小水深（米） Min Depth (m)	3880	最大水深（米） Max Depth (m)	4250
地理实体描述 Feature Description	荆溪海丘位于中沙海台东部，其东、西两侧各有一个水深约 3880 米的峰（图 5-24）。 Jingxi Haiqiu is located to the east of Zhongsha Haitai. Both the east and west sides of the Hill has a peak with depth of 3880 m (Fig.5–24).		
命名由来 Origin of Name	以中国古代文学作品中的短语进行地名的团组化命名。该海底地名的专名取词自唐朝诗人王维的《山中》：“荆溪白石出，天寒红叶稀”。“荆溪”指的是浐水，源出陕西蓝田县西南秦岭山中，北流至长安东北入灞。 A group naming of undersea features after the phrases in ancient Chinese literatures. The specific term of this undersea feature name “Jingxi”, refers to Chanshui river, which originates from the Qinling Mountains in the southwest of Lantian county, Shanxi province and flows north to the northeast of Chang'an and enters the Ba River, is derived from the poetic lines “White pebbles expose in thorn brook shallow, red leaves are strewn on the chill hillside” in the poem *In the Hillside* by Wang Wei (701–761 A.D.), a poet in the Tang Dynasty (618–907 A.D.).		

(a)

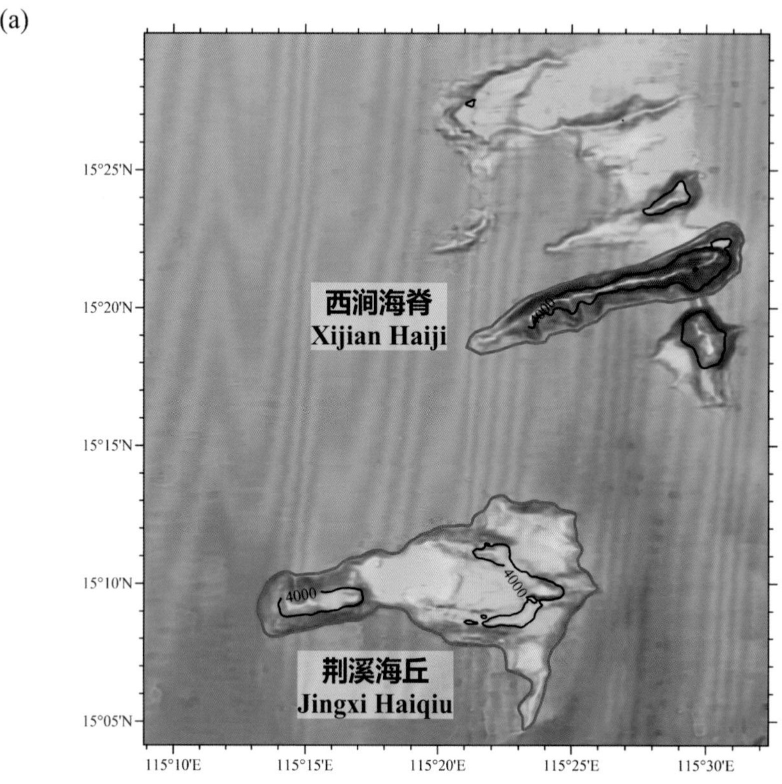

(b)

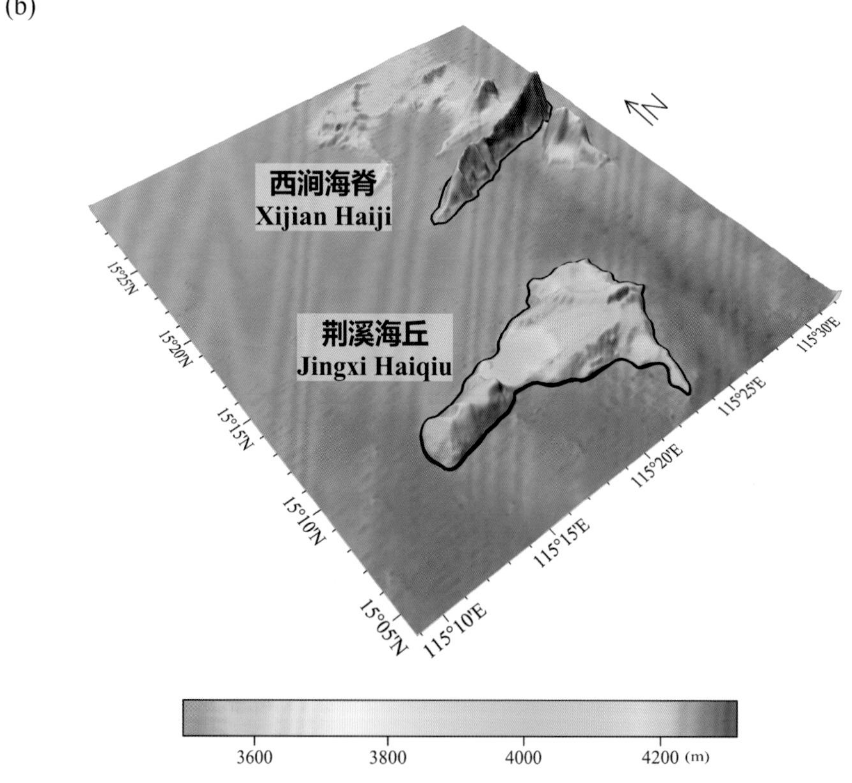

图 5-24　西涧海脊、荆溪海丘

(a) 海底地形图（等深线间隔 500 米）；(b) 三维海底地形图

Fig.5-24　Xijian Haiji, Jingxi Haiqiu

(a) Seafloor topographic map (with contour interval of 500 m)；(b) 3-D seafloor topographic map

5.63 泥蚶海丘

标准名称 Standard Name	泥蚶海丘 Nihan Haiqiu	类别 Generic Term	海丘 Hill
中心点坐标 Center Coordinates	15°21.6'N, 115°03.2'E	规模（千米 × 千米） Dimension（km × km）	15 × 8
最小水深（米） Min Depth (m)	4070	最大水深（米） Max Depth (m)	4300
地理实体描述 Feature Description	泥蚶海丘位于中沙海台东南部，平面形态呈不规则多边形（图 5−25）。 Nihan Haiqiu is located to the southeast of Zhongsha Haitai, with a planform in the shape of an irregular polygon (Fig.5−25).		
命名由来 Origin of Name	以海洋软体动物名进行地名的团组化命名。“泥蚶”别名为泥蚶蜊，是一种生活在海洋里的软体动物。 A group naming of undersea features after the names of marine mollusks. “Nihan” refers to arca granosa, is a species of marine mollusk.		

5.64 双溪海脊

标准名称 Standard Name	双溪海脊 Shuangxi Haiji	类别 Generic Term	海脊 Ridge
中心点坐标 Center Coordinates	15°12.2'N, 114°59.8'E	规模（千米 × 千米） Dimension（km × km）	14 × 4
最小水深（米） Min Depth (m)	3390	最大水深（米） Max Depth (m)	4270
地理实体描述 Feature Description	双溪海脊位于中沙海台东南部，平面形态呈北西—南东向的长条形（图 5−25）。 Shuangxi Haiji is located to the southeast of Zhongsha Haitai, with a planform in the shape of a long strip in the direction of NW−SE (Fig.5−25).		
命名由来 Origin of Name	以中国古代文学作品中的短语进行地名的团组化命名。该海底地名的专名取词自宋朝词人李清照的《武陵春 · 春晚》：“闻说双溪春尚好，也拟泛轻舟”。“双溪”在今浙江省金华市境内，是唐宋时有名的风景秀丽的游览胜地。 A group naming of undersea features after the phrases in ancient Chinese literatures. The specific term of this undersea feature name “Shuangxi”, which is located in Jinhua in Zhejiang province, and is a famous tourist attraction with beautiful scenery, is derived from the poetic lines “I heard the Double Brook is in the prime of spring and ever thought of boating there” in the poem *Late Spring to the Tune of Wu Ling Chun* by Li Qingzhao (1084−1155 A.D.), a poet in the Song Dynasty (960−1279 A.D.).		

(a)

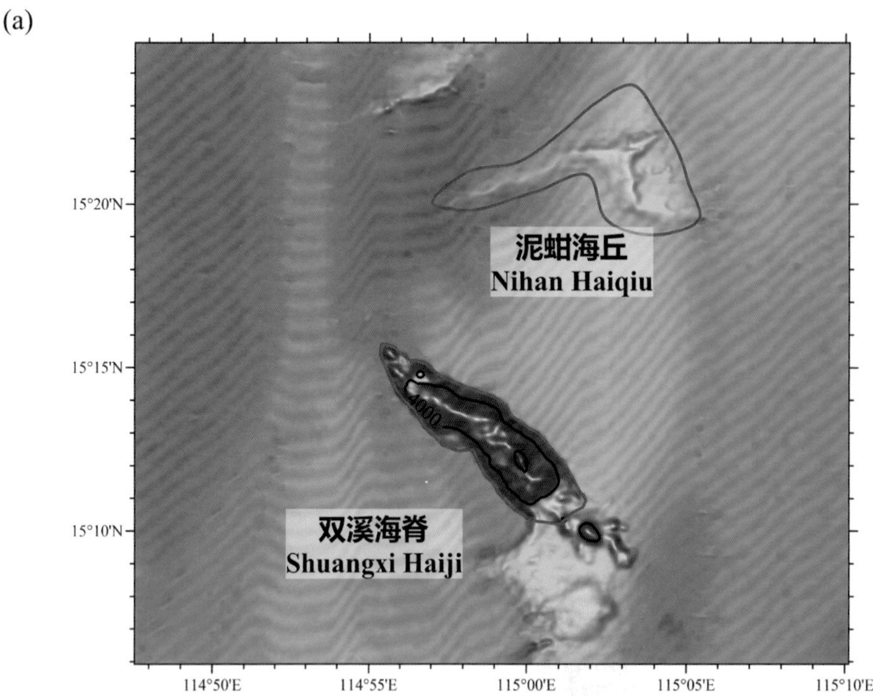

(b)

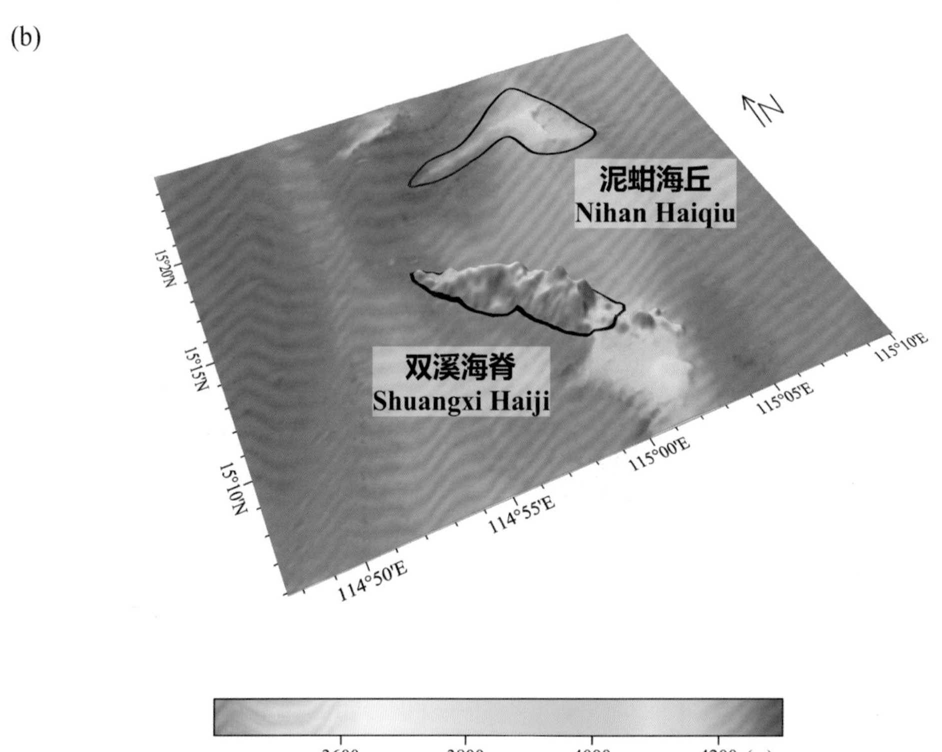

3600 3800 4000 4200 (m)

图 5-25　泥蚶海丘、双溪海脊

(a) 海底地形图（等深线间隔 500 米）；(b) 三维海底地形图

Fig.5-25　Nihan Haiqiu, Shuangxi Haiji

(a) Seafloor topographic map (with contour interval of 500 m)；(b) 3-D seafloor topographic map

6

南海海盆南部和西部海区海底地理实体

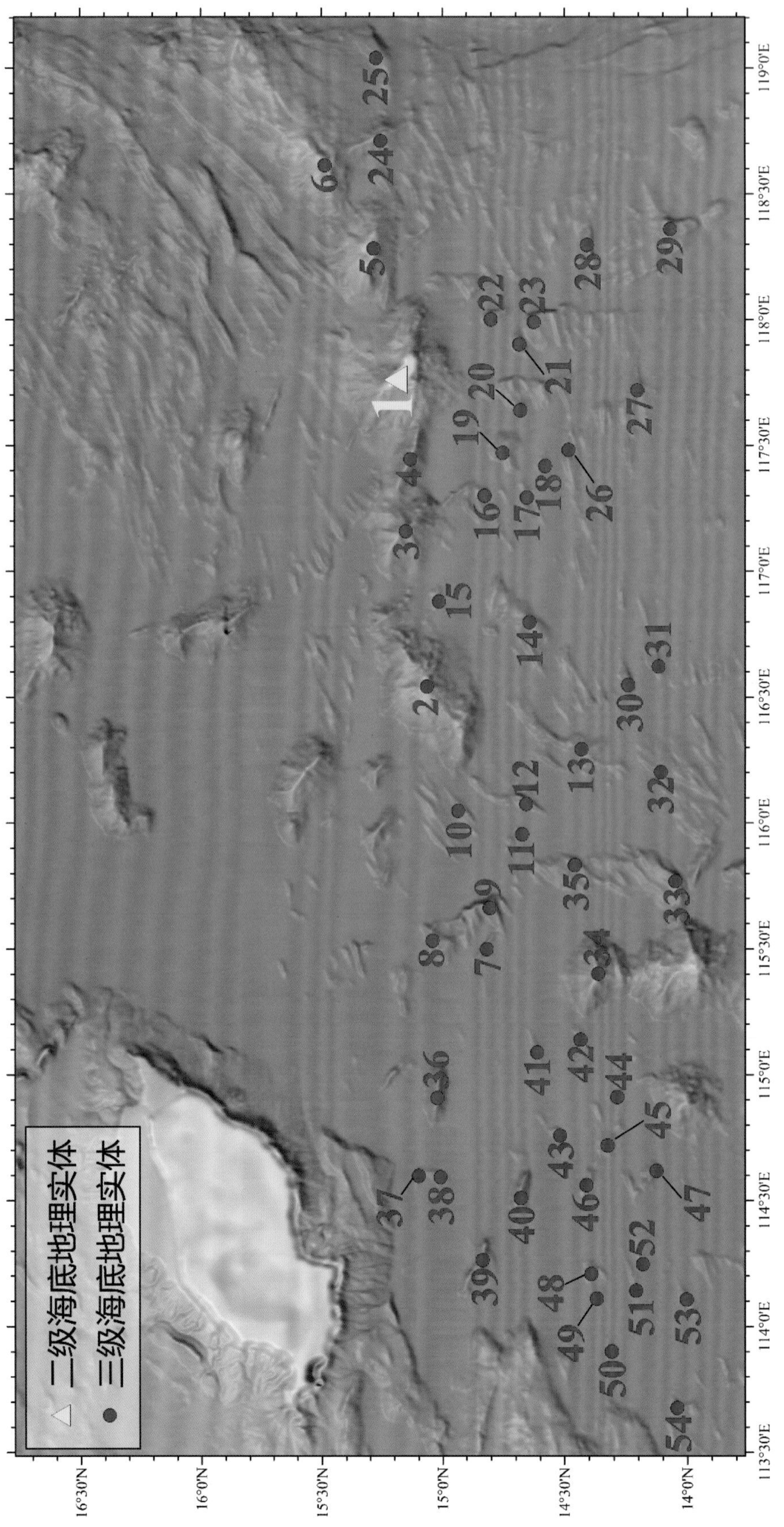

图 6–1　南海海盆南部和西部海区海底地理实体中心点位置示意图，序号含义见表 6–1

Fig.6–1　Location of center coordinates of undersea features in the south and west of Nanhai Haipen, with the meanings of the serial numbers shown in Tab. 6–1

表 6-1 南海海盆南部和西部海区海底地理实体列表

Tab.6-1 List of undersea features in the south and west of Nanhai Haipen

序号 No.	标准名称 Standard Name	汉语拼音 Chinese Phonetic Alphabet	类别 Generic Term	中心点坐标 Center Coordinates		实体等级 Order
				纬度 Latitude	经度 Longitude	
1	珍贝—黄岩海山链 Zhenbei–Huangyan Haishanlian	Zhēnbèi–Huángyán Hǎishānliàn	海山链 Seamount Chain	15°11.6'N	117°45.7'E	2
2	珍贝海山 Zhenbei Haishan	Zhēnbèi Hǎishān	海山 Seamount	15°04.0'N	116°32.5'E	3
3	黄岩西海山 Huangyanxi Haishan	Huángyánxī Hǎishān	海山 Seamount	15°09.1'N	117°09.5'E	3
4	紫贝海山 Zibei Haishan	Zǐbèi Hǎishān	海山 Seamount	15°08.0'N	117°26.7'E	3
5	黄岩东海山 Huangyandong Haishan	Huángyándōng Hǎishān	海山 Seamount	15°17.1'N	118°17.0'E	3
6	贝壳海山 Beike Haishan	Bèiké Hǎishān	海山 Seamount	15°29.2'N	118°36.8'E	3
7	扇贝海脊 Shanbei Haiji	Shànbèi Hǎijǐ	海脊 Ridge	14°49.1'N	115°29.9'E	3
8	明诚海丘 Mingcheng Haiqiu	Míngchéng Hǎiqiū	海丘 Hill	15°02.6'N	115°31.9'E	3
9	易安海山 Yi'an Haishan	Yì'ān Hǎishān	海山 Seamount	14°48.4'N	115°39.7'E	3
10	青蛤海脊 Qingge Haiji	Qīnggé Hǎijǐ	海脊 Ridge	14°56.1'N	116°02.9'E	3
11	紫蛤海丘 Zige Haiqiu	Zǐgé Hǎiqiū	海丘 Hill	14°40.5'N	115°57.3'E	3
12	花螺海脊 Hualuo Haiji	Huāluó Hǎijǐ	海脊 Ridge	14°39.5'N	116°04.5'E	3
13	孟浩然海丘 Menghaoran Haiqiu	Mènghàorán Hǎiqiū	海丘 Hill	14°25.9'N	116°17.6'E	3
14	李白海丘 Libai Haiqiu	Lǐbái Hǎiqiū	海丘 Hill	14°38.7'N	116°47.9'E	3
15	青蛤海丘 Qingge Haiqiu	Qīnggé Hǎiqiū	海丘 Hill	15°00.9'N	116°52.8'E	3
16	深岩海丘 Shenyan Haiqiu	Shēnyán Hǎiqiū	海丘 Hill	14°49.6'N	117°18.0'E	3
17	连岩海丘 Lianyan Haiqiu	Liányán Hǎiqiū	海丘 Hill	14°39.3'N	117°17.6'E	3

续表

序号 No.	标准名称 Standard Name	汉语拼音 Chinese Phonetic Alphabet	类别 Generic Term	中心点坐标 Center Coordinates		实体等级 Order
				纬度 Latitude	经度 Longitude	
18	千岩海丘 Qianyan Haiqiu	Qiānyán Hǎiqiū	海丘 Hill	14°34.8'N	117°25.1'E	3
19	元稹海丘 Yuanzhen Haiqiu	Yuánzhěn Hǎiqiū	海丘 Hill	14°45.2'N	117°28.2'E	3
20	岑参海丘 Censhen Haiqiu	Cénshēn Hǎiqiū	海丘 Hill	14°40.9'N	117°38.3'E	3
21	岩居海丘 Yanju Haiqiu	Yánjū Hǎiqiū	海丘 Hill	14°41.1'N	117°54.1'E	3
22	耕岩海丘 Gengyan Haiqiu	Gēngyán Hǎiqiū	海丘 Hill	14°48.0'N	118°00.0'E	3
23	柳宗元海山 Liuzongyuan Haishan	Liǔzōngyuán Hǎishān	海山 Seamount	14°37.6'N	117°59.4'E	3
24	突岩海山 Tuyan Haishan	Tūyán Hǎishān	海山 Seamount	15°15.3'N	118°42.6'E	3
25	灵岩海山 Lingyan Haishan	Língyán Hǎishān	海山 Seamount	15°16.4'N	119°02.3'E	3
26	杜甫海丘 Dufu Haiqiu	Dùfǔ Hǎiqiū	海丘 Hill	14°29.0'N	117°28.9'E	3
27	孟郊海丘 Mengjiao Haiqiu	Mèngjiāo Hǎiqiū	海丘 Hill	14°12.3'N	117°43.1'E	3
28	王维海丘 Wangwei Haiqiu	Wángwéi Hǎiqiū	海丘 Hill	14°24.5'N	118°17.8'E	3
29	韩愈海山 Hanyu Haishan	Hányù Hǎishān	海山 Seamount	14°04.2'N	118°21.6'E	3
30	瓜螺海脊 Gualuo Haiji	Guāluó Hǎijǐ	海脊 Ridge	14°14.6'N	116°32.9'E	3
31	樱蛤海丘 Yingge Haiqiu	Yīnggé Hǎiqiū	海丘 Hill	14°07.1'N	116°37.3'E	3
32	杜牧海丘 Dumu Haiqiu	Dùmù Hǎiqiū	海丘 Hill	14°06.5'N	116°12.1'E	3
33	秦观海山 Qinguan Haishan	Qínguān Hǎishān	海山 Seamount	14°03.0'N	115°46.1'E	3
34	北岳海山 Beiyue Haishan	Běiyuè Hǎishān	海山 Seamount	14°21.8'N	115°24.1'E	3
35	柳永海丘 Liuyong Haiqiu	Liǔyǒng Hǎiqiū	海丘 Hill	14°27.5'N	115°50.0'E	3
36	钻石海山 Zuanshi Haishan	Zuànshí Hǎishān	海山 Seamount	15°01.4'N	114°54.5'E	3

续表

序号 No.	标准名称 Standard Name	汉语拼音 Chinese Phonetic Alphabet	类别 Generic Term	中心点坐标 Center Coordinates		实体等级 Order
				纬度 Latitude	经度 Longitude	
37	白琥海丘 Baihu Haiqiu	Báihǔ Hǎiqiū	海丘 Hill	15°06.0'N	114°35.9'E	3
38	玄璜海丘 Xuanhuang Haiqiu	Xuánhuáng Hǎiqiū	海丘 Hill	15°00.6'N	114°35.6'E	3
39	蓝宝石海山 Lanbaoshi Haishan	Lánbǎoshí Hǎishān	海山 Seamount	14°50.1'N	114°15.8'E	3
40	红宝石海山 Hongbaoshi Haishan	Hóngbǎoshí Hǎishān	海山 Seamount	14°40.8'N	114°30.6'E	3
41	玉瓶海脊 Yuping Haiji	Yùpíng Hǎijǐ	海脊 Ridge	14°36.8'N	115°05.3'E	3
42	赤璋海丘 Chizhang Haiqiu	Chìzhāng Hǎiqiū	海丘 Hill	14°26.1'N	115°08.3'E	3
43	青圭海丘 Qinggui Haiqiu	Qīngguī Hǎiqiū	海丘 Hill	14°31.1'N	114°45.5'E	3
44	黄琮海丘 Huangcong Haiqiu	Huángcóng Hǎiqiū	海丘 Hill	14°17.2'N	114°54.7'E	3
45	苍璧海丘 Cangbi Haiqiu	Cāngbì Hǎiqiū	海丘 Hill	14°19.5'N	114°43.1'E	3
46	红柱石海丘 Hongzhushi Haiqiu	Hóngzhùshí Hǎiqiū	海丘 Hill	14°24.6'N	114°33.7'E	3
47	孔雀石海丘 Kongqueshi Haiqiu	Kǒngquèshí Hǎiqiū	海丘 Hill	14°07.6'N	114°37.1'E	3
48	绿柱石海丘 Lüzhushi Haiqiu	Lǜzhùshí Hǎiqiū	海丘 Hill	14°23.4'N	114°12.5'E	3
49	理璞海丘 Lipu Haiqiu	Lǐpú Hǎiqiū	海丘 Hill	14°22.1'N	114°06.6'E	3
50	润泽海丘 Runze Haiqiu	Rùnzé Hǎiqiū	海丘 Hill	14°18.5'N	113°54.1'E	3
51	玉璧海脊 Yubi Haiji	Yùbì Hǎijǐ	海脊 Ridge	14°12.6'N	114°08.5'E	3
52	得宝海丘 Debao Haiqiu	Débǎo Hǎiqiū	海丘 Hill	14°10.9'N	114°14.8'E	3
53	石美海丘 Shimei Haiqiu	Shíměi Hǎiqiū	海丘 Hill	14°00.2'N	114°06.4'E	3
54	玉德海脊 Yude Haiji	Yùdé Hǎijǐ	海脊 Ridge	14°02.5'N	113°40.6'E	3

6.1 珍贝—黄岩海山链

标准名称 Standard Name	珍贝—黄岩海山链 Zhenbei–Huangyan Haishanlian	类别 Generic Term	海山链 Seamount Chain
中心点坐标 Center Coordinates	15°11.6′N, 117°45.7′E	规模（千米 × 千米） Dimension（km × km）	40 × 375
最小水深（米） Min Depth (m)	140	最大水深（米） Max Depth (m)	5020
地理实体描述 Feature Description	珍贝—黄岩海山链位于南海海盆的中央深海平原中部，大致沿 15°N 呈北东东—南西西向展布，由珍贝海山、黄岩西海山、紫贝海山、黄岩东海山和贝壳海山等 6 座大小不一的海山呈长条链状排列组成，高耸的海山链与深邃的海盆形成强烈的地形对比，是南海海盆最为壮观的地貌单元（图 6–2）。 Zhenbei–Huangyan Haishanlian is located in the central abyssal plain in Nanhai Haipen and is mainly arranged in the direction of NEE–SWW along 15°N. It consists of six seamounts of various sizes, including Zhenbei Haishan, Huangyanxi Haishan, Zibei Haishan, Huangyandong Haishan, Beike Haishan and so on, which are arranged in the shape of a long chain. The Seamount Chain contrast strikingly with the basins and is the most spectacular geomorphic unit in Nanhai Haipen (Fig.6–2).		
命名由来 Origin of Name	该海山链包含珍贝海山和黄岩岛，因此得名。 The Seamount Chain includes Zhenbei Haishan and Huangyan Dao, so the word “Zhenbei–Huangyan” was used to name the Seamount Chain.		

6.2 珍贝海山

标准名称 Standard Name	珍贝海山 Zhenbei Haishan	类别 Generic Term	海山 Seamount
中心点坐标 Center Coordinates	15°04.0′N, 116°32.5′E	规模（千米 × 千米） Dimension（km × km）	80 × 45
最小水深（米） Min Depth (m)	320	最大水深（米） Max Depth (m)	4540
地理实体描述 Feature Description	珍贝海山属于珍贝—黄岩海山链的一部分，是该海山链最西侧的一座海山，其顶部发育 3 座峰（图 6–2）。 Zhenbei Haishan is a part of Zhenbei–Huangyan Haishanlian and is a seamount on the westernmost side of the seamount chain with three peaks developed on its top (Fig.6–2).		
命名由来 Origin of Name	1986 年，国务院、外交部和中国地名委员会批准了前地矿部第二海洋地质调查大队（现广州海洋地质调查局）对南海 22 个海底地理实体进行的命名，“珍贝海山”是这 22 个海底地理实体名称之一。该海山俯视形态似贝壳，因此得名。 In 1986, the State Council, the Ministry of Foreign Affairs and Chinese Toponymy Committee approved 22 undersea feature names in Nanhai named by former Second Marine Geological Survey Brigade of the Ministry of Geology and Mineral Resources (present Guangzhou Marine Geological Survey of China Geological Survey). Zhenbei Haishan is one of the 22 undersea feature names. The Seamount looks like a shell from above, and “Zhenbei” means rare shell, hence the name.		

6.3 黄岩西海山

标准名称 Standard Name	黄岩西海山 Huangyanxi Haishan	类别 Generic Term	海山 Seamount
中心点坐标 Center Coordinates	15°09.1′N, 117°09.5′E	规模（千米 × 千米） Dimension（km × km）	40 × 32
最小水深（米） Min Depth (m)	480	最大水深（米） Max Depth (m)	4550
地理实体描述 Feature Description	黄岩西海山属于珍贝—黄岩海山链的一部分，其东西两侧分别为紫贝海山和珍贝海山，其山脊线呈北西—南东走向（图 6-2）。 Huangyanxi Haishan is a part of Zhenbei-Huangyan Haishanlian, with Zibei Haishan and Zhenbei Haishan on the east and west respectively. Its ridge lines stretches in the direction of NW-SE (Fig.6-2).		
命名由来 Origin of Name	该海山位于黄岩岛以西，因此得名。 The Seamount is located to the west of Huangyan Dao, and “Xi” means west in Chinese, so the word “Huangyanxi” was used to name the Seamount.		

6.4 紫贝海山

标准名称 Standard Name	紫贝海山 Zibei Haishan	类别 Generic Term	海山 Seamount
中心点坐标 Center Coordinates	15°08.0′N, 117°26.7′E	规模（千米 × 千米） Dimension（km × km）	35 × 15
最小水深（米） Min Depth (m)	1800	最大水深（米） Max Depth (m)	4540
地理实体描述 Feature Description	紫贝海山属于珍贝—黄岩海山链的一部分，是海山链中最小的一座海山。该海山的东西两侧分别为黄岩岛和紫贝海山，其山脊线呈北西西—南东东走向（图 6-2）。 Zibei Haishan is a part of Zhenbei-Huangyan Haishanlian and is the smallest seamount of the Seamount Chain. The east and west sides of this Seamount are Huangyan Dao and Zibei Haishan respectively, and its ridge lines stretches in the direction NWW-SEE (Fig.6-2).		
命名由来 Origin of Name	该海山位于珍贝—黄岩海山链，因此以一种贝壳（紫贝）命名。 The feature is located in Zhenbei-Huangyan Haishanlian, and “Zibei” a kind of shell (arabic cowry shell), so the word “Zibei” was used to name the Seamount.		

6.5 黄岩东海山

标准名称 Standard Name	黄岩东海山 Huangyandong Haishan	类别 Generic Term	海山 Seamount
中心点坐标 Center Coordinates	15°17.1′N, 118°17.0′E	规模（千米 × 千米） Dimension（km × km）	58 × 32
最小水深（米） Min Depth (m)	860	最大水深（米） Max Depth (m)	4380
地理实体描述 Feature Description	黄岩东海山属于珍贝—黄岩海山链的一部分，其东西两侧分别为贝壳海山和黄岩岛，其山脊线呈北西西—南东东走向（图 6−2）。 Huangyandong Haishan is a part of Zhenbei−Huangyan Haishanlian, with Beike Haishan and Huangyan Dao on the east and west side respectively. Its ridge line stretches in the direction of NWW−SEE (Fig.6−2).		
命名由来 Origin of Name	该海山位于黄岩岛以东，因此得名。 The Seamount is located on the east of Huangyan Dao, and “Dong” means east in Chinese, so the word “Huangyandong” was used to name the Seamount.		

6.6 贝壳海山

标准名称 Standard Name	贝壳海山 Beike Haishan	类别 Generic Term	海山 Seamount
中心点坐标 Center Coordinates	15°29.2′N, 118°36.8′E	规模（千米 × 千米） Dimension（km × km）	110 × 45
最小水深（米） Min Depth (m)	1260	最大水深（米） Max Depth (m)	4950
地理实体描述 Feature Description	贝壳海山是珍贝—黄岩海山链中最东侧的一座海山，同时也是最大的一座海山，其东、西两侧分别为马尼拉海沟和黄岩东海山（图 6−2）。 Beike Haishan is the easternmost seamount in the Zhenbei−Huangyan Haishanlian, and also the largest one, with Manila Trench and Huangyandong Haishan on its east and west sides respectively (Fig.6−2).		
命名由来 Origin of Name	该海山俯视形态似贝壳，因此得名。 The Seamount looks like a shell from above, and “Beike” means shell in Chinese, so the word “Beike” was used to name the Seamount.		

(a)

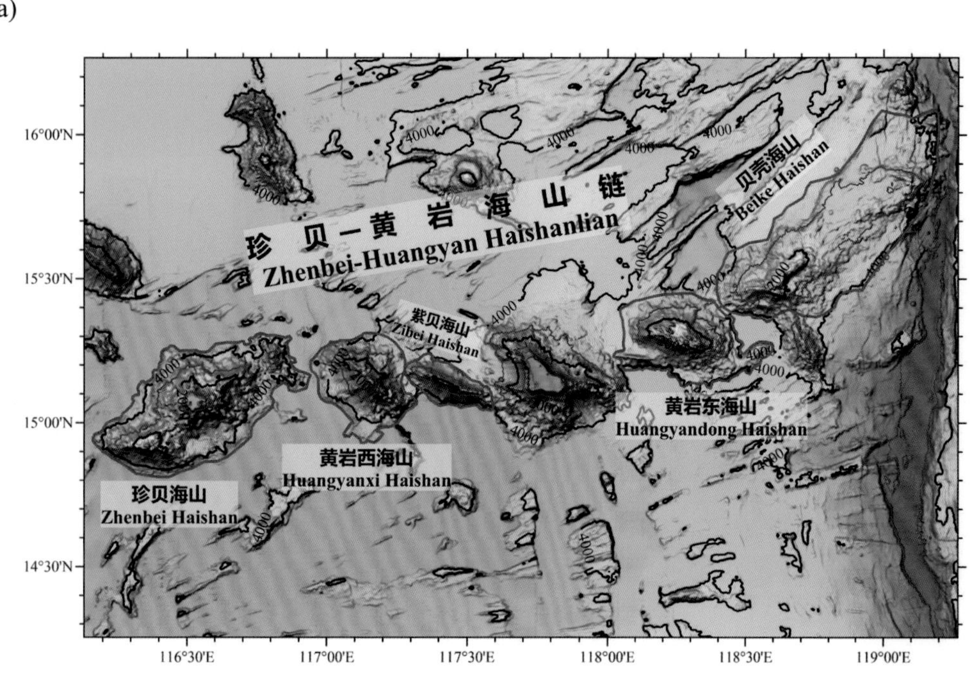

(b)

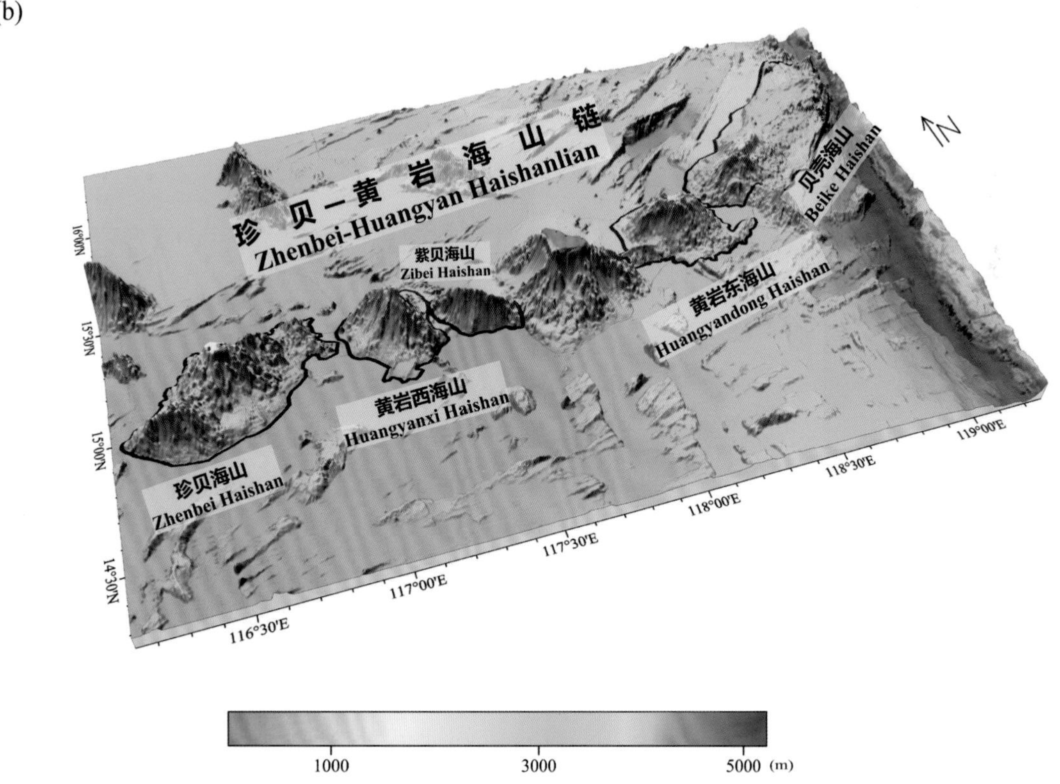

图 6-2　珍贝—黄岩海山链、珍贝海山、黄岩西海山、紫贝海山、黄岩东海山、贝壳海山

(a) 海底地形图（等深线间隔 1000 米）；(b) 三维海底地形图

Fig.6-2　Zhenbei-Huangyan Haishanlian, Zhenbei Haishan, Huangyanxi Haishan, Zibei Haishan, Huangyandong Haishan, Beike Haishan

(a) Seafloor topographic map (with contour interval of 1000 m)；(b) 3-D seafloor topographic map

6.7 扇贝海脊

标准名称 Standard Name	扇贝海脊 Shanbei Haiji	类别 Generic Term	海脊 Ridge
中心点坐标 Center Coordinates	14°49.1'N, 115°29.9'E	规模（千米 × 千米） Dimension（km × km）	17 × 3
最小水深（米） Min Depth (m)	3660	最大水深（米） Max Depth (m)	4300
地理实体描述 Feature Description	扇贝海脊位于南海海盆南部，其脊线呈北西—南东走向（图 6–3）。 Shanbei Haiji is located in the south of Nanhai Haipen, with its ridge line stretches in the direction of NW–SE (Fig.6–3).		
命名由来 Origin of Name	以海洋软体动物名进行地名的团组化命名。“扇贝”是一种生活在海洋里的软体动物。 A group naming of undersea features after the names of marine mollusks. “Shanbei” refers to scallops, is a species of marine mollusk.		

6.8 明诚海丘

标准名称 Standard Name	明诚海丘 Mingcheng Haiqiu	类别 Generic Term	海丘 Hill
中心点坐标 Center Coordinates	15°02.6′N, 115°31.9′E	规模（千米 × 千米） Dimension（km × km）	21 × 14
最小水深（米） Min Depth (m)	3414	最大水深（米） Max Depth (m)	4312
地理实体描述 Feature Description	明诚海丘位于南海海盆南部，平面形态呈不规则多边形（图 6–3）。 Mingcheng Haiqiu is located in the south of Nanhai Haipen, with a planform in the shape of an irregular polygon (Fig.6–3).		
命名由来 Origin of Name	以唐宋文人的名字进行地名的团组化命名。该海丘以宋代著名文学家赵明诚命名，纪念他在中国文学史上的重要成就。他的 30 卷巨著《金石录》长期以来被誉为中国书学发展的重要作品。 A group naming of undersea features after the names of famous men of letter of the Tang and Song Dynasties. The Hill is named after Zhao Mingcheng (1081–1129 A.D.), a famous literati during the Song Dynasty (960–1279 A.D.), to commemorate his outstanding achievement to Chinese literature history. His 30-volume magnum opus “Jin Shi Lu” has long been hailed as an important work in the development of Chinese epigraphy.		

6.9 易安海山

标准名称 Standard Name	易安海山 Yi’an Haishan	类别 Generic Term	海山 Seamount
中心点坐标 Center Coordinates	14°48.4′N, 115°39.7′E	规模（千米 × 千米） Dimension（km × km）	28 × 17
最小水深（米） Min Depth (m)	2604	最大水深（米） Max Depth (m)	4328
地理实体描述 Feature Description	易安海山位于南海海盆南部，由多条北东东—南西西向的脊状地形构成（图 6−3）。 Yi’an Haishan is located in the south of Nanhai Haipen and is composed of multiple ridge-shaped topographical units in the direction of NEE−SWW (Fig.6−3).		
命名由来 Origin of Name	以唐宋文人的名字进行地名的团组化命名。该海山以宋代著名女词人李清照，号易安居士命名，纪念她在中国诗词史上的重要成就。她是中国历史上最伟大的女词人之一。 A group naming of undersea features after the names of famous men of letter of the Tang and Song Dynasties. The Seamount is named after Li Qingzhao (1084−1155 A.D.), who was also named Yi’an Jushi, a famous female lyricist during the Song Dynasty (960−1279 A.D.), to commemorate his outstanding achievement to Chinese poem and lyrics. She was one of the greatest female lyricists in Chinese history.		

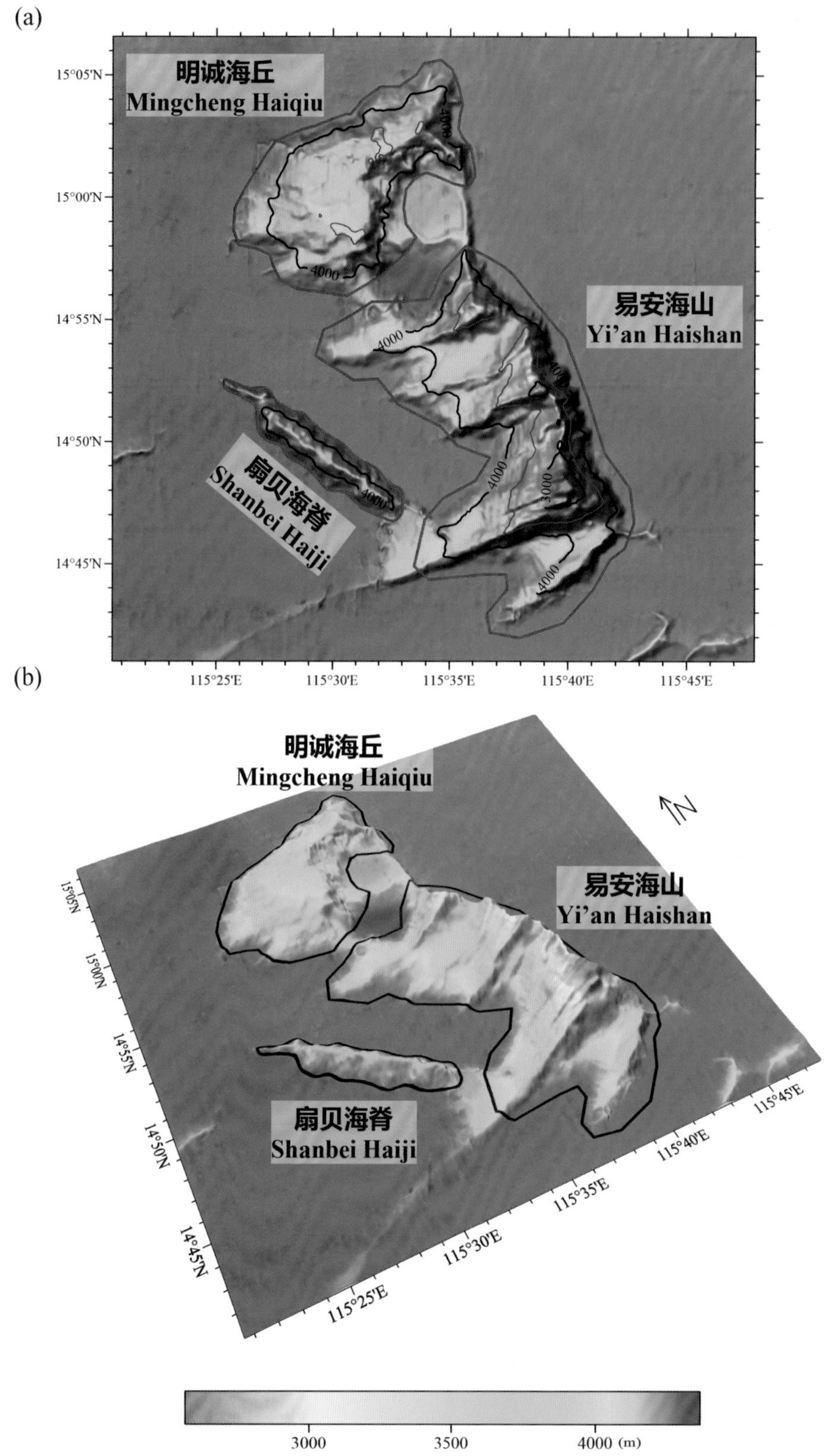

图 6-3　扇贝海脊、明诚海丘、易安海山

(a) 海底地形图（等深线间隔 500 米）；(b) 三维海底地形图

Fig.6-3　Shanbei Haiji, Mingcheng Haiqiu, Yi'an Haishan

(a) Seafloor topographic map (with contour interval of 500 m)；(b) 3-D seafloor topographic map

6.10 青蛤海脊

标准名称 Standard Name	青蛤海脊 Qingge Haiji	类别 Generic Term	海脊 Ridge
中心点坐标 Center Coordinates	14°56.1'N, 116°02.9'E	规模（千米 × 千米） Dimension（km × km）	28 × 6
最小水深（米） Min Depth (m)	3840	最大水深（米） Max Depth (m)	4250
地理实体描述 Feature Description	青蛤海脊位于南海海盆南部，珍贝海山西侧，其脊线呈北东—南西走向（图 6–4）。 Qingge Haiji is located in the south of Nanhai Haipen and is to the west of Zhenbei Haishan. Its ridge line stretches in the direction of NE–SW (Fig.6–4).		
命名由来 Origin of Name	以海洋软体动物名进行地名的团组化命名。"青蛤"是一种生活在海洋里的软体动物。 A group naming of undersea features after the names of marine mollusks. "Qingge" refers to cyclina sinensis, is a species of marine mollusk.		

6.11 紫蛤海丘

标准名称 Standard Name	紫蛤海丘 Zige Haiqiu	类别 Generic Term	海丘 Hill
中心点坐标 Center Coordinates	14°40.5'N, 115°57.3'E	规模（千米 × 千米） Dimension（km × km）	8 × 4
最小水深（米） Min Depth (m)	4030	最大水深（米） Max Depth (m)	4250
地理实体描述 Feature Description	紫蛤海丘位于南海海盆南部，珍贝海山西侧，平面形态呈菱形（图 6–4）。 Zige Haiqiu is located to the south of Nanhai Haipen and is on the west of Zhenbei Haishan, with a planform in the shape of a diamond (Fig.6–4).		
命名由来 Origin of Name	以海洋软体动物名进行地名的团组化命名。"紫蛤"学名长型紫云蛤，是一种生活在海洋里的软体动物。 A group naming of undersea features after the names of marine mollusks. "Zige" refers to psammotaea elongate, is a species of marine mollusk.		

6.12 花螺海脊

标准名称 Standard Name	花螺海脊 Hualuo Haiji	类别 Generic Term	海脊 Ridge
中心点坐标 Center Coordinates	14°39.5'N, 116°04.5'E	规模（千米 × 千米） Dimension（km × km）	40 × 6
最小水深（米） Min Depth (m)	3400	最大水深（米） Max Depth (m)	4335
地理实体描述 Feature Description	花螺海脊位于南海海盆南部，珍贝海山西南侧，其脊线呈北北东—南南西走向（图 6–4）。 Hualuo Haiji is located in the south of Nanhai Haipen and is on the southwest side of Zhenbei Haishan, with a planform in the shape of a long strip in the direction of NNE–SSW (Fig.6–4).		
命名由来 Origin of Name	以海洋软体动物名进行地名的团组化命名。"花螺"俗称东风螺、海猪螺，是一种生活在海洋里的软体动物。 A group naming of undersea features after the names of marine mollusks. "Hualuo" refers to babylonia formorae, is a kind of marine mollusk.		

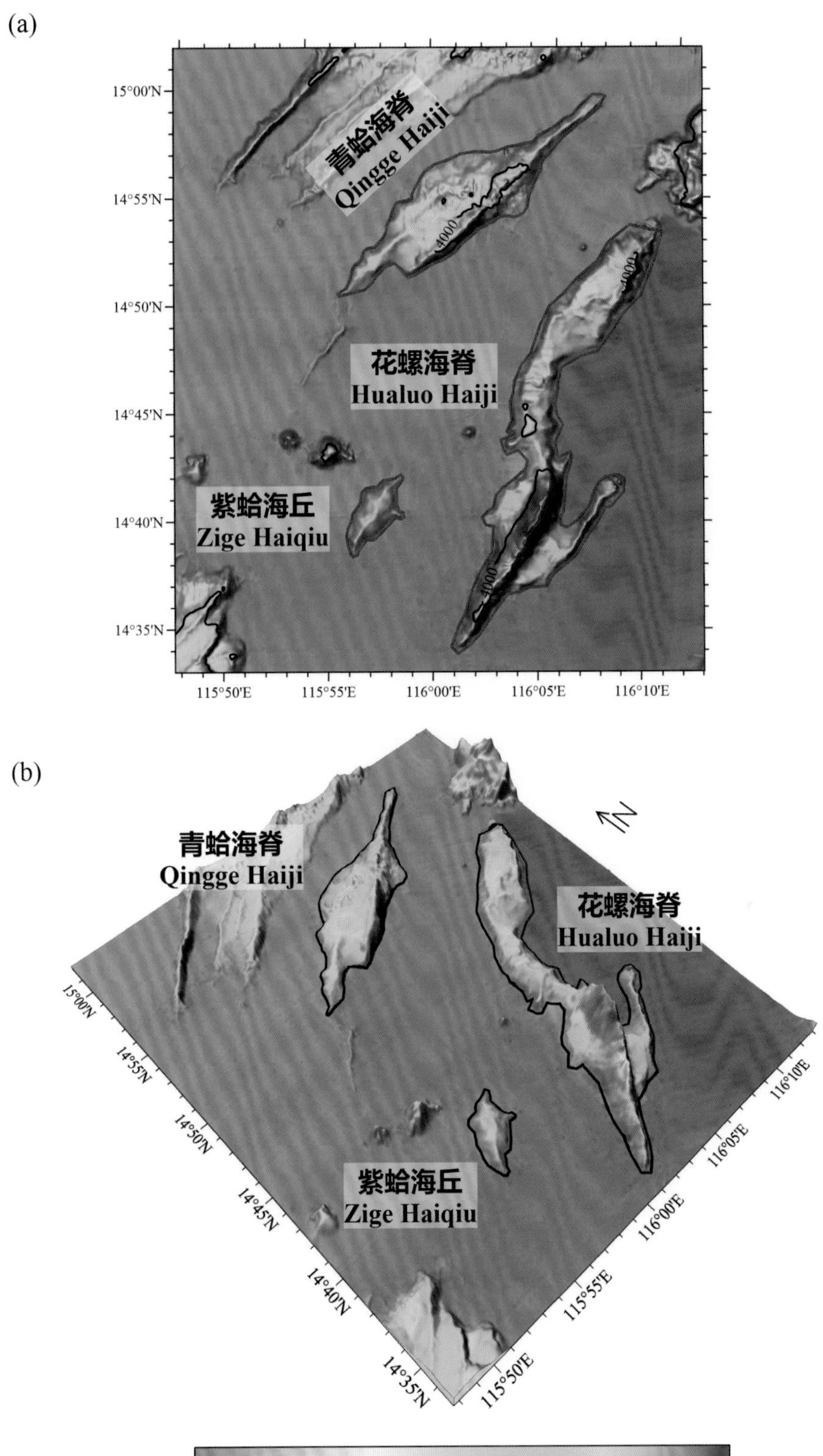

图 6-4 青蛤海脊、紫蛤海丘、花螺海脊

(a) 海底地形图（等深线间隔 500 米）；(b) 三维海底地形图

Fig.6-4 Qingge Haiji, Zige Haiqiu, Hualuo Haiji

(a) Seafloor topographic map (with contour interval of 500 m); (b) 3-D seafloor topographic map

6.13 孟浩然海丘

标准名称 Standard Name	孟浩然海丘 Menghaoran Haiqiu	类别 Generic Term	海丘 Hill
中心点坐标 Center Coordinates	14°25.9′N, 116°17.6′E	规模（千米 × 千米） Dimension（km × km）	51 × 10
最小水深（米） Min Depth (m)	3630	最大水深（米） Max Depth (m)	4388
地理实体描述 Feature Description	孟浩然海丘位于南海海盆东南部，珍贝海山南侧，平面形态呈北东—南西向的长条形，其顶部发育有多座峰（图 6-5）。 Menghaoran Haiqiu is located in the southeast of Hainan Haipen and is on the south of Zhenbei Haishan, with a planform in the shape of a long strip in the direction of NE−SW and multiple peaks have developed on its top (Fig.6−5).		
命名由来 Origin of Name	以唐宋文人的名字进行地名的团组化命名。该海丘以唐朝著名诗人孟浩然命名，纪念他在中国文学史上的重要成就。 A group naming of undersea features after the names of famous men of letter of the Tang and Song Dynasties. The Hill is named after Meng Haoran (689−740 A.D.), a prominent Chinese poet in the Tang Dynasty (618−907 A.D.), to commemorate his outstanding achievement to Chinese literature history.		

6.14 李白海丘

标准名称 Standard Name	李白海丘 Libai Haiqiu	类别 Generic Term	海丘 Hill
中心点坐标 Center Coordinates	14°38.7′N, 116°47.9′E	规模（千米 × 千米） Dimension（km × km）	37 × 24
最小水深（米） Min Depth (m)	3640	最大水深（米） Max Depth (m)	4610
地理实体描述 Feature Description	李白海丘位于南海海盆南部，珍贝海山东南侧，平面形态呈北东—南西向的长条形，其顶部发育多座峰（图 6-5）。 Libai Haiqiu is located in the south of Hainan Haipen and is on the southeast of Zhenbei Haishan, with a planform in the shape of a long strip in the direction of NE−SW and multiple peaks have developed on its top (Fig.6−5).		
命名由来 Origin of Name	以唐宋文人的名字进行地名的团组化命名。该海丘以唐朝著名诗人李白命名，纪念他在中国文学史上的重要成就。 A group naming of undersea features after the names of famous men of letter of the Tang and Song Dynasties. The Hill is named after Li Bai (701−762 A.D.), a famous poet in the Tang Dynasty (618−907 A.D.), to commemorate his outstanding achievement to Chinese literature history.		

(a)

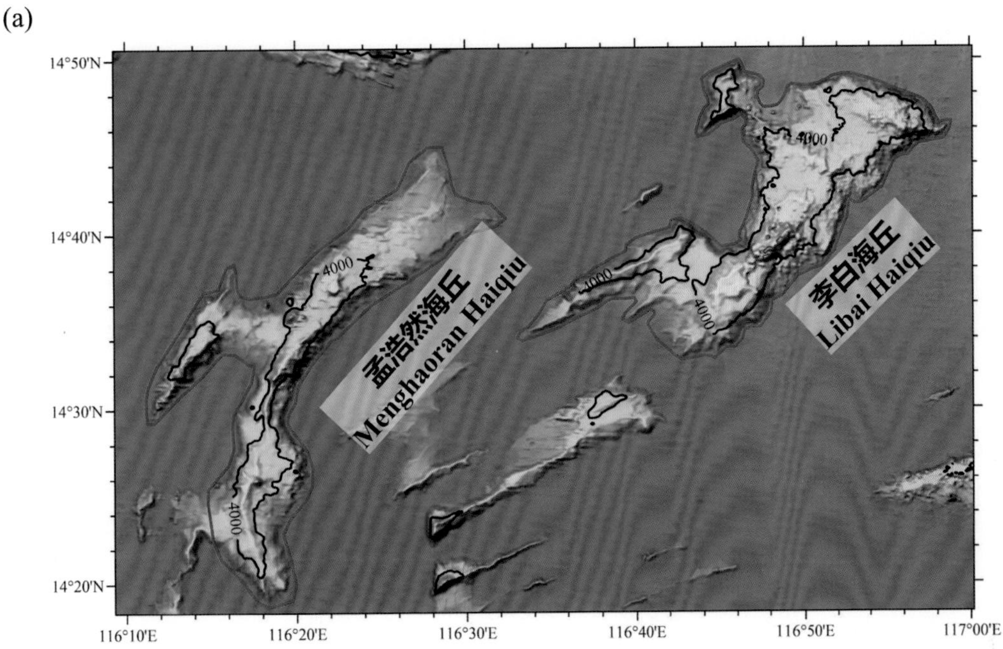

(b)

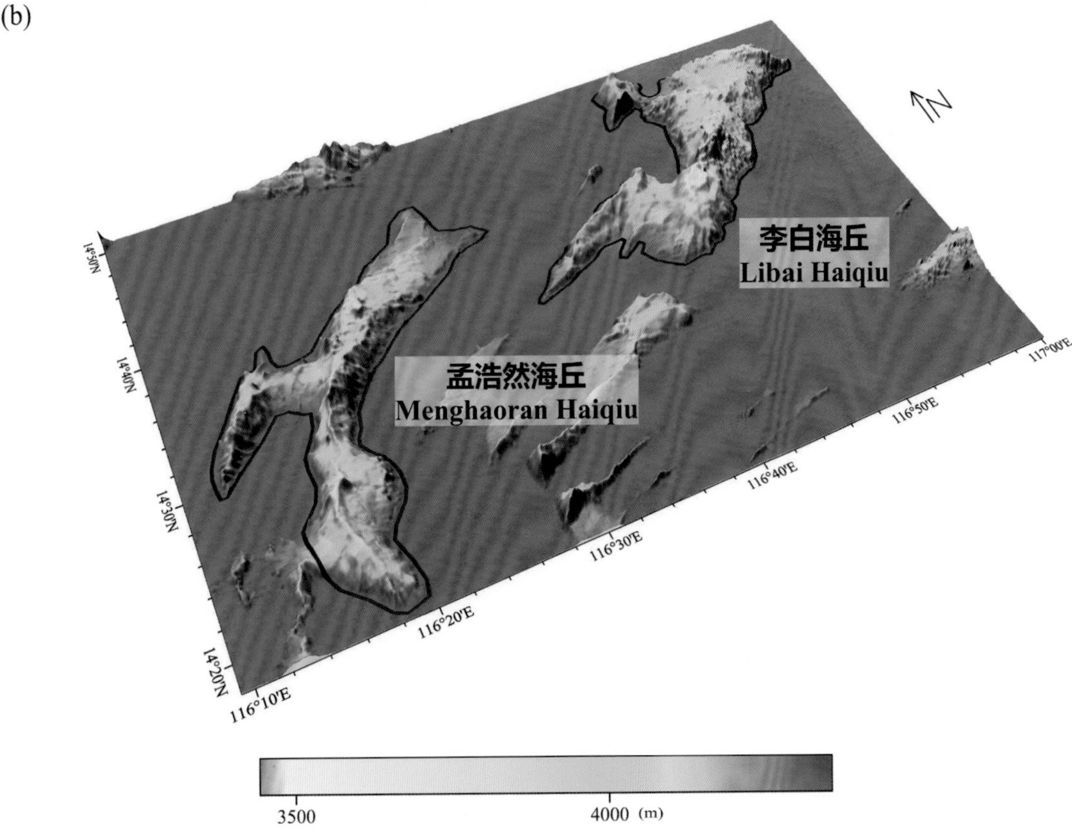

图 6-5　孟浩然海丘、李白海丘

(a) 海底地形图（等深线间隔 500 米）；(b) 三维海底地形图

Fig.6-5　Menghaoran Haiqiu, Libai Haiqiu

(a) Seafloor topographic map (with contour interval of 500 m)；(b) 3-D seafloor topographic map

6.15 青蛤海丘

标准名称 Standard Name	青蛤海丘 Qingge Haiqiu	类别 Generic Term	海丘 Hill
中心点坐标 Center Coordinates	15°00.9'N, 116°52.8'E	规模（千米 × 千米） Dimension（km × km）	10 × 5
最小水深（米） Min Depth (m)	3980	最大水深（米） Max Depth (m)	4270
地理实体描述 Feature Description	青蛤海丘位于南海海盆南部，珍贝海山与黄岩西海山之间，平面形态呈东—西向的多边形（图 6-6）。 Qingge Haiqiu is located in the south of Nanhai Haipen and is between Zhenbei Haishan and Huangyanxi Haishan, with a polygon-shaped planform in the direction of E–W (Fig.6–6).		
命名由来 Origin of Name	以海洋软体动物名进行地名的团组化命名。“青蛤”是一种生活在海洋里的软体动物。 A group naming of undersea features after the names of marine mollusks. “Qingge” refers to cyclina sinensis, is a species of marine mollusk.		

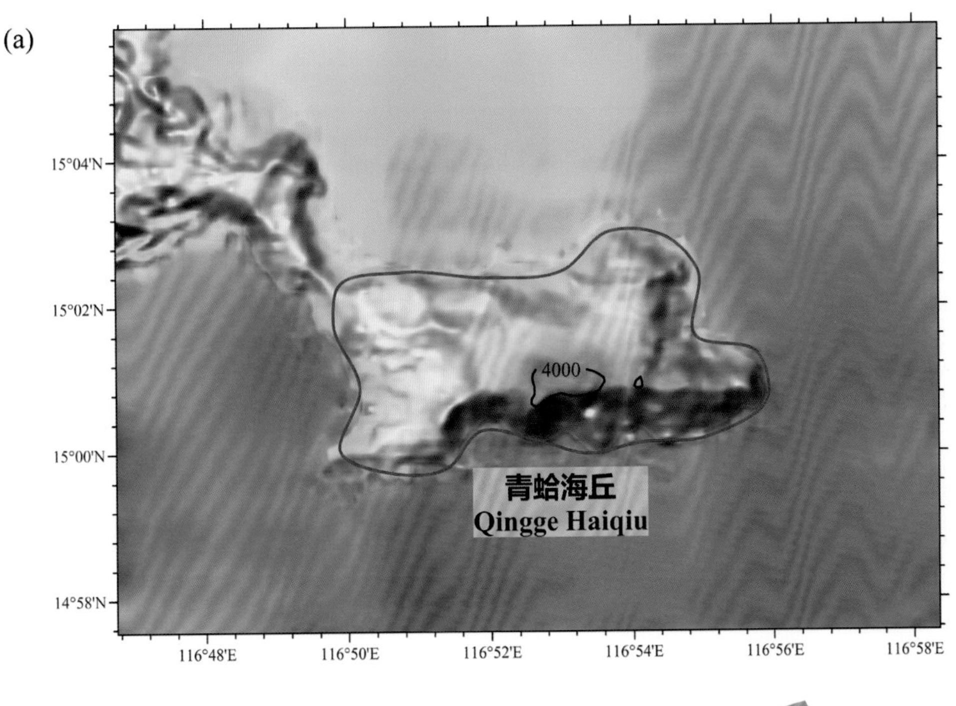

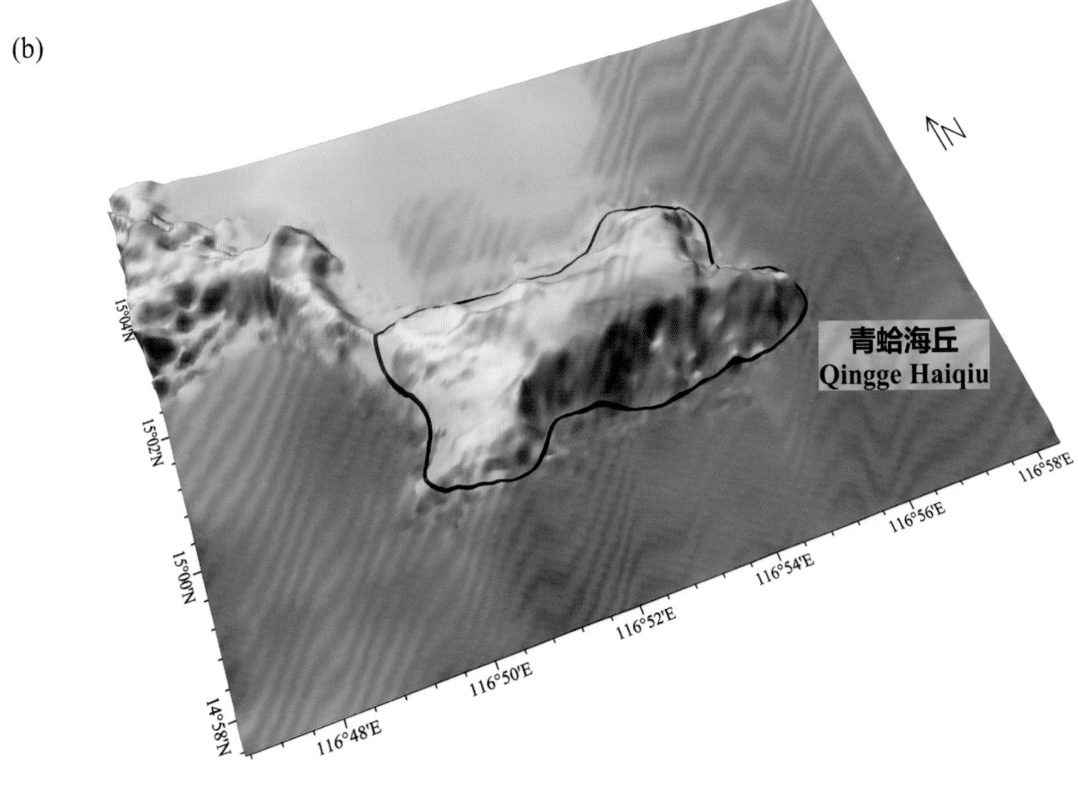

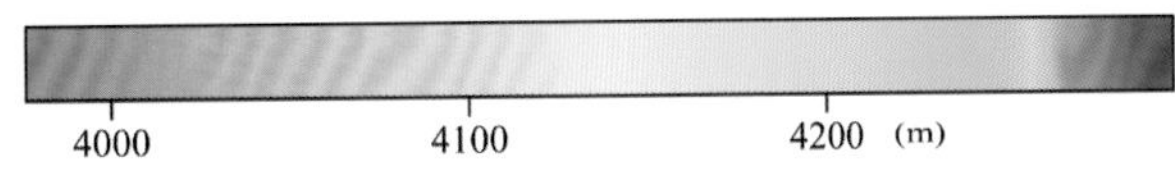

图 6-6 青蛤海丘

(a) 海底地形图（等深线间隔 500 米）；(b) 三维海底地形图

Fig.6-6 Qingge Haiqiu

(a) Seafloor topographic map (with contour interval of 500 m); (b) 3-D seafloor topographic map

6.16 深岩海丘

标准名称 Standard Name	深岩海丘 Shenyan Haiqiu	类别 Generic Term	海丘 Hill
中心点坐标 Center Coordinates	14°49.6'N, 117°18.0'E	规模（千米 × 千米） Dimension（km × km）	21 × 18
最小水深（米） Min Depth (m)	3800	最大水深（米） Max Depth (m)	4240
地理实体描述 Feature Description	深岩海丘位于南海海盆南部，黄岩西海山南侧，平面形态似一弯曲的箭头（图 6−7）。 Shenyan Haiqiu is located in the south of Nanhai Haipen and is on the south of Huangyanxi Haishan, with a planform in the shape of a bent arrow (Fig.6−7).		
命名由来 Origin of Name	以中国古代文学作品中的短语进行地名的团组化命名。该海底地名的专名取词自唐朝诗人杜甫的《柏学士茅屋》：“碧山学士焚银鱼，白马却走深岩居”。“深岩”即崖边水洼曲处。 A group naming of undersea features after the phrases in ancient Chinese literatures. The specific term of this undersea feature name “Shenyan”, means a pool on the edge of the cliff, is derived from the poetic lines “In remote mountain, the scholar burned fish-shaped silver official medal, with a white horse, he went to live in seclusion near a pool on the edge of the cliff” in the poem *Scholar Bo's Thatched Cottage* by Du Fu (712−770 A.D.), a poet in the Tang Dynasty (618−907 A.D.).		

6.17 连岩海丘

标准名称 Standard Name	连岩海丘 Lianyan Haiqiu	类别 Generic Term	海丘 Hill
中心点坐标 Center Coordinates	14°39.3'N, 117°17.6'E	规模（千米 × 千米） Dimension（km × km）	9 × 4
最小水深（米） Min Depth (m)	4060	最大水深（米） Max Depth (m)	4330
地理实体描述 Feature Description	连岩海丘位于南海海盆南部，黄岩西海山南侧，其山脊线呈北东—南西走向（图 6−7）。 Lianyan Haiqiu is located in the south of Nanhai Haipen and is to the south of Huangyanxi Haishan, with its ridge line stretches in the direction of NE−SW (Fig.6−7).		
命名由来 Origin of Name	以中国古代文学作品中的短语进行地名的团组化命名。该海底地名的专名取词自南北朝时期文学家王褒的《和从弟佑山家诗》：“散云非一色，连岩异众峰”。“连岩”即接连不断的岩石。 A group naming of undersea features after the phrases in ancient Chinese literatures. The specific term of this undersea feature name “Lianyan”, means continuous rocks, is derived from the poetic lines “Clouds scatter in different colors, the continuous rocks stand in contrast with the mountain peaks” in the poem *A Poem Composed in the Rhythm of Youshan's Poem* by Wang Bao (513−576 A.D.), a litterateur in the Southern and Northern Dynasties (420−589 A.D.).		

6.18 千岩海丘

标准名称 Standard Name	千岩海丘 Qianyan Haiqiu	类别 Generic Term	海丘 Hill
中心点坐标 Center Coordinates	14°34.8'N, 117°25.1'E	规模（千米 × 千米） Dimension（km × km）	11 × 6
最小水深（米） Min Depth (m)	4010	最大水深（米） Max Depth (m)	4330
地理实体描述 Feature Description	千岩海丘位于南海海盆南部，紫贝海山南侧，其山脊线呈北东—南西走向（图 6−7）。 Qianyan Haiqiu is located in the south of Nanhai Haipen and is to the south of Zibei Haishan, with its ridge line stretches in the direction of NE−SW (Fig.6−7).		
命名由来 Origin of Name	以中国古代文学作品中的短语进行地名的团组化命名。该海底地名的专名取词自南北朝时期文学家刘义庆的《世说新语·言语》：“千岩竞秀，万壑争流”。“千岩”形容许多陡峭的山。 A group naming of undersea features after the phrases in ancient Chinese literatures. The specific term of this undersea feature name “Qianyan”, refers to the many steep mountains, is derived from the poetic lines “Many steep mountains seem to compete with each other in beauty, and many streams run in competition with each other.” in the novel collection *New Saying of the Day* by Liu Yiqing (403−444 A.D.), a litterateur in the Southern and Northern Dynasties (420−589 A.D.).		

6.19 元稹海丘

标准名称 Standard Name	元稹海丘 Yuanzhen Haiqiu	类别 Generic Term	海丘 Hill
中心点坐标 Center Coordinates	14°45.2′N, 117°28.2′E	规模（千米 × 千米） Dimension（km × km）	20 × 16
最小水深（米） Min Depth (m)	3680	最大水深（米） Max Depth (m)	4550
地理实体描述 Feature Description	元稹海丘位于南海海盆南部，紫贝海山南侧，平面形态似蘑菇状（图 6−7）。 Yuanzhen Haiqiu is located in the south of Nanhai Haipen and is to the south of Zibei Haishan with a mushroom-shaped planform (Fig.6−7).		
命名由来 Origin of Name	以唐宋文人的名字进行地名的团组化命名。该海丘以唐朝著名诗人元稹命名，纪念他在中国诗词史上的重要成就。 A group naming of undersea features after the names of famous men of letter of the Tang and Song Dynasties. The Hill is named after Yuan Zhen (779−831 A.D.), a famous poet in the Tang Dynasty (618−907 A.D.), to commemorate his outstanding achievement to Chinese poet and lyrics history.		

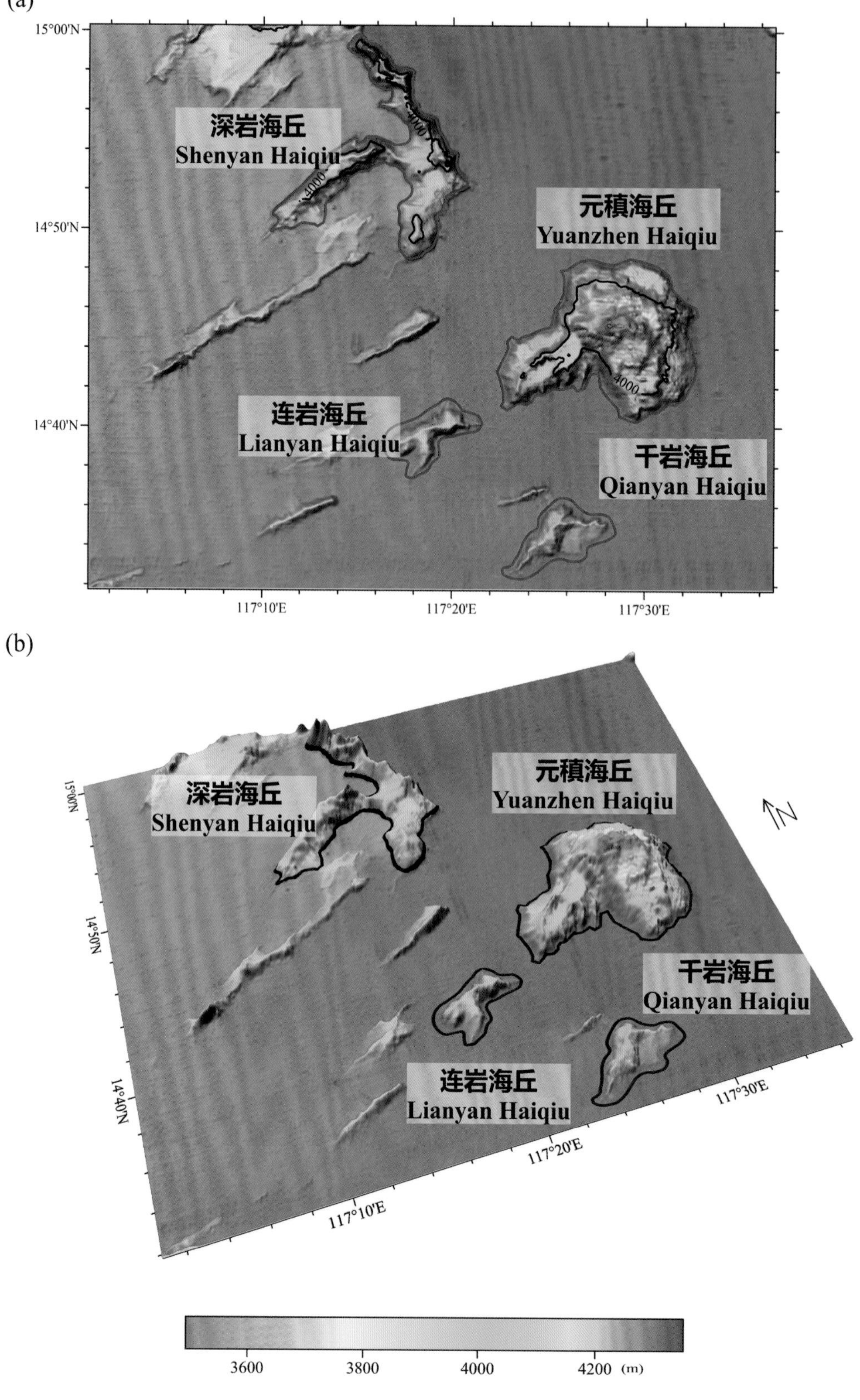

图 6-7　深岩海丘、连岩海丘、千岩海丘、元積海丘

(a) 海底地形图（等深线间隔 500 米）；(b) 三维海底地形图

Fig.6-7　Shenyan Haiqiu, Lianyan Haiqiu, Qianyan Haiqiu, Yuanzhen Haiqiu

(a) Seafloor topographic map (with contour interval of 500 m); (b) 3-D seafloor topographic map

6.20 岑参海丘

标准名称 Standard Name	岑参海丘 Censhen Haiqiu	类别 Generic Term	海丘 Hill
中心点坐标 Center Coordinates	14°40.9′N, 117°38.3′E	规模（千米 × 千米） Dimension（km × km）	40 × 20
最小水深（米） Min Depth (m)	3690	最大水深（米） Max Depth (m)	4560
地理实体描述 Feature Description	岑参海丘位于南海海盆南部，黄岩岛南侧，由多条北东东—南西西向的脊状地形构成，平面形态似梳子，其顶部发育多座峰（图 6–8）。 Censhen Haiqiu is located in the south of Nanhai Haipen and is to the south of Huangyan Dao. It is composed of multiple ridge-shaped topographical units in the direction of NEE–SWW with a planform in the shape of a comb. Multiple peaks have developed on its top (Fig.6–8).		
命名由来 Origin of Name	以唐宋文人的名字进行地名的团组化命名。该海丘以唐朝著名诗人岑参命名，纪念他在中国诗词史上的重要成就。 A group naming of undersea features after the names of famous men of letter of the Tang and Song Dynasties. The Hill is named after Cen Shen (718–769 A.D.), a famous poet in the Tang Dynasty (618–907 A.D.), to commemorate his outstanding achievement to Chinese poet and lyrics history.		

6.21 岩居海丘

标准名称 Standard Name	岩居海丘 Yanju Haiqiu	类别 Generic Term	海丘 Hill
中心点坐标 Center Coordinates	14°41.1'N, 117°54.1'E	规模（千米 × 千米） Dimension（km × km）	16 × 6
最小水深（米） Min Depth (m)	3450	最大水深（米） Max Depth (m)	4280
地理实体描述 Feature Description	岩居海丘位于南海海盆南部，黄岩岛南侧，平面形态呈北东东—南西西向的长条形（图 6–8）。 Yanju Haiqiu is located in the south of Nanhai Haipen and is to the south of Huangyan Dao with a planform in the shape of a long strip in the direction of NEE–SWW (Fig.6–8).		
命名由来 Origin of Name	以中国古代文学作品中的短语进行地名的团组化命名。该海底地名的专名取词自西汉时期思想家刘安主编的《淮南子·人间训》："岩居谷饮"。"岩居"意为住在深山洞穴中，指隐居。 A group naming of undersea features after the phrases in ancient Chinese literatures. The specific term of this undersea feature name "Yanju", refers to living in cave, implying living in seclusion, is derived from the poetic lines "living in cave and dining in valley." in the philosophical work *Book of Prince of Huainan compiled* by Liu An (179 B.C.–122 B.C.), a thinker in the Western Han Dynasty (202 B.C.–8 A.D.).		

6.22 耕岩海丘

标准名称 Standard Name	耕岩海丘 Gengyan Haiqiu	类别 Generic Term	海丘 Hill
中心点坐标 Center Coordinates	14°48.0′N, 118°00.0′E	规模（千米 × 千米） Dimension（km × km）	15 × 6
最小水深（米） Min Depth (m)	3302	最大水深（米） Max Depth (m)	4232
地理实体描述 Feature Description	耕岩海丘位于南海海盆南部，黄岩岛南侧，平面形态呈水滴形（图 6−8）。 Gengyan Haiqiu is located in the south of Nanhai Haipen and is to the south of Huangyan Dao, with a planform in the shape of a water drop (Fig.6−8).		
命名由来 Origin of Name	以中国古代文学作品中的短语进行地名的团组化命名。该海底地名的专名取词自唐朝诗人杜甫的《寄张十二山人》："耕岩非谷口，结草即河滨"。"耕岩"即耕种于岩石之下，指隐居。 A group naming of undersea features after the phrases in ancient Chinese literatures. The specific term of this undersea feature name "Gengyan", means farming under a rock, implying living in seclusion, is derived from the poetic lines "Farming under a rock, and living in a riverside thatched cottage in seclusion" in the poem *To a Helmet Named Zhang Shi'er* by Du Fu (712−770 A.D.), a poet in the Tang Dynasty (618−907 A.D.).		

6.23 柳宗元海山

标准名称 Standard Name	柳宗元海山 Liuzongyuan Haishan	类别 Generic Term	海山 Seamount
中心点坐标 Center Coordinates	14°37.6′N, 117°59.4′E	规模（千米 × 千米） Dimension（km × km）	27 × 20
最小水深（米） Min Depth (m)	3390	最大水深（米） Max Depth (m)	4510
地理实体描述 Feature Description	柳宗元海山位于南海海盆南部、黄岩岛南侧，由多条北东东—南西西向的脊状地形构成，平面形态似梳子，其顶部发育多座峰（图 6−8）。 Liuzongyuan Haishan is located in the south of Nanhai Haipen and is to the south of Huangyan Dao. It is composed of multiple ridge-shaped topographical units in the direction of NEE−SWW, with a planform in the shape of a comb. Multiple peaks have developed on its top (Fig.6−8).		
命名由来 Origin of Name	以唐宋文人的名字进行地名的团组化命名。该海山以唐朝著名文学家柳宗元命名，纪念他在中国文学史上的重要成就。 A group naming of undersea features after the names of famous men of letter of the Tang and Song Dynasties. The Seamount is named after Liu Zongyuan (773−819 A.D.), a famous litterateur in the Tang Dynasty (618−907 A.D.), to commemorate his outstanding achievements to Chinese literature history.		

(a)

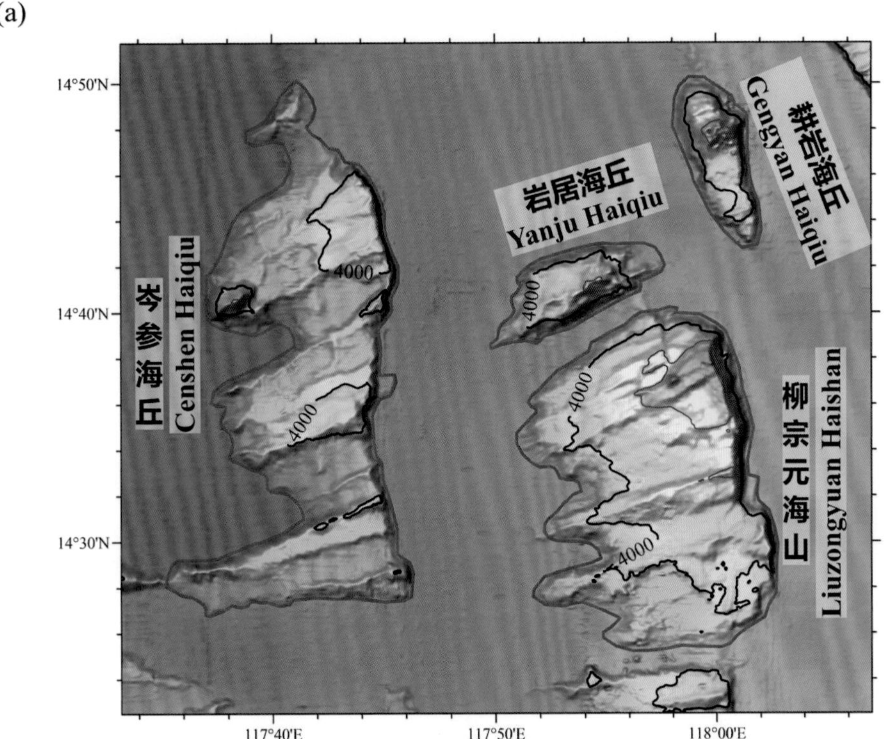

(b)

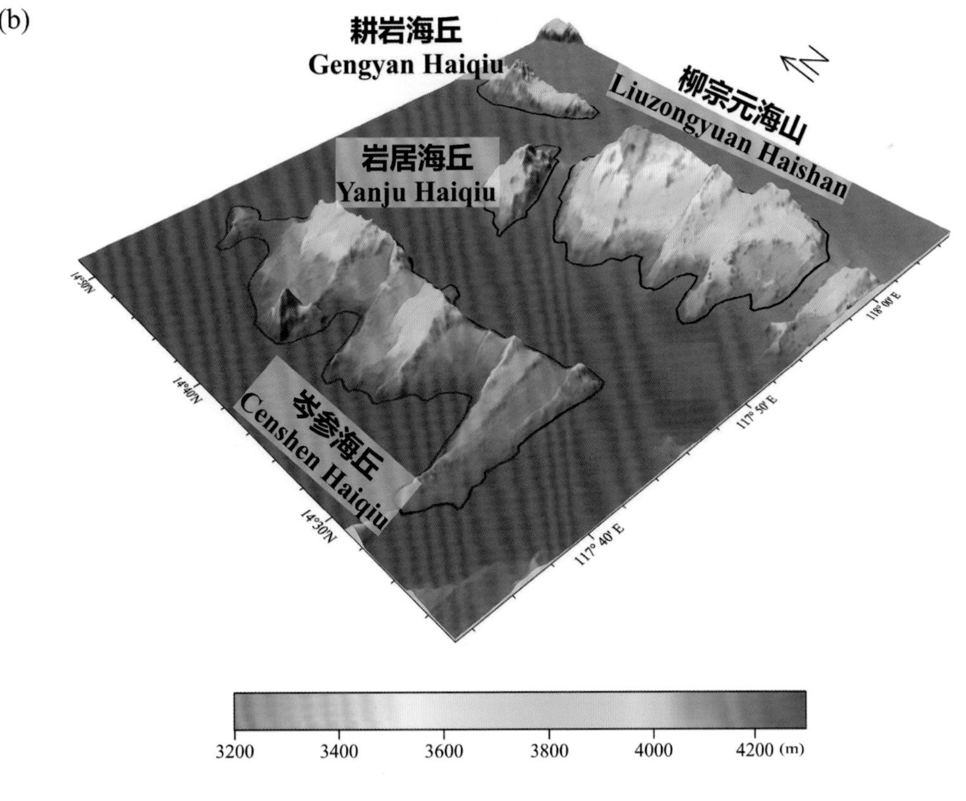

图 6-8　岑参海丘、岩居海丘、耕岩海丘、柳宗元海山

(a) 海底地形图（等深线间隔 500 米）；(b) 三维海底地形图

Fig.6-8　Censhen Haiqiu, Yanju Haiqiu, Gengyan Haiqiu, Liuzongyuan Haishan

(a) Seafloor topographic map (with contour interval of 500 m); (b) 3-D seafloor topographic map

6.24 突岩海山

标准名称 Standard Name	突岩海山 Tuyan Haishan	类别 Generic Term	海山 Seamount
中心点坐标 Center Coordinates	15°15.3'N, 118°42.6'E	规模（千米 × 千米） Dimension（km × km）	35 × 18
最小水深（米） Min Depth (m)	2050	最大水深（米） Max Depth (m)	4160
地理实体描述 Feature Description	突岩海山位于南海海盆南部，贝壳海山南侧，平面形态呈菱形，其山脊线程北北西—南南东走向（图 6–9）。 Tuyan Haishan is located in the south of Nanhai Haipen and is on the south of Beike Haishan, with a diamond-shaped planform. Its ridge line stretches in the direction of NNW–SSE (Fig.6–9).		
命名由来 Origin of Name	以中国古代文学作品中的短语进行地名的团组化命名。该海底地名的专名取词自清朝文学家郑板桥的《题画兰》："身在千山顶上头，突岩深缝妙香稠"。"突岩"即突出的岩石。 A group naming of undersea features after the phrases in ancient Chinese literatures. The specific term of this undersea feature name "Tuyan", means protruding rock, is derived from the poetic lines "The orchids grow on the top of numerous mountains, with rich fragrance penetrating into deep crevices of protruding rocks" in the poem *Poem Inscribed for Painted Orchid* by Zheng Banqiao (1693–1766 A.D.), a litterateur in the Qing Dynasty (1636–1912 A.D.).		

6.25 灵岩海山

标准名称 Standard Name	灵岩海山 Lingyan Haishan	类别 Generic Term	海山 Seamount
中心点坐标 Center Coordinates	15°16.4'N, 119°02.3'E	规模（千米 × 千米） Dimension（km × km）	26 × 10
最小水深（米） Min Depth (m)	3810	最大水深（米） Max Depth (m)	5180
地理实体描述 Feature Description	灵岩海山位于南海海盆南部，贝壳海山南侧，平面形态呈北东—南西向的长条形，顶部发育多座峰（图 6–9）。 Lingyan Haishan is located in the south of Nanhai Haipen and is to the south of Beike Haishan with a planform in the shape of a long strip in the direction of NE–SW. Multiple peaks developed on its top (Fig.6–9).		
命名由来 Origin of Name	以中国古代文学作品中的短语进行地名的团组化命名。该海底地名的专名取词自唐朝诗人李绅《姑苏台杂句》："灵岩香径掩禅扉，秋草荒凉遍落晖"。"灵岩"即灵验的石头。 A group naming of undersea features after the phrases in ancient Chinese literatures. The specific term of this undersea feature name "Lingyan", means inspiring rocks, is derived from the poetic lines "The fragrant path among the inspiring rocks leads to a closed gate of temple, the desolated grasses of autumn look bleak in the afterglow" in the poem *The Verses on Gusu Platform* by Li Shen (772–846 A.D.), a poet in the Tang Dynasty (618–907 A.D.).		

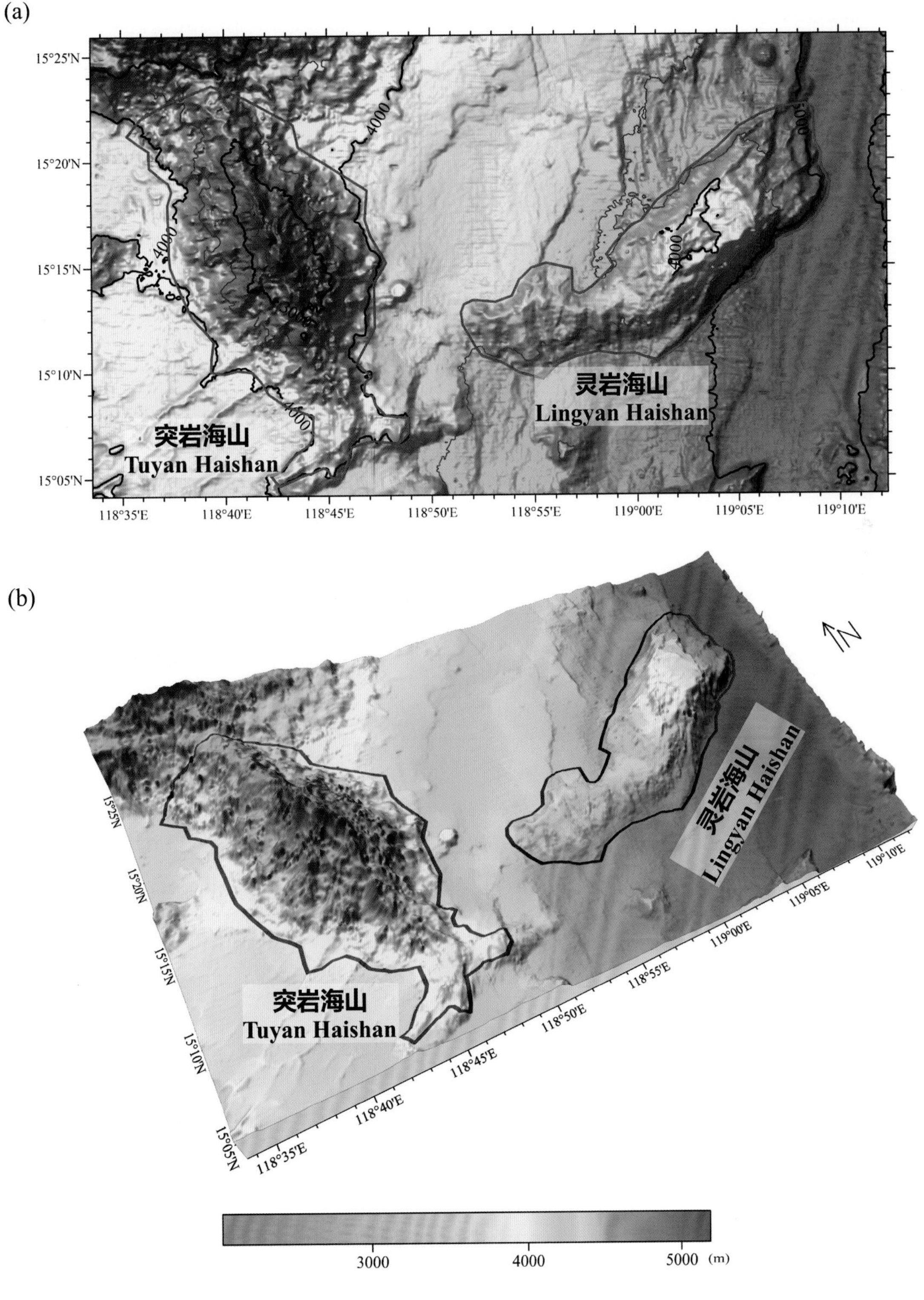

图 6-9 突岩海山、灵岩海山

(a) 海底地形图（等深线间隔 500 米）；(b) 三维海底地形图

Fig.6-9 Tuyan Haishan, Lingyan Haishan

(a) Seafloor topographic map (with contour interval of 500 m); (b) 3-D seafloor topographic map

6.26 杜甫海丘

标准名称 Standard Name	杜甫海丘 Dufu Haiqiu	类别 Generic Term	海丘 Hill
中心点坐标 Center Coordinates	14°29.0′N, 117°28.9′E	规模（千米 × 千米） Dimension（km × km）	60 × 14
最小水深（米） Min Depth (m)	3940	最大水深（米） Max Depth (m)	4550
地理实体描述 Feature Description	杜甫海丘位于南海海盆南部，平面形态呈北东东—南西西向的长条形，其顶部发育多座峰（图 6-10）。 Dufu Haiqiu is located in the south of Nanhai Haipen, with a planform in the shape of a long strip in the direction of NEE−SWW. Multiple peaks developed on its top (Fig.6−10).		
命名由来 Origin of Name	以唐宋文人的名字进行地名的团组化命名。该海丘以唐朝著名诗人杜甫命名，纪念他在中国文学史上的重要成就。 A group naming of undersea features after the names of famous men of letter of the Tang and Song Dynasties. The Hill is named after Du Fu (712−770 A.D.), a famous poet in the Tang Dynasty (618−907 A.D.), to commemorate his outstanding achievement to Chinese literature history.		

6.27 孟郊海丘

标准名称 Standard Name	孟郊海丘 Mengjiao Haiqiu	类别 Generic Term	海丘 Hill
中心点坐标 Center Coordinates	14°12.3'N, 117°43.1'E	规模（千米 × 千米） Dimension（km × km）	55 × 15
最小水深（米） Min Depth (m)	4170	最大水深（米） Max Depth (m)	4520
地理实体描述 Feature Description	孟郊海丘位于南海海盆南部，平面形态呈东—西向的长条形。海丘的两侧窄，中部宽，顶部发育多座峰（图 6-10）。 Mengjiao Haiqiu is located in the south of Nanhai Haipen with a planform in the shape of a long strip in the direction of E−W. It is narrow on the both sides and is wide in the middle, and has multiple peaks developed on its top (Fig.6−10).		
命名由来 Origin of Name	以唐宋文人的名字进行地名的团组化命名。该海丘以唐朝著名诗人孟郊命名，纪念他在中国文学史上的重要成就。 A group naming of undersea features after the names of famous men of letter of the Tang and Song Dynasties. The Hill is named after Meng Jiao (751−814 A.D.), a famous poet in the Tang Dynasty (618−907 A.D.), to commemorate his outstanding achievement to Chinese literature history.		

(a)

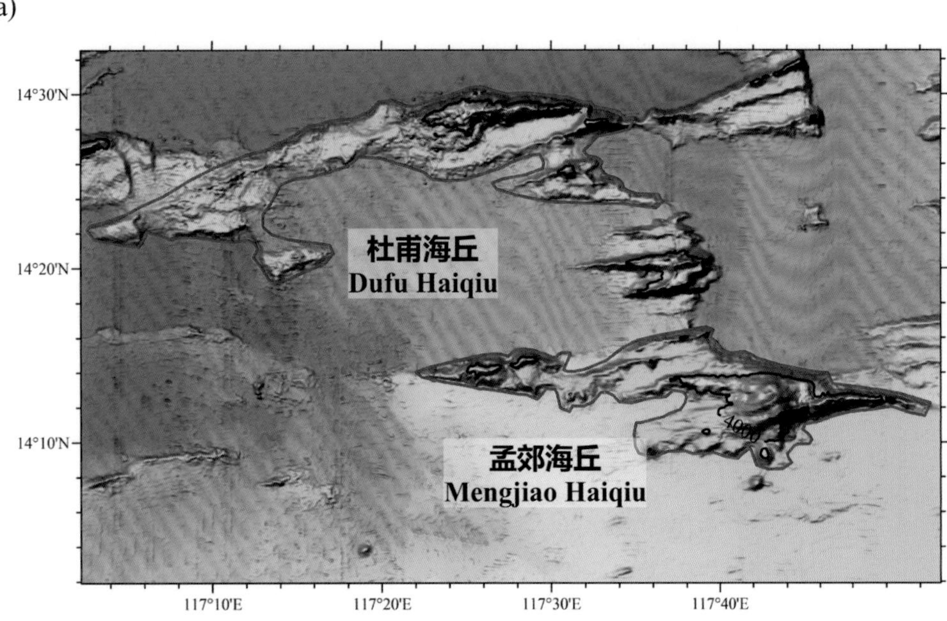

(b)

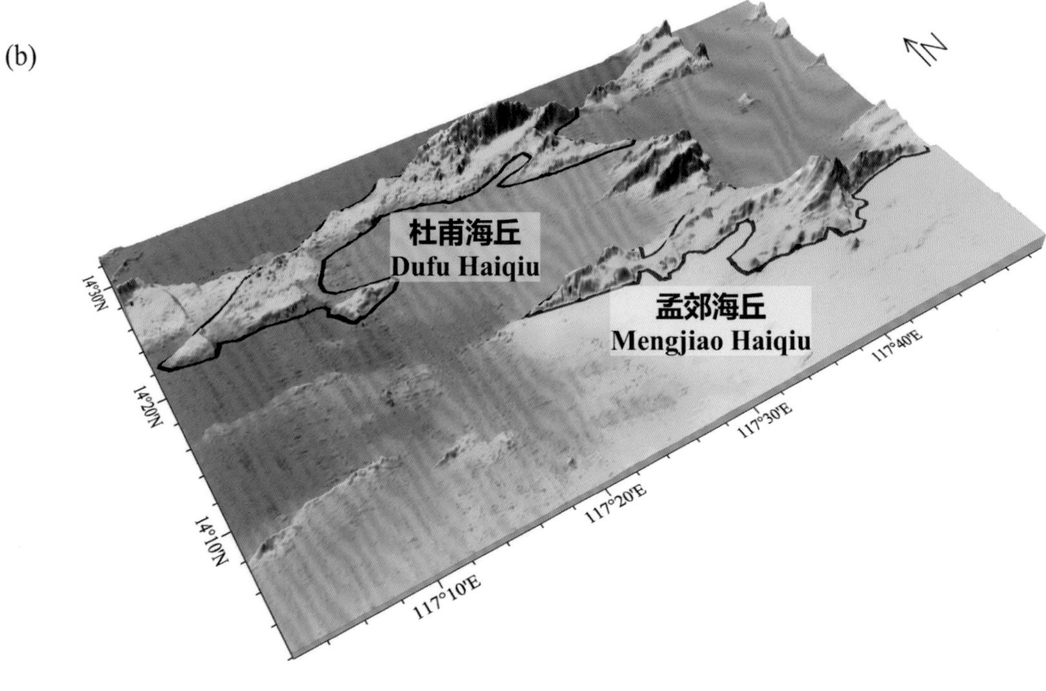

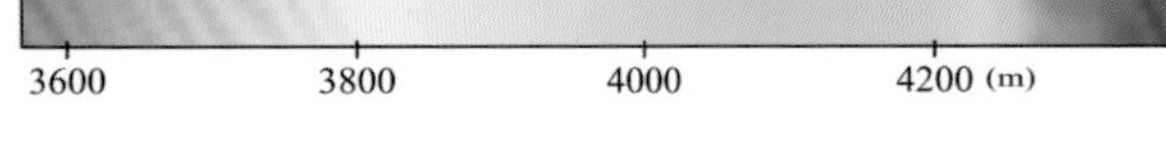

图 6-10　杜甫海丘、孟郊海丘

(a) 海底地形图（等深线间隔 500 米）；(b) 三维海底地形图

Fig.6-10　Dufu Haiqiu, Mengjiao Haiqiu

(a) Seafloor topographic map (with contour interval of 500 m)；(b) 3-D seafloor topographic map

6.28 王维海丘

标准名称 Standard Name	王维海丘 Wangwei Haiqiu	类别 Generic Term	海丘 Hill
中心点坐标 Center Coordinates	14°24.5′N, 118°17.8′E	规模（千米 × 千米） Dimension（km × km）	30 × 12
最小水深（米） Min Depth (m)	3470	最大水深（米） Max Depth (m)	4410
地理实体描述 Feature Description	王维海丘位于南海海盆东南部，平面形态呈北东东—南西西向的长条形（图 6–11）。 Wangwei Haiqiu is located in the southeast of Nanhai Haipen, with a planform in the shape of a long strip in the direction of NEE–SWW (Fig.6–11).		
命名由来 Origin of Name	以唐宋文人的名字进行地名的团组化命名。该海丘以唐朝著名诗人王维命名，纪念他在中国文学史上的重要成就。 A group naming of undersea features after the names of famous men of letter of the Tang and Song Dynasties. The Hill is named after Wang Wei (701–761 A.D.), a famous poet in the Tang Dynasty (618–907 A.D.), to commemorate his outstanding achievement to Chinese literature history.		

6.29 韩愈海山

标准名称 Standard Name	韩愈海山 Hanyu Haishan	类别 Generic Term	海山 Seamount
中心点坐标 Center Coordinates	14°04.2'N, 118°21.6'E	规模（千米 × 千米） Dimension（km × km）	45 × 22
最小水深（米） Min Depth (m)	2950	最大水深（米） Max Depth (m)	4370
地理实体描述 Feature Description	韩愈海丘位于南海海盆东南部，平面形态呈三角形，其顶部发育塌陷火山口（图 6–11）。 Hanyu Haiqiu is located in the southeast of Nanhai Haipen, with a planform in the shape of a triangle. A sunken crater developed on its top (Fig.6–11).		
命名由来 Origin of Name	以唐宋文人的名字进行地名的团组化命名。该海山以唐朝著名诗人韩愈命名，纪念他在中国文学史上的重要成就。 A group naming of undersea features after the names of famous men of letter of the Tang and Song Dynasties. The Seamount is named after Han Yu (768–824 A.D.), a famous poet in the Tang Dynasty (618–907 A.D.), to commemorate his outstanding achievement to Chinese literature history.		

(a)

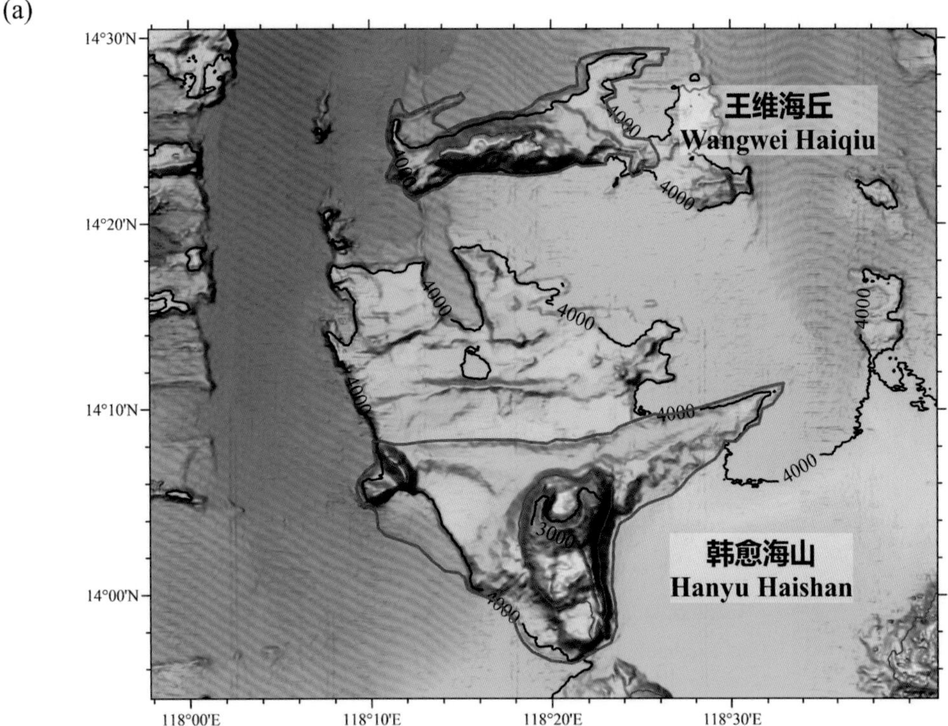

(b)

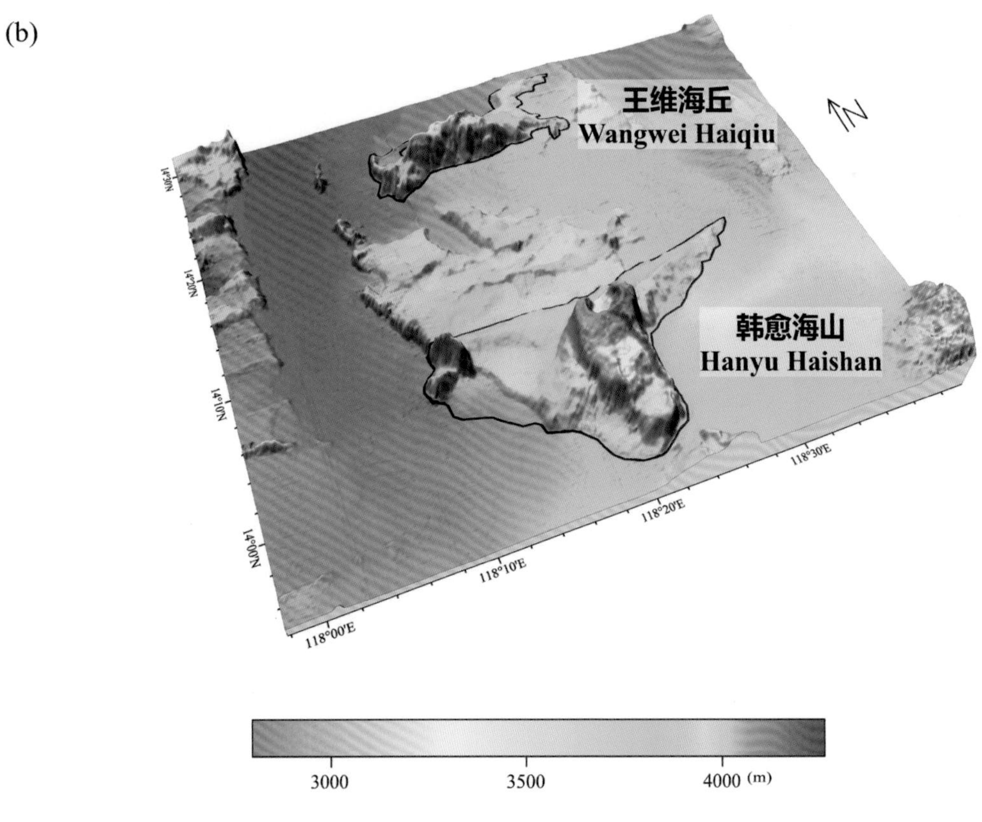

图 6-11 王维海丘、韩愈海山

(a) 海底地形图（等深线间隔 500 米）；(b) 三维海底地形图

Fig.6-11 Wangwei Haiqiu, Hanyu Haishan

(a) Seafloor topographic map (with contour interval of 500 m); (b) 3-D seafloor topographic map

6.30 瓜螺海脊

标准名称 Standard Name	瓜螺海脊 Gualuo Haiji	类别 Generic Term	海脊 Ridge
中心点坐标 Center Coordinates	14°14.6'N, 116°32.9'E	规模（千米 × 千米） Dimension（km × km）	17 × 5
最小水深（米） Min Depth (m)	3640	最大水深（米） Max Depth (m)	4330
地理实体描述 Feature Description	瓜螺海脊位于南海海盆南部，平面形态呈北东东—南西西向的长条形（图 6−12）。 Gualuo Haiji is located in the south of Nanhai Haipen, with a planform in the shape of a long strip in the direction of NEE−SWW (Fig.6−12).		
命名由来 Origin of Name	以海洋软体动物名进行地名的团组化命名。“瓜螺”是一种生活在海洋里的软体动物。 A group naming of undersea features after the names of marine mollusks. “Gualuo” refers to cymbium melo, is a species of marine mollusk.		

6.31 樱蛤海丘

标准名称 Standard Name	樱蛤海丘 Yingge Haiqiu	类别 Generic Term	海丘 Hill
中心点坐标 Center Coordinates	14°07.1'N, 116°37.3'E	规模（千米 × 千米） Dimension（km × km）	19 × 6
最小水深（米） Min Depth (m)	3810	最大水深（米） Max Depth (m)	4340
地理实体描述 Feature Description	樱蛤海丘位于南海海盆南部，平面形态呈北东东—南西西向的长条形（图 6−12）。 Yingge Haiqiu is located in the south of Nanhai Haipen, with a planform in the shape of a long strip in the direction of NEE−SWW (Fig.6−12).		
命名由来 Origin of Name	以海洋软体动物名进行地名的团组化命名。“樱蛤”是一种生活在海洋里的软体动物。 A group naming of undersea features after the names of marine mollusks. “Yingge” refers to tellina, is a species of marine mollusk.		

(a)

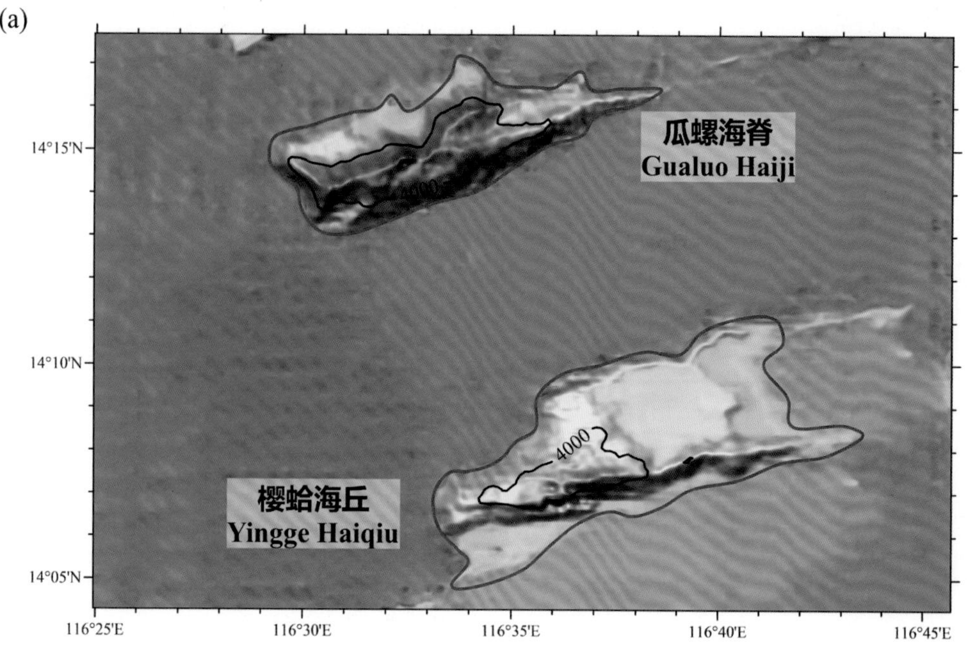

(b)

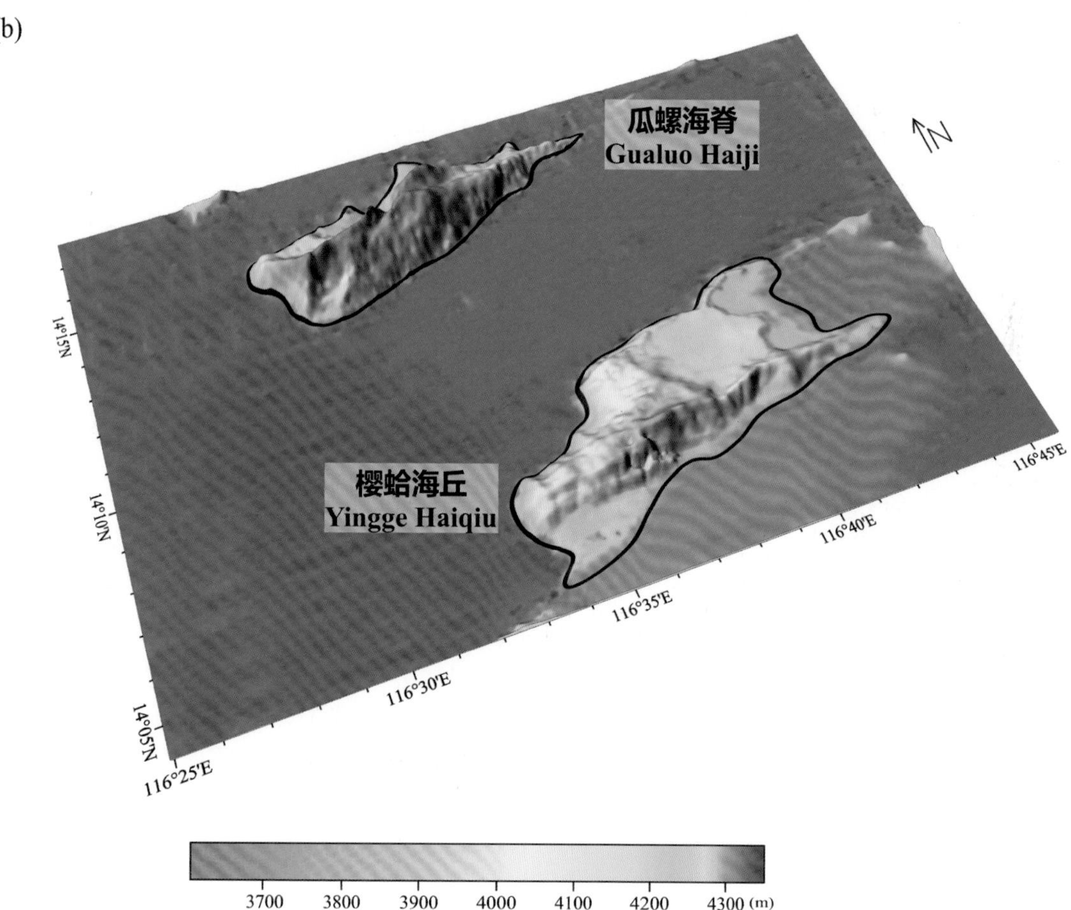

图 6-12　瓜螺海脊、樱蛤海丘

(a) 海底地形图（等深线间隔 500 米）；(b) 三维海底地形图

Fig.6-12　Gualuo Haiji, Yingge Haiqiu

(a) Seafloor topographic map (with contour interval of 500 m); (b) 3-D seafloor topographic map

6.32 杜牧海丘

标准名称 Standard Name	杜牧海丘 Dumu Haiqiu	类别 Generic Term	海丘 Hill
中心点坐标 Center Coordinates	14°06.5'N, 116°12.1'E	规模（千米 × 千米） Dimension（km × km）	46 × 18
最小水深（米） Min Depth (m)	3940	最大水深（米） Max Depth (m)	4650
地理实体描述 Feature Description	杜牧海丘位于南海海盆南部，平面形态呈北东—南西向的扇形，其顶部发育多座峰（图 6-13）。 Dumu Haiqiu is located in the south of Nanhai Haipen, with a planform in the shape of a fan in the direction of NE–SW, Multiple peaks developed on its top (Fig.6–13).		
命名由来 Origin of Name	以唐宋文人的名字进行地名的团组化命名。该海丘以唐朝著名诗人杜牧命名，纪念他在中国文学史上的重要成就。 A group naming of undersea features after the names of famous men of letter of the Tang and Song Dynasties. The Hill is named after Du Mu (803–852 A.D.), a famous poet in the Tang Dynasty (618–907 A.D.), to commemorate his outstanding achievement to Chinese literature history.		

6.33 秦观海山

标准名称 Standard Name	秦观海山 Qinguan Haishan	类别 Generic Term	海山 Seamount
中心点坐标 Center Coordinates	14°03.0'N, 115°46.1'E	规模（千米 × 千米） Dimension（km × km）	40 × 16
最小水深（米） Min Depth (m)	3260	最大水深（米） Max Depth (m)	4580
地理实体描述 Feature Description	秦观海山位于南海海盆南部，与中南海山东侧相邻，平面形态呈北东—南西向的多边形，其顶部发育多座峰（图 6-13）。 Qinguan Haishan is located in the south of Nanhai Haipen and is adjacent to the east side of Zhongnan Haishan, with a planform in the shape of a polygon in the direction of NE–SW. Multiple peaks developed on its top (Fig.6–13).		
命名由来 Origin of Name	以唐宋文人的名字进行地名的团组化命名。该海山以北宋著名诗人秦观命名，纪念他在中国文学史上的重要成就。 A group naming of undersea features after the names of famous men of letter of the Tang and Song Dynasties. The Seamount is named after Qin Guan (1049–1100 A.D.), a famous poet in the Song Dynasty (960–1279 A.D.), to commemorate his outstanding achievement to Chinese literature history.		

(a)

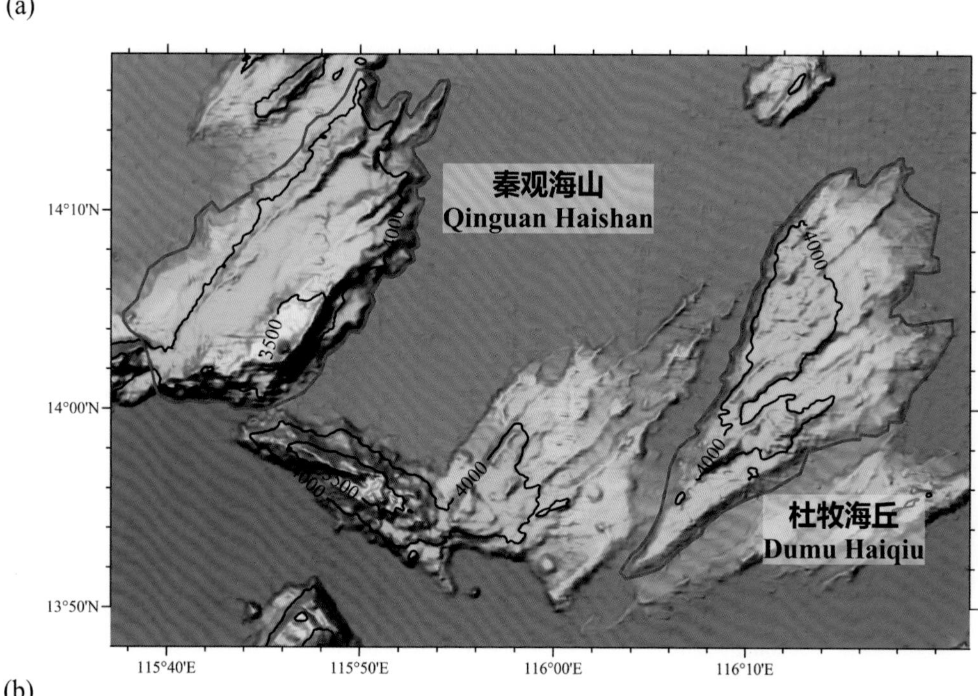

(b)

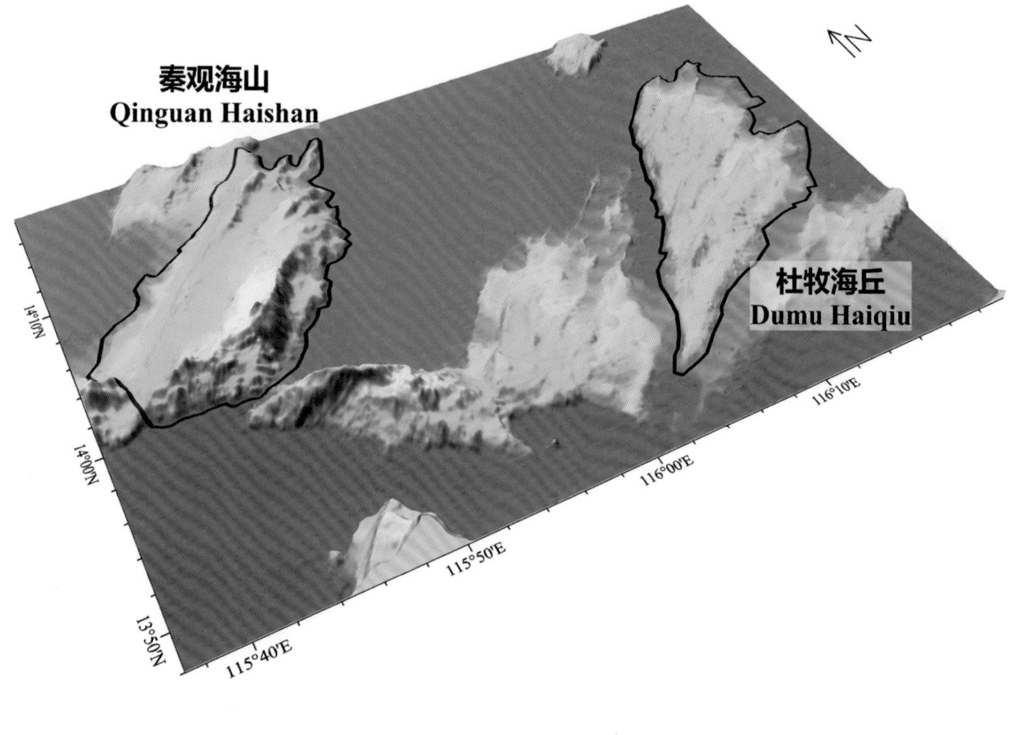

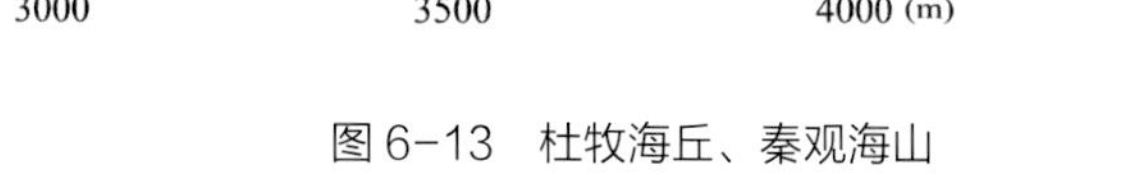

图 6-13　杜牧海丘、秦观海山

(a) 海底地形图（等深线间隔 500 米）；(b) 三维海底地形图

Fig.6-13　Dumu Haiqiu, Qinguan Haishan

(a) Seafloor topographic map (with contour interval of 500 m)；(b) 3-D seafloor topographic map

6.34 北岳海山

标准名称 Standard Name	北岳海山 Beiyue Haishan	类别 Generic Term	海山 Seamount
中心点坐标 Center Coordinates	14°21.8'N, 115°24.1'E	规模（千米 × 千米） Dimension（km × km）	36 × 32
最小水深（米） Min Depth (m)	810	最大水深（米） Max Depth (m)	4690
地理实体描述 Feature Description	北岳海山位于南海海盆南部，平面形态近似圆形，其顶部发育多个峰（图 6–14）。 Beiyue Haishan is located in the south of Nanhai Haipen, with a nearly circular planform, and has multiple peaks developed on its top (Fig.6–14).		
命名由来 Origin of Name	该海山位于中南海山以北，山体规模大，因此得名。 The Seamount is located on the north of Zhongnan Haishan with a large scale, and “Bei” means north, “Yue” means massive mountain, so the word “Beiyue” was used to name the Seamount.		

6.35 柳永海丘

标准名称 Standard Name	柳永海丘 Liuyong Haiqiu	类别 Generic Term	海丘 Hill
中心点坐标 Center Coordinates	14°27.5'N, 115°50.0'E	规模（千米 × 千米） Dimension（km × km）	32 × 15
最小水深（米） Min Depth (m)	3860	最大水深（米） Max Depth (m)	4600
地理实体描述 Feature Description	柳永海丘位于南海海盆南部，北岳海山东侧，由多条北东—南西向的脊状地形构成（图 6–14）。 Liuyong Haiqiu is located in the south of Nanhai Haipen and to the east of Beiyue Haishan. The Hill composed of multiple ridge-shaped topographical units in the direction of NE–SW (Fig.6–14).		
命名由来 Origin of Name	以唐宋文人的名字进行地名的团组化命名。该海丘以北宋著名词人柳永命名，纪念他在中国文学史上的重要成就。 A group naming of undersea features after the names of famous men of letter of the Tang and Song Dynasties. The Hill is named after Liu Yong (984–1053 A.D.), a famous lyricist in the Song Dynasty (960–1279 A.D.), to commemorate his outstanding achievement to Chinese literature history.		

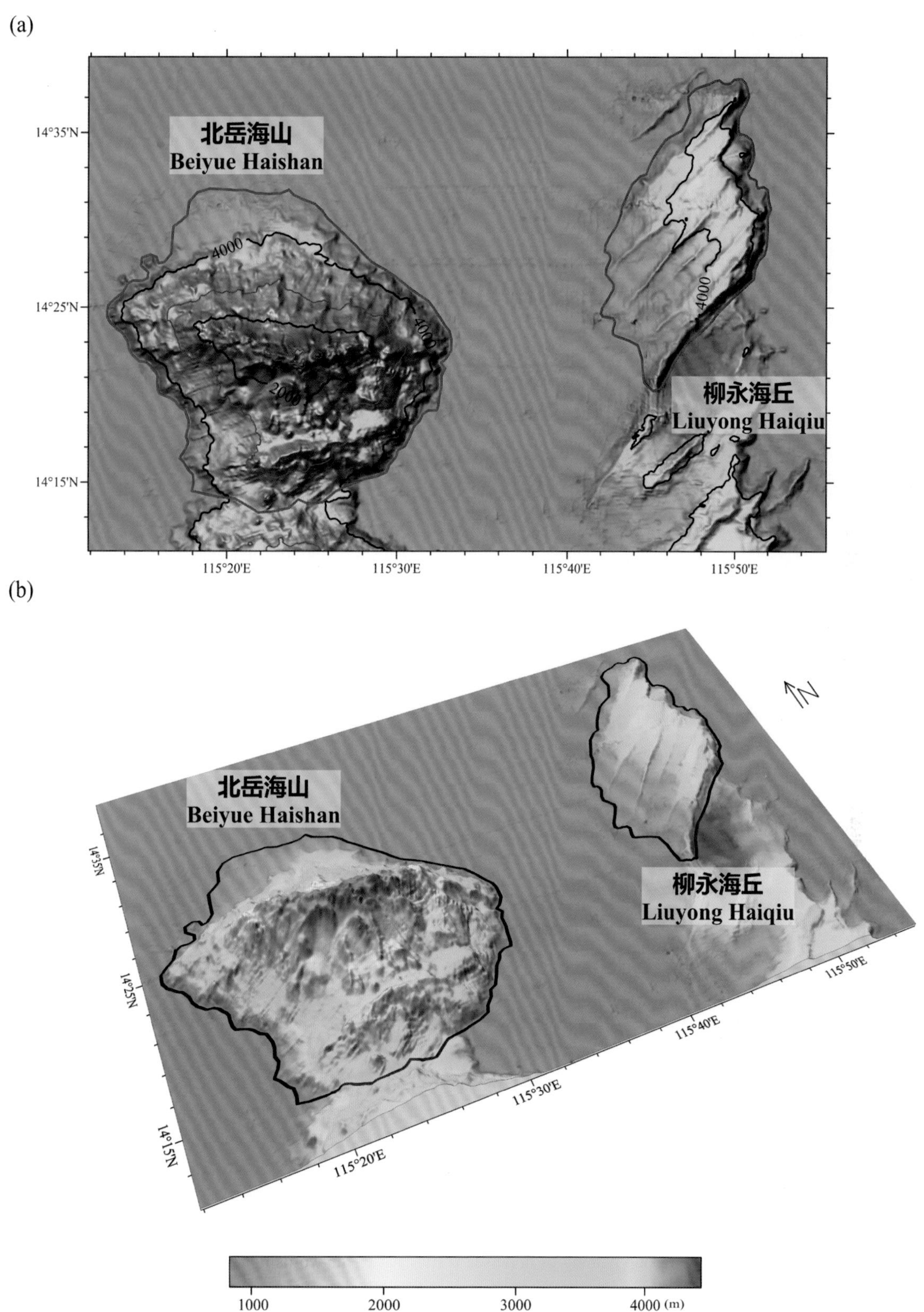

图 6-14　北岳海山、柳永海丘

(a) 海底地形图（等深线间隔 1000 米）；(b) 三维海底地形图

Fig.6-14　Beiyue Haishan, Liuyong Haiqiu

(a) Seafloor topographic map (with contour interval of 1000 m)；(b) 3-D seafloor topographic map

6.36 钻石海山

标准名称 Standard Name	钻石海山 Zuanshi Haishan	类别 Generic Term	海山 Seamount
中心点坐标 Center Coordinates	15°01.4'N, 114°54.5'E	规模（千米 × 千米） Dimension（km × km）	30 × 10
最小水深（米） Min Depth (m)	2370	最大水深（米） Max Depth (m)	4500
地理实体描述 Feature Description	钻石海山位于南海海盆西部，平面形态呈东—西向的菱形（图 6−15）。 Zuanshi Haishan is located in the west of Nanhai Haipen with a planform in the shape of a diamond in the direction of E−W (Fig.6−15).		
命名由来 Origin of Name	以宝石、矿物的名称进行地名的团组化命名。“钻石”是指经过琢磨的金刚石，是自然界中天然存在的最坚硬的物质，也是贵重的观赏宝石和佩戴饰物。 A group naming of undersea features after the names of gems and minerals. The specific term of the undersea feature name “Zuanshi”, is a polished diamond and is the hardest natural object in nature, as well as a kind of valuable ornamental gems and jewelries.		

6.37 白琥海丘

标准名称 Standard Name	白琥海丘 Baihu Haiqiu	类别 Generic Term	海丘 Hill
中心点坐标 Center Coordinates	15°06.0'N, 114°35.9'E	规模（千米 × 千米） Dimension（km × km）	10 × 3
最小水深（米） Min Depth (m)	3390	最大水深（米） Max Depth (m)	4300
地理实体描述 Feature Description	白琥海丘位于南海海盆西部，中沙海台南侧，平面形态呈北西西—南东东向的菱形（图 6−15）。 Baihu Haiqiu is located in the west of Nanhai Haipen and to the south of Zhongsha Haitai, with a planform in the shape of a diamond in the direction of NWW−SEE (Fig.6−15).		
命名由来 Origin of Name	以与“玉”相关的词进行地名的团组化命名。“白琥”是指雕成虎形的白玉，古代祭祀西方时用之。 A group naming of undersea features after the words related to jade (“Yu” in Chinese). “Baihu” is a kind white jade carved in the shape of tiger and was used for sacrifice to the west in the ancient time.		

6.38 玄璜海丘

标准名称 Standard Name	玄璜海丘 Xuanhuang Haiqiu	类别 Generic Term	海丘 Hill
中心点坐标 Center Coordinates	15°00.6'N, 114°35.6'E	规模（千米 × 千米） Dimension（km × km）	8 × 13
最小水深（米） Min Depth (m)	3910	最大水深（米） Max Depth (m)	4310
地理实体描述 Feature Description	玄璜海丘位于南海海盆西部，中沙海台南侧，平面形态呈近南—北向的菱形（图 6-15）。 Xuanhuang Haiqiu is located in the west of Nanhai Haipen and to the south of Zhongsha Haitai, with a planform in the shape of a diamond nearly in the direction of S-N (Fig.6-15).		
命名由来 Origin of Name	以与“玉”相关的词进行地名的团组化命名。“玄璜”是指用黑玉制作的半圆形瑞玉。 A group naming of undersea features after the words related to jade (“Yu” in Chinese). “Xuanhuang” is a semi-circular ritual jade made of dark jade.		

(a)

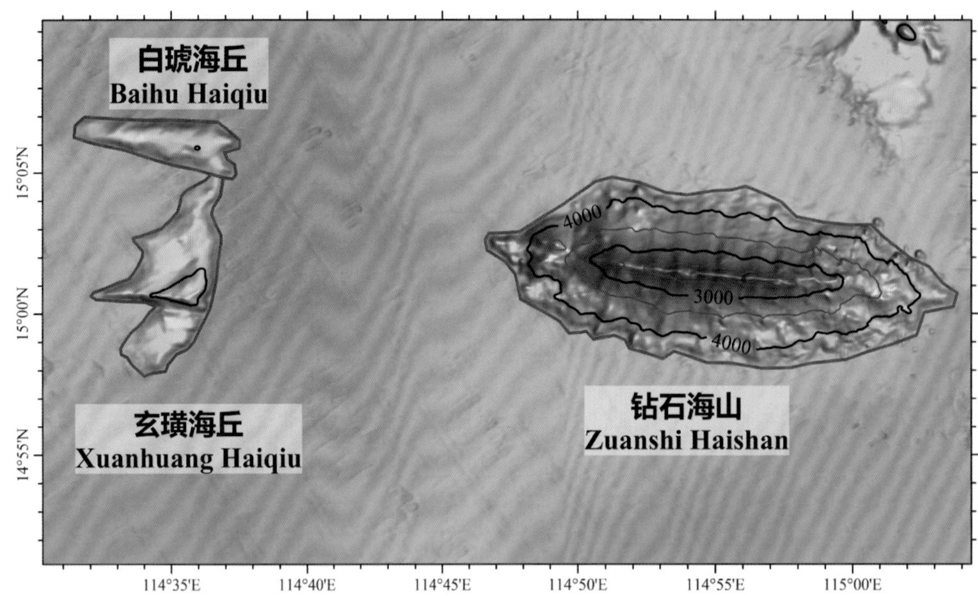

(b)

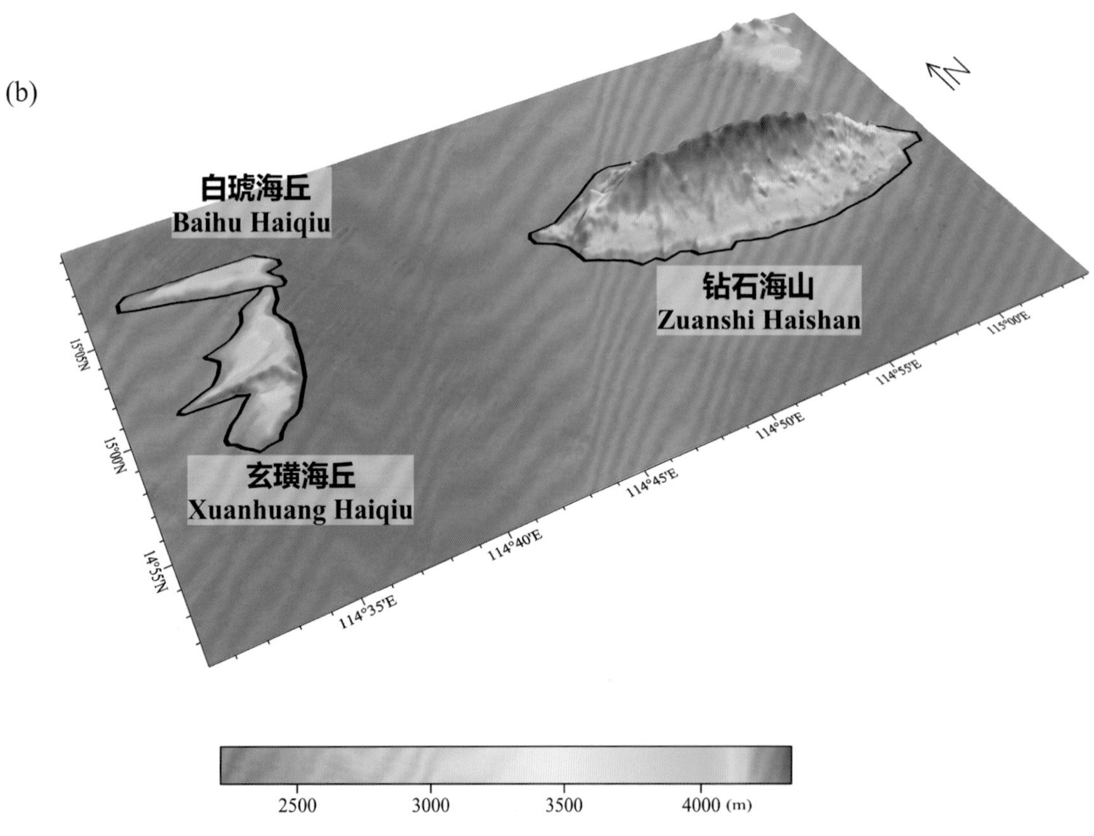

图 6-15　钻石海山、白琥海丘、玄璜海丘

(a) 海底地形图（等深线间隔 500 米）；(b) 三维海底地形图

Fig.6-15　Zuanshi Haishan, Baihu Haiqiu, Xuanhuang Haiqiu

(a) Seafloor topographic map (with contour interval of 500 m)；(b) 3-D seafloor topographic map

6.39 蓝宝石海山

标准名称 Standard Name	蓝宝石海山 Lanbaoshi Haishan	类别 Generic Term	海山 Seamount
中心点坐标 Center Coordinates	14°50.1'N, 114°15.8'E	规模（千米 × 千米） Dimension（km × km）	30 × 8
最小水深（米） Min Depth (m)	2960	最大水深（米） Max Depth (m)	4530
地理实体描述 Feature Description	蓝宝石海山位于南海海盆西部，中沙海台南侧，平面形态呈北西—南东向的长条形（图 6–16）。 Lanbaoshi Haishan is located in the west of Nanhai Haipen and to the south of Zhongsha Haitai, with a planform in the shape of a long strip in the direction of NW–SE (Fig.6–16).		
命名由来 Origin of Name	以宝石、矿物的名称进行地名的团组化命名。“蓝宝石”是指是刚玉宝石中除红宝石之外，其他颜色刚玉宝石的通称。 A group naming of undersea features after the names of gems and minerals. “Lanbaoshi” means sapphire, is the general term for other colors of corundum jewels than ruby.		

6.40 红宝石海山

标准名称 Standard Name	红宝石海山 Hongbaoshi Haishan	类别 Generic Term	海山 Seamount
中心点坐标 Center Coordinates	14°40.8'N, 114°30.6'E	规模（千米 × 千米） Dimension（km × km）	20 × 8
最小水深（米） Min Depth (m)	2830	最大水深（米） Max Depth (m)	4520
地理实体描述 Feature Description	红宝石海山位于南海海盆西部，中沙海台南侧，平面形态呈北西西—南东东向的长条形（图 6–16）。 Hongbaoshi Haishan is located in the west of Nanhai Haipen and to the south of Zhongsha Haitai with a planform in the shape of a long strip in the direction of NWW–SEE (Fig.6–16).		
命名由来 Origin of Name	以宝石、矿物的名称进行地名的团组化命名。“红宝石”是指颜色呈红色的刚玉。 A group naming of undersea features after the names of gems and minerals. “Hongbaoshi” means ruby, is a red-colored corundum.		

(a)

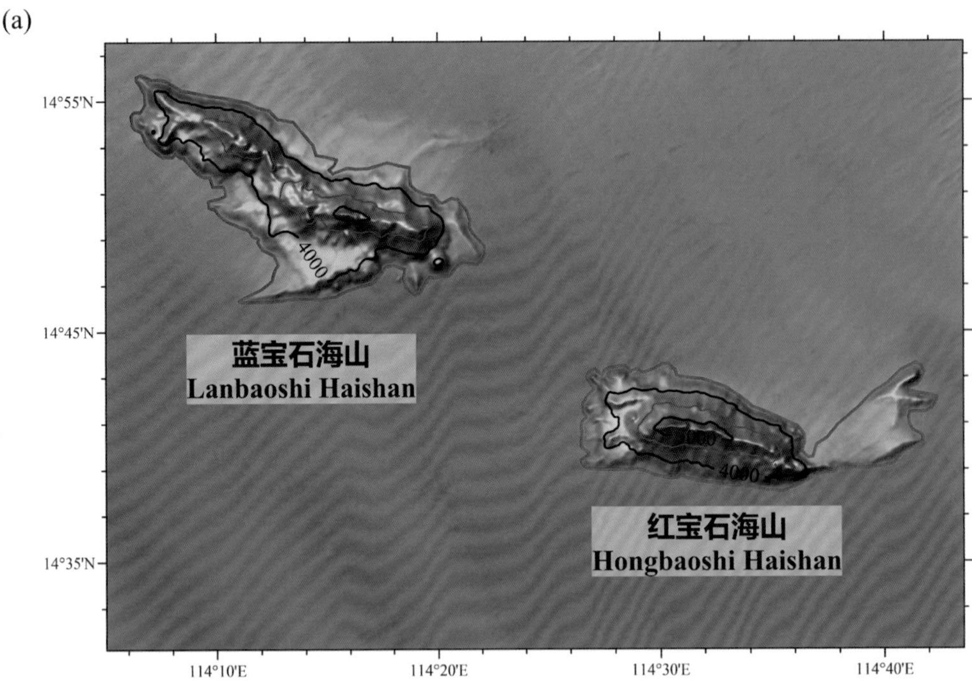

(b)

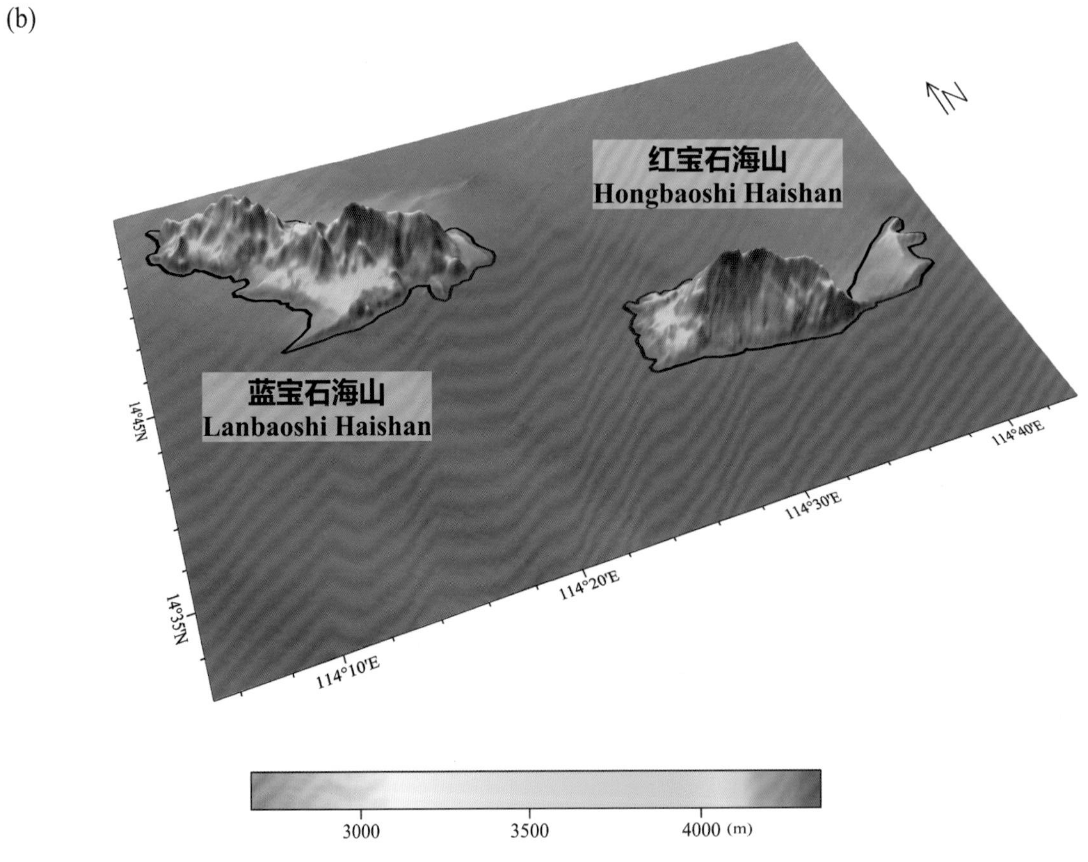

图 6-16　蓝宝石海山、红宝石海山

(a) 海底地形图（等深线间隔 200 米）；(b) 三维海底地形图

Fig.6-16　Lanbaoshi Haishan, Hongbaoshi Haishan

(a) Seafloor topographic map (with contour interval of 200 m)；(b) 3-D seafloor topographic map

6.41 玉瓶海脊

标准名称 Standard Name	玉瓶海脊 Yuping Haiji	类别 Generic Term	海脊 Ridge
中心点坐标 Center Coordinates	14°36.8'N, 115°05.3'E	规模（千米 × 千米） Dimension（km × km）	16 × 4
最小水深（米） Min Depth (m)	4000	最大水深（米） Max Depth (m)	4350
地理实体描述 Feature Description	玉瓶海脊位于南海海盆西部，北岳海山西北侧，平面形态呈北东—南西向的长条形（图 6–17）。 Yuping Haiji is located in the west of Nanhai Haipen and to the northwest of Beiyue Haishan, with a planform in the shape of a long strip in the direction of NE–SW (Fig.6–17).		
命名由来 Origin of Name	以与"玉"相关的词进行地名的团组化命名。"玉瓶"是瓷瓶的美称。 A group naming of undersea features after the words related to jade ("Yu" in Chinese). "Yuping" means porcelain vase.		

6.42 赤璋海丘

标准名称 Standard Name	赤璋海丘 Chizhang Haiqiu	类别 Generic Term	海丘 Hill
中心点坐标 Center Coordinates	14°26.1'N, 115°08.3'E	规模（千米 × 千米） Dimension（km × km）	12 × 5
最小水深（米） Min Depth (m)	3970	最大水深（米） Max Depth (m)	4340
地理实体描述 Feature Description	赤璋海丘位于南海海盆西部，北岳海山西侧，平面形态呈北东—南西向的长条形（图 6–17）。 Chizhang Haiqiu is located in the west of Nanhai Haipen and to the west of Beiyue Haishan, with planform in the shape of a long strip in the direction of NE–SW (Fig.6–17).		
命名由来 Origin of Name	以与"玉"相关的词进行地名的团组化命名。"赤璋"是中国古代天子举行祭祀活动时，专门用于敬献给南方之神的规范化玉制礼器。 A group naming of undersea features after the words related to jade ("Yu" in Chinese). "Chizhang" is a kind of standard sacrificial vessels made of jade and was dedicated for sacrificing the God of the South during the sacrificial activities held by the ancient emperors.		

(a)

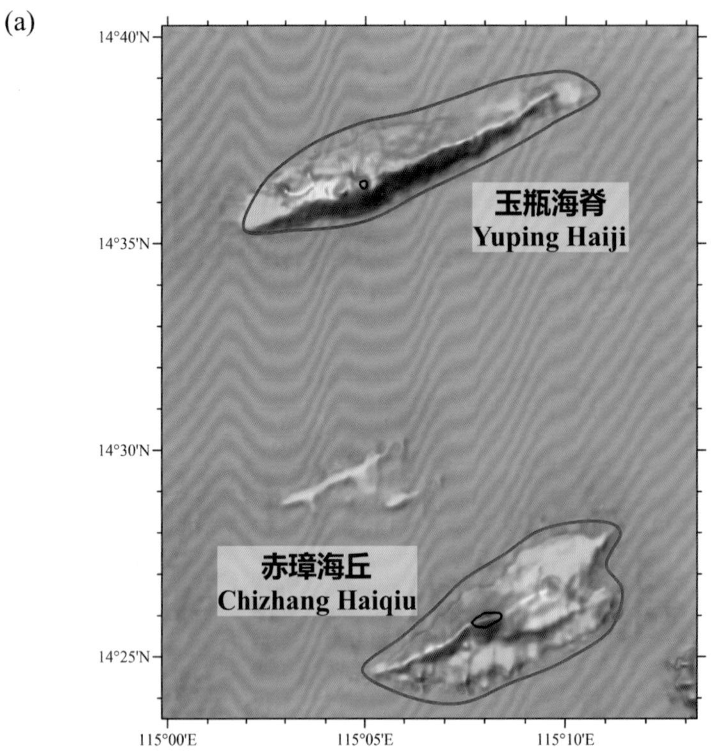

(b)

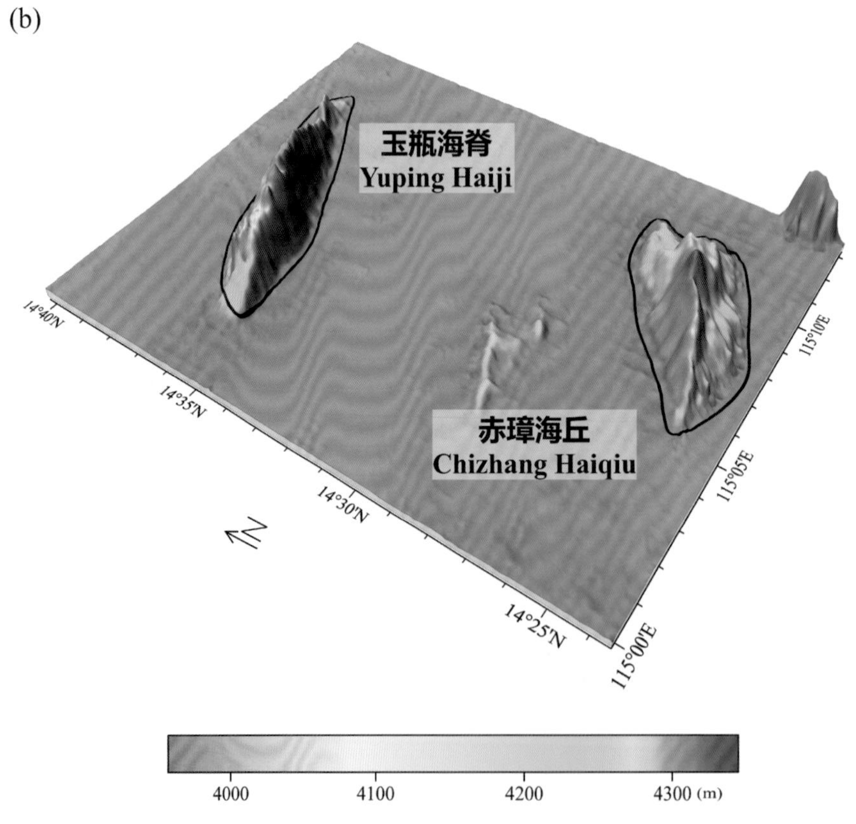

图 6-17　玉瓶海脊、赤璋海丘

(a) 海底地形图（等深线间隔 500 米）；(b) 三维海底地形图

Fig.6-17　Yuping Haiji, Chizhang Haiqiu

(a) Seafloor topographic map (with contour interval of 500 m)；(b) 3-D seafloor topographic map

6.43 青圭海丘

标准名称 Standard Name	青圭海丘 Qinggui Haiqiu	类别 Generic Term	海丘 Hill
中心点坐标 Center Coordinates	14°31.1'N, 114°45.5'E	规模（千米 × 千米） Dimension（km × km）	16 × 5
最小水深（米） Min Depth (m)	3960	最大水深（米） Max Depth (m)	4330
地理实体描述 Feature Description	青圭海丘位于南海海盆西部，中沙海台南侧，平面形态呈北东—南西向的长条形，其顶部发育两座峰（图 6−18）。 Qinggui Haiqiu is located in the west of Nanhai Haipen and to the south side of Zhongsha Haitai, with a planform in the shape of a long strip in the direction of NE−SW, and has two peaks developed on its top (Fig.6−18).		
命名由来 Origin of Name	以与“玉”相关的词进行地名的团组化命名。“青圭”是中国古代青玉制作的礼器，上圆（或剑头形）下方。 A group naming of undersea features after the words related to jade (“Yu” in Chinese). “Qinggui” is a sacrificial vessels made of sapphire in ancient China, with a round upper part (or in the shape of head of sword) and square lower part.		

6.44 黄琮海丘

标准名称 Standard Name	黄琮海丘 Huangcong Haiqiu	类别 Generic Term	海丘 Hill
中心点坐标 Center Coordinates	14°17.2'N, 114°54.7'E	规模（千米 × 千米） Dimension（km × km）	4 × 5
最小水深（米） Min Depth (m)	3950	最大水深（米） Max Depth (m)	4330
地理实体描述 Feature Description	黄琮海丘位于南海海盆西部，龙北海山北侧，平面形态呈圆形（图 6−18）。 Huangcong Haiqiu is located in the west of Nanhai Haipen and to the north of Longbei Haishan with a planform in the shape of a circle (Fig.6−18).		
命名由来 Origin of Name	以与“玉”相关的词进行地名的团组化命名。“黄琮”是我国古代祭祀所用之玉制礼器，专为祭祀皇地祇而用。 A group naming of undersea features after the words related to jade (“Yu” in Chinese). “Huangcong” is a kind of sacrificial jade vessel dedicated for sacrificing gnome in ancient China.		

6.45 苍璧海丘

标准名称 Standard Name	苍璧海丘 Cangbi Haiqiu	类别 Generic Term	海丘 Hill
中心点坐标 Center Coordinates	14°19.5'N, 114°43.1'E	规模（千米 × 千米） Dimension（km × km）	20 × 10
最小水深（米） Min Depth (m)	3720	最大水深（米） Max Depth (m)	4350
地理实体描述 Feature Description	苍璧海丘位于南海海盆西部，龙北海山西北侧，平面形态呈北西—南东向的长条形，其顶部发育多座峰（图 6−18）。 Cangbi Haiqiu is located in the west of Nanhai Haipen and to the northwest of Longbei Haishan, with a planform in the shape of a long stripe in the direction of NW−SE, and has multiple peaks developed on its top (Fig.6−18).		
命名由来 Origin of Name	以与"玉"相关的词进行地名的团组化命名。"苍璧"即灰白色的玉璧，是我国古代祭祀所用之玉制礼器。 A group naming of undersea features after the words related to jade ("Yu" in Chinese). "Cangbi" means greyish white jade, is a sacrificial jade vessel in ancient China.		

6.46 红柱石海丘

标准名称 Standard Name	红柱石海丘 Hongzhushi Haiqiu	类别 Generic Term	海丘 Hill
中心点坐标 Center Coordinates	14°24.6′N, 114°33.7′E	规模（千米 × 千米） Dimension（km × km）	22 × 5
最小水深（米） Min Depth (m)	3940	最大水深（米） Max Depth (m)	4550
地理实体描述 Feature Description	红柱石海丘位于南海海盆西部，龙北海山西北侧，平面形态呈北西—南东向的长条形，其顶部发育多座峰（图 6−18）。 Hongzhushi Haiqiu is located in the west of Nanhai Haipen and is to the northwest of Longbei Haishan, with a planform in the shape of a long strip in the direction of NW−SE, Multiple peaks developed on its top (Fig.6−18).		
命名由来 Origin of Name	以宝石、矿物的名称进行地名的团组化命名。"红柱石"是一种铝硅酸盐矿物，是一种优质的耐火材料。 A group naming of undersea features after the names of gems and minerals. "Hongzhushi" means andalusite, is a kind of silicate minerals and is a high quality fireproofing materials.		

6.47 孔雀石海丘

标准名称 Standard Name	孔雀石海丘 Kongqueshi Haiqiu	类别 Generic Term	海丘 Hill
中心点坐标 Center Coordinates	14°07.6′N, 114°37.1′E	规模（千米 × 千米） Dimension（km × km）	14 × 7
最小水深（米） Min Depth (m)	3920	最大水深（米） Max Depth (m)	4560
地理实体描述 Feature Description	孔雀石海丘位于南海海盆西部，龙北海山西北侧，平面形态呈北西西—南东东向的多边形，其顶部发育多座峰（图 6–18）。 Kongqueshi Haiqiu is located in the west of Nanhai Haipen and is to the northwest of Longbei Haishan, with a planform in the shape of a polygon in the direction of NWW–SEE, and has multiple peaks developed on its top (Fig.6–18).		
命名由来 Origin of Name	以宝石、矿物的名称进行地名的团组化命名。“孔雀石”是一种含铜的碳酸盐矿物，呈现浓绿、翠绿的光泽。 A group naming of undersea features after the names of gems and minerals. “Kongqueshi” means malachite, is a kind of carbonate minerals containing copper and is emerald green or deep green in color.		

(a)

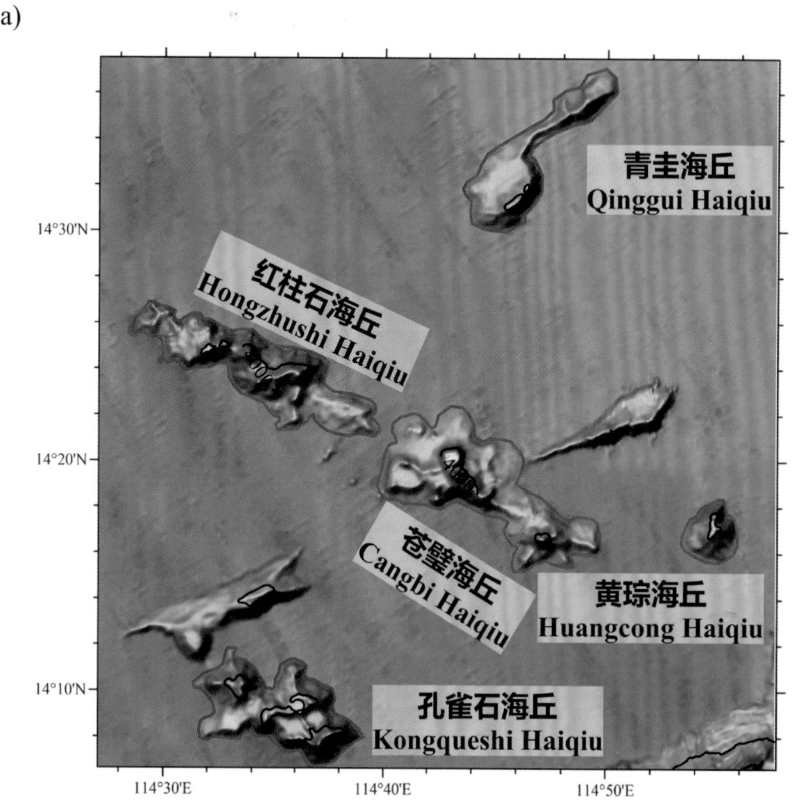

(b)

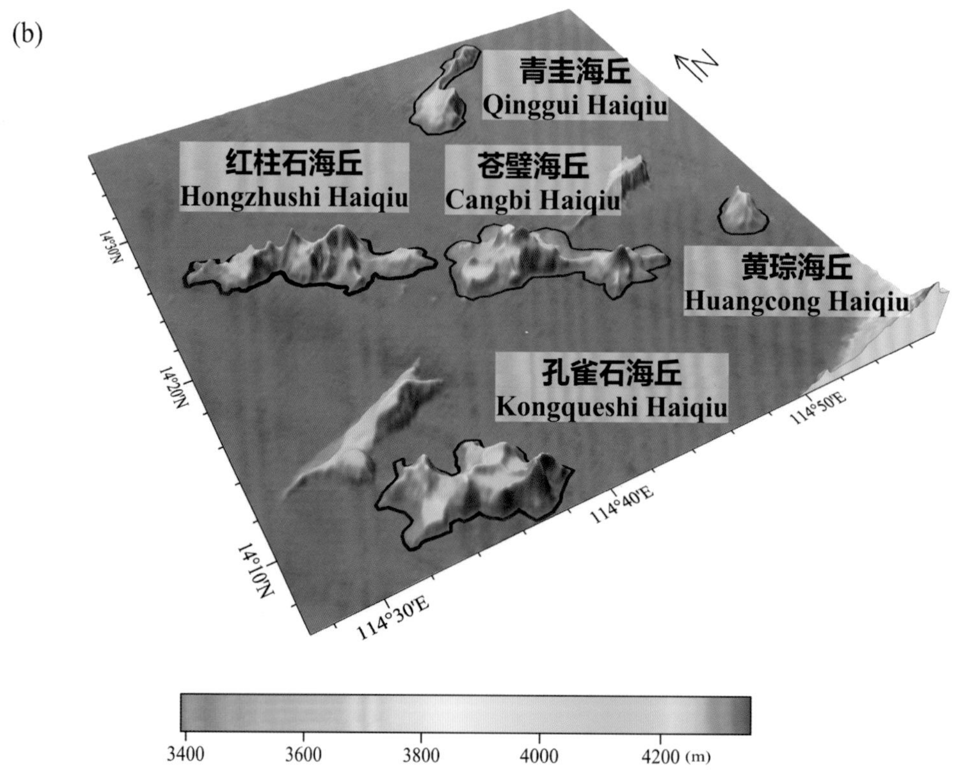

图 6-18 青圭海丘、黄琮海丘、苍璧海丘、红柱石海丘、孔雀石海丘

(a) 海底地形图（等深线间隔 500 米）；(b) 三维海底地形图

Fig.6-18 Qinggui Haiqiu, Huangcong Haiqiu, Cangbi Haiqiu, Hongzhushi Haiqiu, Kongqueshi Haiqiu

(a) Seafloor topographic map (with contour interval of 500 m); (b) 3-D seafloor topographic map

6.48 绿柱石海丘

标准名称 Standard Name	绿柱石海丘 Lüzhushi Haiqiu	类别 Generic Term	海丘 Hill
中心点坐标 Center Coordinates	14°23.4′N, 114°12.5′E	规模（千米 × 千米） Dimension（km × km）	14 × 10
最小水深（米） Min Depth (m)	3880	最大水深（米） Max Depth (m)	4540
地理实体描述 Feature Description	绿柱石海丘位于南海海盆西部，盆西海岭东侧，平面形态呈不规则多边形（图 6–19）。 Lüzhushi Haiqiu is located in the west of Nanhai Haipen and is to the east of Penxi Hailing, with a planform in the shape of an irregular polygon (Fig.6–19).		
命名由来 Origin of Name	以宝石、矿物的名称进行地名的团组化命名。"绿柱石"是一种含铜的碳酸盐矿物，呈现浓绿、翠绿的光泽。 A group naming of undersea features after the names of gems and minerals. "Lüzhushi" means beryl, is a kind of carbonate minerals containing copper and is emerald green or deep green in color.		

6.49 理璞海丘

标准名称 Standard Name	理璞海丘 Lipu Haiqiu	类别 Generic Term	海丘 Hill
中心点坐标 Center Coordinates	14°22.1'N, 114°06.6'E	规模（千米 × 千米） Dimension（km × km）	7 × 8
最小水深（米） Min Depth (m)	4040	最大水深（米） Max Depth (m)	4330
地理实体描述 Feature Description	理璞海丘位于南海海盆西部，盆西海岭东侧，平面形态呈不规则多边形（图 6–19）。 Lipu Haiqiu is located in the west of Nanhai Haipen and to the east of Penxi Hailing, with a planform in the shape of in irregular polygon (Fig.6–19).		
命名由来 Origin of Name	以与"玉"相关的词进行地名的团组化命名。"理璞"即加工璞玉，取词自《韩非子・和氏》："王乃使玉人理其璞而得宝焉"。 A group naming of undersea features after the words related to jade ("Yu" in Chinese). The specific term of this undersea feature name "Lipu", means processed jade, is derived from the quote "The king then had the jade processed and obtained a priceless treasure" in *Han Fei Zi · He Shi*.		

6.50 润泽海丘

标准名称 Standard Name	润泽海丘 Runze Haiqiu	类别 Generic Term	海丘 Hill
中心点坐标 Center Coordinates	14°18.5'N, 113°54.1'E	规模（千米 × 千米） Dimension（km × km）	10 × 3
最小水深（米） Min Depth (m)	4020	最大水深（米） Max Depth (m)	4340
地理实体描述 Feature Description	润泽海丘位于南海海盆西部，盆西海岭东侧，平面形态呈北东—南西向的长条形（图 6-19）。 Runze Haiqiu is located in the west of Nanhai Haipen and to the east of Penxi Hailing, with a planform in the shape of a long strip in the direction of NE-SW (Fig.6-19).		
命名由来 Origin of Name	以与"玉"相关的词进行地名的团组化命名。"润泽"指玉石的温润、油性，取词自汉代的汉文字学家许慎的《说文解字》："润泽以温，仁之方也"。 A group naming of undersea features after the words related to jade ("Yu" in Chinese). The specific term of this undersea feature name "Runze", means moisturized and not dry, implying the jade is fine and oily, is derived from the quote "The jade is gentle, moisturized and lustrous, implying of its benevolence and generosity" in *Origin of Chinese Characters* by Xu Shen (58-147 A.D.), a philologist in the Han Dynasty (202 B.C.-220 A.D.)		

6.51 玉璧海脊

标准名称 Standard Name	玉璧海脊 Yubi Haiji	类别 Generic Term	海脊 Ridge
中心点坐标 Center Coordinates	14°12.6'N, 114°08.5'E	规模（千米 × 千米） Dimension（km × km）	28 × 5
最小水深（米） Min Depth (m)	3950	最大水深（米） Max Depth (m)	4350
地理实体描述 Feature Description	玉璧海脊位于南海海盆西部，盆西海岭东侧，平面形态呈北东—南西向的长条形。海脊的两侧窄，中部宽（图 6-19）。 Yubi Haiji is located in the west of Nanhai Haipen and to the east side of Penxi Hailing with a planform in the shape of a long strip in the direction of NE-SW. The Ridge is narrow on the both ends and wide in the middle (Fig.6-19).		
命名由来 Origin of Name	以与"玉"相关的词进行地名的团组化命名。"玉璧"是一种中央有穿孔的扁平状圆形玉器，为我国传统的玉制礼器之一。 A group naming of undersea features after the words related to jade ("Yu" in Chinese). "Yubi" is a kind of flat circular jade ware with a hole in the center and is one of the traditional sacrificial jade ware in China.		

6.52 得宝海丘

标准名称 Standard Name	得宝海丘 Debao Haiqiu	类别 Generic Term	海丘 Hill
中心点坐标 Center Coordinates	14°10.9'N, 114°14.8'E	规模（千米 × 千米） Dimension（km × km）	10 × 3
最小水深（米） Min Depth (m)	4130	最大水深（米） Max Depth (m)	4340
地理实体描述 Feature Description	得宝海丘位于南海海盆西部，盆西海岭东侧，平面形态呈北东—南西向的长条形（图 6–19）。 Debao Haiqiu is located in the west of Nanhai Haipen and to the east of Penxi Hailing, with a planform in the shape of a long strip in the direction of NE–SW (Fig.6–19).		
命名由来 Origin of Name	以与"玉"相关的词进行地名的团组化命名。"得宝"即得到宝玉，取词自《韩非子·和氏》："王乃使玉人理其璞而得宝焉"。 A group naming of undersea features after the words related to jade ("Yu" in Chinese). The specific term of this undersea feature name "Debao", means obtaining priceless treasure, is derived from the lines "The king had the jade processed and obtained a priceless treasure" in *Han Fei Zi · He Shi*.		

(a)

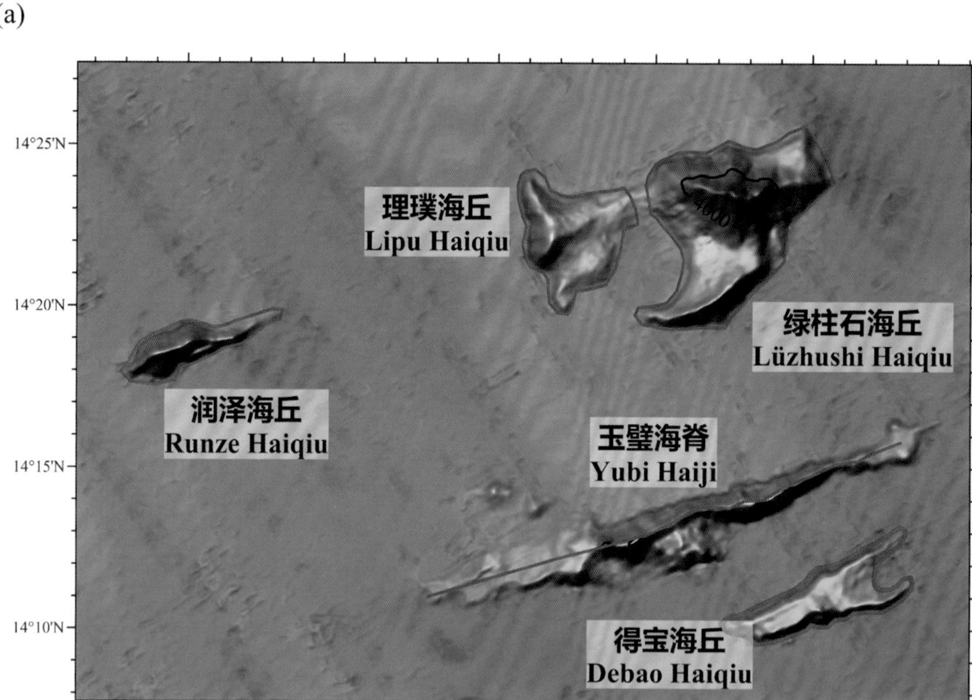

(b)

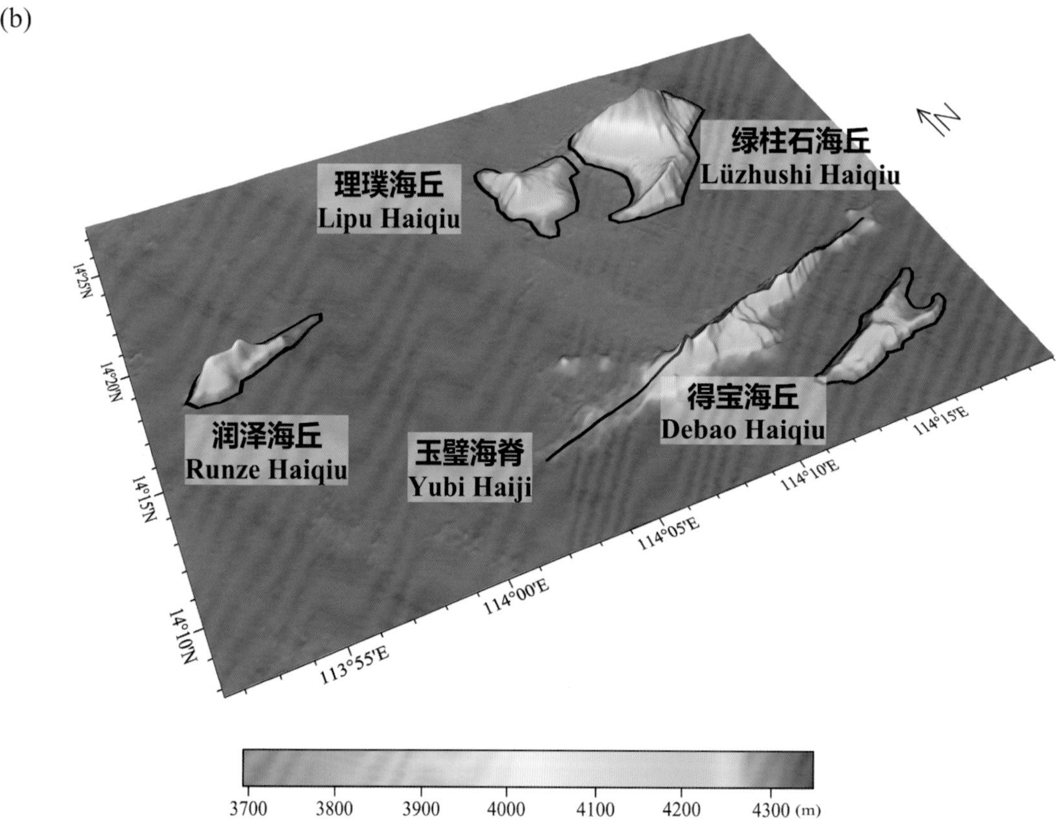

图 6-19　绿柱石海丘、理璞海丘、润泽海丘、玉璧海脊、得宝海丘

(a) 海底地形图（等深线间隔 500 米）；(b) 三维海底地形图

Fig.6-19　Lüzhushi Haiqiu, Lipu Haiqiu, Runze Haiqiu, Yubi Haiji, Debao Haiqiu

(a) Seafloor topographic map (with contour interval of 500 m); (b) 3-D seafloor topographic map

6.53 石美海丘

标准名称 Standard Name	石美海丘 Shimei Haiqiu	类别 Generic Term	海丘 Hill
中心点坐标 Center Coordinates	14°00.2'N, 114°06.4'E	规模（千米 × 千米） Dimension（km × km）	20 × 8
最小水深（米） Min Depth (m)	3930	最大水深（米） Max Depth (m)	4340
地理实体描述 Feature Description	石美海丘位于南海海盆西部，盆西海岭东侧，平面形态呈北东—南西向的长条形（图 6-20）。 Shimei Haiqiu is located in the west of Nanhai Haipen and to the east of Penxi Hailing, with a planform in the shape of a long strip in the direction of NE–SW (Fig.6–20).		
命名由来 Origin of Name	以与"玉"相关的词进行地名的团组化命名。"石美"即美丽的石头，暗指玉石。 A group naming of undersea features after the words related to jade ("Yu" in Chinese). "Shimei", meaning beautiful stones, allusions to jade.		

6.54 玉德海脊

标准名称 Standard Name	玉德海脊 Yude Haiji	类别 Generic Term	海脊 Ridge
中心点坐标 Center Coordinates	14°02.5'N, 113°40.6'E	规模（千米 × 千米） Dimension（km × km）	14 × 12
最小水深（米） Min Depth (m)	3690	最大水深（米） Max Depth (m)	4320
地理实体描述 Feature Description	玉德海脊位于南海海盆西部，盆西海岭东侧，平面形态呈南—北向的长条形，其顶部发育多座峰（图 6-20）。 Yude Haiji is located in the west of Nanhai Haipen and to the east of Penxi Hailing with a planform in the shape of a long strip in the direction of S–N. Multiple peaks have developed on its peak (Fig.6–20).		
命名由来 Origin of Name	以与"玉"相关的词进行地名的团组化命名。"玉德"即玉石代表的品德。 A group naming of undersea features after the words related to jade ("Yu" in Chinese). "Yude" literally means the virtue of a jade.		

(a)

(b)

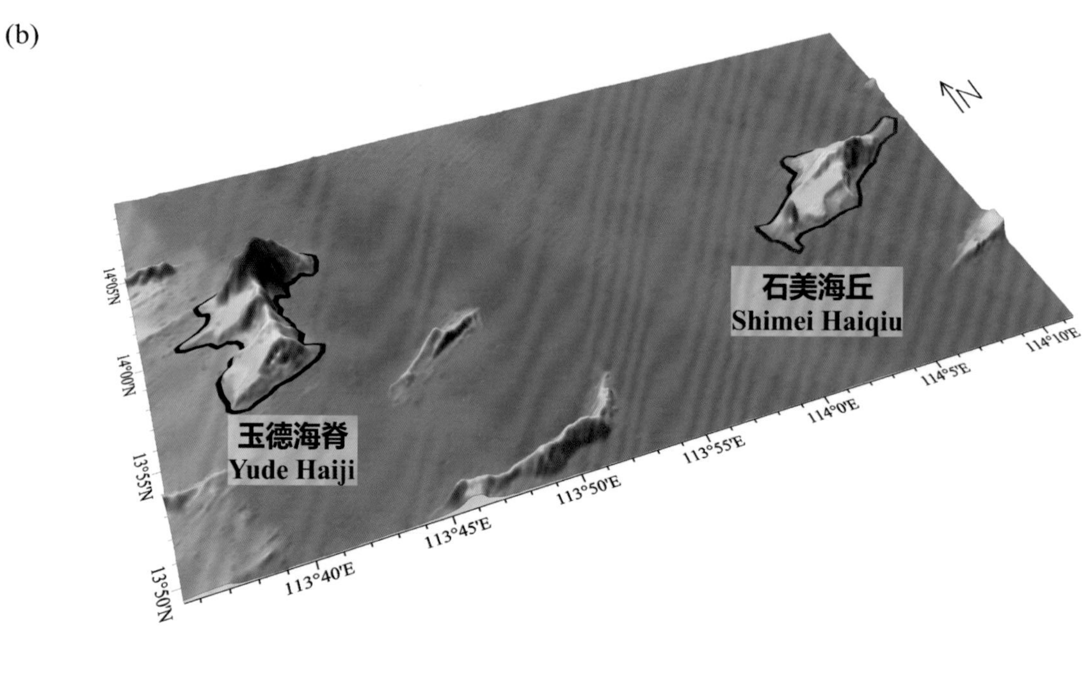

图 6-20 石美海丘、玉德海脊

(a) 海底地形图（等深线间隔 500 米）；(b) 三维海底地形图

Fig.6-20 Shimei Haiqiu, Yude Haiji

(a) Seafloor topographic map (with contour interval of 500 m)；(b) 3-D seafloor topographic map

索 引
Index

续表

续表

续表

续表

续表

续表

续表

续表

续表